HARMONIC BALANCE FINITE ELEMENT METHOD

HARMONIC BALANCE FINITE ELEMENT METHOD

APPLICATIONS IN NONLINEAR ELECTROMAGNETICS AND POWER SYSTEMS

Junwei Lu, Xiaojun Zhao and Sotoshi Yamada

WILEY

IEEE PRESS

This edition first published 2016

Registered Office
John Wiley & Sons Singapore Pte. Ltd., 1 Fusionopolis Walk, #07-01 Solaris South Tower, Singapore 138628.

For details of our global editorial offices, for customer services and for information about how to apply for permission to reuse the copyright material in this book please see our website at www.wiley.com.

Library of Congress Cataloging-in-Publication Data

Names: Lu, Junwei, author. | Zhao, Xiaojun, (Electrical engineer), author. | Yamada, Sotoshi, author.
Title: Harmonic balance finite element method : applications in nonlinear electromagnetics and power systems / Junwei Lu, Xiaojun Zhao, and Sotoshi Yamada.
Description: Solaris South Tower, Singapore : John Wiley & Sons, Inc., [2016] | Includes bibliographical references and index.
Identifiers: LCCN 2016009676| ISBN 9781118975763 (cloth) | ISBN 9781118975787 (epub)
Subjects: LCSH: Electric power systems–Mathematical models. | Harmonics (Electric waves)–Mathematics. | Finite element method.
Classification: LCC TK3226 .L757 2016 | DDC 621.3101/51825–dc23
LC record available at https://lccn.loc.gov/2016009676

Set in 10.5/13pt Times by SPi Global, Pondicherry, India
Printed and bound in Singapore by Markono Print Media Pte Ltd

1 2016

This book is dedicated to my wife Michelle, without her support I would never complete this book, and in memory to my parents.

– Junwei Lu

This book is dedicated to my wife Weichun Cui, since she has helped me a lot during the writing of this book. I also would like to express my gratitude to my beloved parents, who have always supported me.

– Xiaojun Zhao

Contents

Preface

In writing this book on the *Harmonic Balance Finite Element Method (HBFEM): Applications in Nonlinear Electromagnetics and Power Systems*, two major objectives were borne in my mind. Firstly, the book intends to teach postgraduate students and design engineers how to define quasi-static nonlinear electromagnetic (EM) field and harmonic problems, build EM simulation models, and solve EM problems by using the HBFEM. Secondly, this book will delve into a field of challenging innovations pertinent to a large readership, ranging from students and academics to engineers and seasoned professionals.

The art of HBFEM is to use Computational Electromagnetics (CEMs) with harmonic balance theories, and CEM technologies (with IEEE Standard 1597.1 and IEEE Standard 1597.2) to analyze or investigate nonlinear EM field and harmonic problems in electrical and electronic engineering and electrical power systems. CEM technologies have been significantly developed in the last three decades, and many commercially available software packages are widely used by students, academics and professional engineers for research and product design. However, it takes untrained engineers or users several months to understand how to use those packages properly, due to a lack of knowledge on CEMs and EM modeling, and computer simulation techniques. This is particularly true for the harmonic analysis technique, which has not been fully presented in any CEM textbook or used in any commercially available packages. Although a number of CEM-related books are available, these books are normally written for experts rather than students and design engineers. Some of these books only cover one or a few areas of CEMs, and many common CEM techniques and real-world harmonic problems are not introduced. This book attempts to combine the fundamental elements of nonlinear EM, harmonic balance theories, CEM techniques and HBFEM approaches, rather than providing a comprehensive treatment of each area.

This book covers broad areas of harmonic problems in electrical and electronic engineering and power systems, and includes the basic concepts of CEMs, nonlinear EM field and harmonic problems, IEEE Standards 1597.1 and 1597.2, and various numerical analysis methods. In particular, it covers some of the methods that are very useful in solving harmonic-related problems – such as the HBFEM – that are not mentioned in any other numerical calculation books or commercial software packages. In relation to computational technology, this book introduces high-performance parallel computation, cloud computing, and visualization techniques. It covers application problems from component level to system level, from low-frequency to high-frequency, and from electronics to power systems.

This book is divided into six chapters and three appendices. Chapter 1 provides a short introduction to the HBFEM used for solving various harmonic problems in nonlinear electromagnetic field and power systems. This chapter will also discuss definitions of CEM techniques and the various methods used for nonlinear EM problem solving. It also describes high-performance computation, visualization and optimization techniques for EMs, and CEM standards and validation (IEEE Standard 1597.1 and IEEE Standard 1597.2, 2010).

Chapter 2 highlights some fundamental EM theory used in nonlinear EM fields, harmonic problems in transformer power supplies, DC-biased phenomenon in High Voltage Direct Current (HVDC) power transformers, harmonic problems in geomagnetic disturbances (GMDS), geomagnetic induced current (GIC), harmonic problems in distributed energy resource (DER) systems and microgrids, and future smart grids with electric vehicles (EV) and vehicle to grid (V2G).

Chapter 3 covers: the fundamental theory of harmonic balance methods used in nonlinear circuit problems; CEM for nonlinear EM field and harmonic problems; basic concepts of HBFEM used in nonlinear magnetic field analysis; HBFEM for electric circuits and magnetic field coupled problems; HBFEM for three-phase electric circuits coupled with magnetic field; and HBFEM for DC-biased HVDC power transformers.

Chapter 4 investigates HBFEM and its applications in nonlinear magnetic fields and harmonic problems. Several case study problems are presented, such as: HBFEM for a nonlinear magnetic field with current driven (inductor and single phase transformer); HBFEM for a nonlinear magnetic field with voltage-driven (switch mode power supply transformer); three-phase magnetic tripler transformer (electric circuit and magnetic field coupled problems); three-phase high speed motor based on frequency tripler using HBFEM; DC-biased 3D asymmetrical magnetic structure transformer using HBFEM.

Chapter 5 is devoted to the advanced numerical approaches of HBFEM. These include: the decomposed algorithm of HBFEM; HBFEM with a fixed-point technique; hysteresis model based on a neural network and consuming function; and analysis of hysteretic characteristics under sinusoidal and DC bias excitation, parallel computing techniques for multi-frequency domain problem.

Chapter 6 discusses: three-phase power supply transformer model; magnetically controlled shunt reactors (MCSR); computation taking account of hysteresis effects based on fixed-point reluctance; harmonics analysis in HVDC transformers (three phase model) with geo-magnetics and geomagnetic induced current (GIC); HBFEM used for low-voltage network transformers in renewable energy and microgrid grid systems with distributed energy resource (DER); and electric vehicle (EV) charging systems and vehicle to grid (V2G).

There are three appendices included in this book: MATLAB Program 1 (magnetic circuit analysis of a single phase transformer) and MATLAB Program 2 (main program for 2D magnetic field analysis in current driven); and Fortran program 3 (3D Asymmetrical magnetic structure transformer using HBFEM).

Junwei Lu

About the Companion Website

Don't forget to visit the companion website for this book:

www.wiley.com/go/lu/HBFEM

There you will find valuable material designed to enhance your learning, including:

- HBFEM program codes
- Explanations

Scan this QR code to visit the companion website

1

Introduction to Harmonic Balance Finite Element Method (HBFEM)

1.1 Harmonic Problems in Power Systems

The harmonics problem in power systems is not a new problem. It has existed since the early 1900s – as long as AC power itself has been available. The earliest harmonic distortion issues were associated with third harmonic currents produced by saturated iron in machines and transformers, or so-called ferromagnetic loads. Later, arcing loads, like lighting and electric arc furnaces, were also shown to produce harmonic distortion. The final type, electronic loads, burst onto the power scene in the 1970s and 1980s, and has represented the fastest growing category ever since [1].

Since power system harmonic distortion is mainly caused by non-linear loads and power electronics used in the electrical power system [2, 3], the presence of non-linear loads and the increasing number of distributed generation power systems in electrical grids contributes to changing the characteristics of voltage and current waveforms in power systems (which differ from pure sinusoidal constant amplitude signals). The impact of non-linear loads and power electronics used in electrical power systems has been increasing during the last decade.

Such electrical loads, which introduce non-sinusoidal current consumption patterns (current harmonics), can be found in power electronics [4], such as: DC/AC inverters; switch mode power supplies; rectification front-ends in motor drives; electronic ballasts for discharge lamps; personal computers or electrical appliances; high-voltage DC (HVDC) power systems; impulse transformers; magnetic induction devices; and various

Harmonic Balance Finite Element Method: Applications in Nonlinear Electromagnetics and Power Systems, First Edition. Junwei Lu, Xiaojun Zhao and Sotoshi Yamada.

Companion website: www.wiley.com/go/lu/HBFEM

electric machines. In addition, the harmonics can be generated in distributed renewable energy systems, geomagnetic disturbances (GMDs) and geomagnetic induced currents (GICs) [5, 6].

Harmonics in power systems means the existence of signals, superimposed on the fundamental signal, whose frequencies are integer numbers of the fundamental frequency. The presence of harmonics in the voltage or current waveform leads to a distorted signal for the voltage or current, and the signal becomes non-sinusoidal. Thus, the study of power system harmonics is an important subject for electrical engineers. Electricity supply authorities normally abrogate responsibility on harmonic matters by introducing standards or recommendations for the limitation of voltage harmonic levels at the points of common coupling between consumers.

1.1.1 Harmonic Phenomena in Power Systems

A better understanding of power system harmonic phenomena can be achieved by consideration of some fundamental concepts, especially the nature of non-linear loads, and the interaction of harmonic currents and voltages within the power system. By definition, harmonic (or non-linear) loads are those devices that naturally produce a non-sinusoidal current when energized by a sinusoidal voltage source. As shown in Figure 1.1, each "waveform" represents the variation in instantaneous current over time for two different loads each energized from a sinusoidal voltage source. This pattern is repeated continuously, as long as the device is energized, creating a set of largely-identical waveforms that adhere to a common time period. Both current waveforms were produced by turning on some type of load device. In the case of the current on the left, this device was probably an electric motor or resistance heater. The current on the right could have been produced by an electronic variable-speed drive, for example. The devices could be single- or three-phase, but only one phase current waveform is shown for illustration. The other phases would be similar.

A French mathematician, Jean Fourier, discovered a special characteristic of periodic waveforms in the early 19th century. The method describing the non-sinusoidal

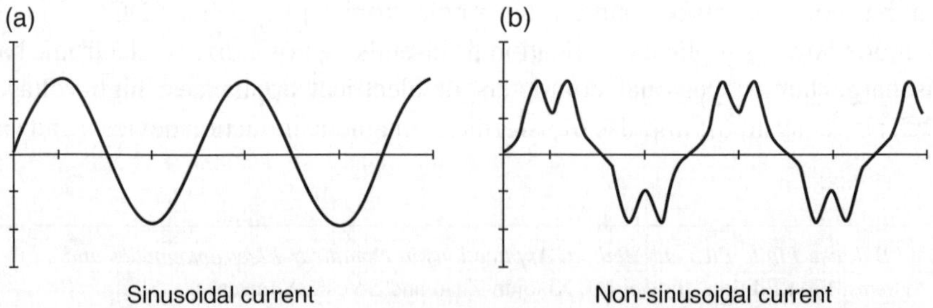

Figure 1.1 (a) Sine wave. (b) Distorted waveform or non-sinusoidal

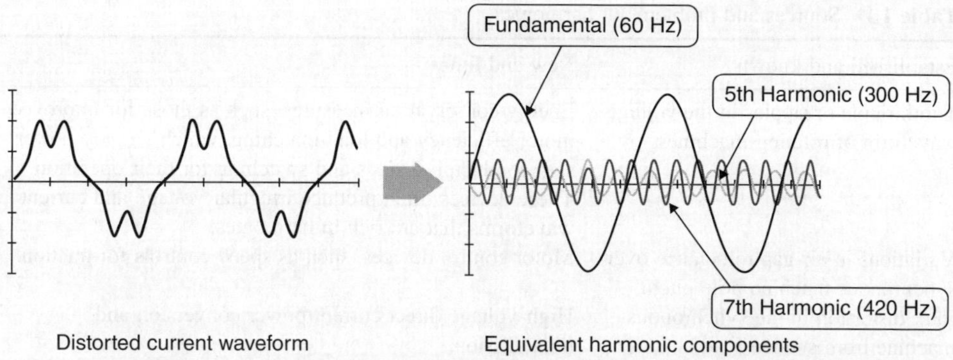

Figure 1.2 Distorted waveform and number of harmonics by Fourier series

waveform is called its Fourier Series. The Fourier theorem breaks down a periodic wave into its component frequencies. Periodic waveforms are those waveforms comprised of identical values that repeat in the same time interval, as shown in Figure 1.2. Fourier discovered that periodic waveforms can be represented by a series of sinusoids summed together. The frequency of these sinusoids is an integer multiple of the frequency represented by the fundamental periodic waveform.

The distorted (non-linear) waveform, however, deserves further scrutiny. This waveform meets the continuous, periodic requirement established by Fourier. It can be described, therefore, by a series of sinusoids. This example waveform is represented by only three harmonic components, but some real-world waveforms (square wave, for example) require hundreds of sinusoidal components to describe them fully. The magnitude of these sinusoids decreases with increasing frequency, often allowing the power engineer to ignore the effect of components above the 50th harmonic.

1.1.2 Sources and Problems of Harmonics in Power Systems

Harmonic sources generated in power systems can be divided into two categories: established and known; and new and future. Table 1.1 presents sources and problems of harmonics. Harmonic problems in power systems can be traced to a number of factors [3], such as: (a) the substantial increase of non-linear loads resulting from new technologies such as silicon-controlled rectifiers (SCRs), power transistors, and microprocessor controls, which create load-generated harmonics throughout the system; and (b) a change in equipment design philosophy.

In the past, equipment designs tended to be under-rated or over-designed. Nowadays, in order to be competitive, power devices and equipment are more critically designed and, in the case of iron-core devices, their operating points are more focused on non-linear regions. Operation in these regions results in a sharp rise in harmonics.

Table 1.1 Sources and problems of harmonics

Established and known	New and future
Tooth ripple or ripples in the voltage waveform of rotating machines.	Energy conservation measures, such as those for improved motor efficiency and load-matching, which employ power semiconductor devices and switching for their operation. These devices often produce irregular voltage and current waveforms that are rich in harmonics.
Variations in air-gap reluctance over synchronous machine pole pitch.	Motor control devices, such as speed controls for traction.
Flux distortion in the synchronous machine from sudden load changes.	High-voltage direct current power conversion and transmission.
Non-sinusoidal distribution of the flux in the air gap of synchronous machines.	Interconnection of wind and solar power converters with distribution systems.
Transformer magnetizing currents.	Static var compensators which have largely replaced synchronous condensers as continuously variable-var sources.
Network non-linearities from loads such as rectifiers, inverters, welders, arc furnaces, voltage controllers, frequency converters, etc.	The development and potential use of electric vehicles that require a significant amount of power rectification for battery charging.
N/A	The potential use of direct energy conversion devices, such as magneto-hydrodynamics, storage batteries, and fuel cells that require DC/AC power converters.
N/A	Cyclo-converters used for low-speed high-torque machines.
N/A	Pulse-burst-modulated heating elements for large furnaces.

1.1.3 Total Harmonic Distortion (THD)

The reduced impedance at the peak voltage results in a large, sudden rise in current flow until the impedance is suddenly increased, resulting in a sudden drop in current. Because the voltage and current waveforms are no longer related, they are said to be "non-linear". These non-sinusoidal current pulses introduce unanticipated reflective currents back into the power distribution system, and the currents operate at frequencies other than the fundamental 50/60 Hz. Ideally, voltage and current waveforms are perfect sinusoids. However, because of the increased non-linear load and power electronic devices based on switch mode power supplies and motor drives, these waveforms quite often become distorted. This deviation from a perfect sine wave can be represented by harmonics – sinusoidal components having a frequency that is an integral multiple of the fundamental frequency, as shown in Figure 1.3. Thus, a non-sinusoidal wave has distortion and harmonics. To quantify the distortion, the term total harmonic distortion (THD) is used, and this expresses the distortion as a percentage of the fundamental voltage and current waveforms.

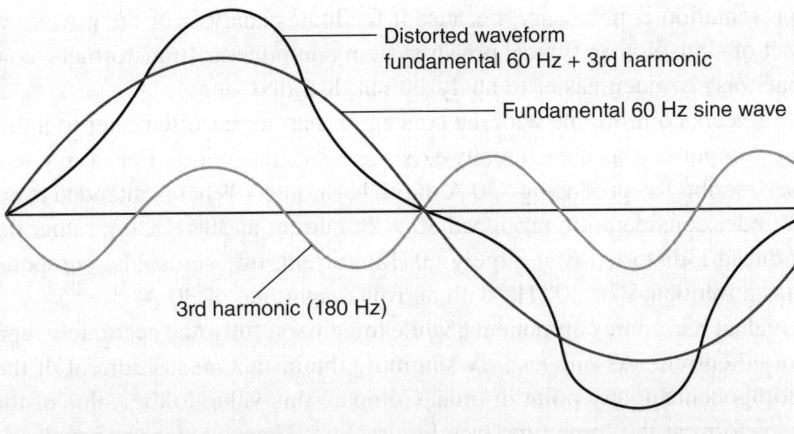

Figure 1.3 Harmonic distortion of the electrical current waveform, where the distorted waveform is composed of fundamental and 3rd harmonics

Harmonics have frequencies that are integer multiples of the waveform's fundamental frequency. For example, given a 60 Hz fundamental waveform, the 2nd, 3rd, 4th and 5th harmonic components will be at 120 Hz, 180 Hz, 240 Hz and 300 Hz, respectively. Thus, harmonic distortion is the degree to which a waveform deviates from its pure sinusoidal values as a result of the summation of all these harmonic elements. The ideal sine wave has zero harmonic components. In that case, there is nothing to distort this perfect wave. Total harmonic distortion, or THD, is the summation of all harmonic components of the voltage or current waveform, compared against the fundamental component of the voltage or current wave:

$$THD = \frac{\sqrt{\left(V_2^2 + V_3^2 + V_4^2 + \cdots V_n^2\right)}}{V_1} \tag{1-1}$$

The formula above (Equation 1-1) shows the calculation for THD on a voltage signal. The end result is a percentage comparing the harmonic components to the fundamental component of a signal. The higher the percentage, the more distortion that is present on the mains signal. The concept that a distorted waveform (including a square wave) can be represented by a series of sinusoids is difficult for many engineers, but it is absolutely essential for understanding the harmonic analysis and mitigation to follow. It is important for the power engineer to keep the following facts in mind:

- The equivalent harmonic components are just a representation – the instantaneous current as described by the distorted waveform is what is actually flowing on the wire.

- This representation is necessary, because it facilitates analysis of the power system. The effect of sinusoids on typical power system components (transformers, conductors, capacitors) is much easier to analyze than distorted signals.
- Power engineers comfortable with the concept of harmonics often refer to individual harmonic components as if each really exists as a separate entity. For example, a load might be described as producing "30 A of 5th harmonic." What is intended is not that the load under consideration produced 30 A of current at 300 Hz, but rather that the load produced a distorted (but largely 60 Hz) current, one sinusoidal component of which has a frequency of 300 Hz with an rms magnitude of 30 A.
- The equivalent harmonic components, while imaginary, fully and accurately represent the distorted current. As one test, try summing the instantaneous current of the harmonic components at any point in time. Compare this value to the value of the distorted waveform at the same time (see Figure 1.3). These values are equal.

The current drawn by non-linear loads passes through all of the impedance between the system source and load. This current produces harmonic voltages for each harmonic as it flows through the system impedance. The sum of these harmonic voltages produces a distorted voltage when combined with the fundamental. The voltage distortion magnitude is dependent on the source impedance and the harmonic voltages produced. Figure 1.4 illustrates how the distorted voltage is created. As illustrated, non-linear loads are typically modeled as a source of harmonic current.

With low source impedance, the voltage distortion will be low for a given level of harmonic current. If the harmonic current increases, however, system impedance changes due to the harmonic resonance (discussed below) can significantly increase voltage distortion.

IEEE Std. 519-1992, which is titled *IEEE Recommended Practices and Requirements for Harmonic Control in Electric Power Systems*, is the main document for harmonics in North America. This standard serves as an excellent tutorial on harmonics. The most

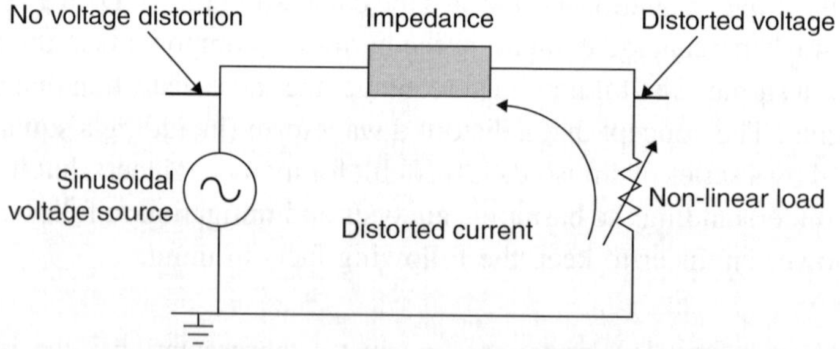

Figure 1.4 Creation of distorted current

important part of this document to the industrial user is Chapter 10 ("Recommended Practices for Individual Consumers" [7]).

The electric consumption is a significant part of the total energy consumption and, consequently, the complete chain of generation, transportation and usage of electricity should be optimized. The usage of electrical energy is often optimized by controlling the output of electrical equipment towards the desired value. Advances in power electronic (PE) energy conversion have led to an optimization of electrical equipment. Practical examples of PE-controlled energy conversion are dimmable halogen lighting, low- and high-pressurized discharge lights, AC drives for induction machines (IM), and so on. In addition to the advantages of PE in terms of energy optimization, a lot of PE is also used for DC power supply, such as IT equipment, DC arcing or electrolysis [8, 9].

Harmonics are a distortion of the normal electrical current waveform, generally transmitted by non-linear loads. Switch-mode power supplies (SMPS), variable speed motors and drives, photocopiers, personal computers, laser printers, fax machines, battery chargers and UPSs are examples of non-linear loads. Single-phase non-linear loads are prevalent in modern office buildings, while three-phase, non-linear loads are widespread in factories and industrial plants, and in DC-biased power transformers in HVDC power systems [10].

The study of these harmonics problems is normally focused on the electrical circuit level. A large number of articles and reports have been published in this area. However, the harmonics problem in the component level (or electromagnetic fields) has not been fully investigated, due to a lack of understanding of the characteristics of non-linear electromagnetic fields and a lack of theory and methodology dealing with harmonics generated from non-linear electromagnetic fields. Only a very limited number of papers and reports related to HBFEM used in solving the harmonic problems in electromagnetic field [11, 12]. Detailed HBFEM theory development and various application problem-solving examples are presented in later chapters.

1.2 Definitions of Computational Electromagnetics and IEEE Standards 1597.1 and 1597.2

1.2.1 "The Building Block" of the Computational Electromagnetics Model [13, 14]

The objective of computational electromagnetics (CEM) is to create a representation of real-life problems that can be examined and analyzed by computer resources, as an alternative to building a system, exciting it, and measuring the generated fields. Once the problem has been defined, the important physical characteristics must be identified. All CEM models can be broken into three parts: the source of EM energy, the geometry of the model components, and the remaining problem space. The following elements of a physical CEM model should be taken into account during the simulation:

1.2.1.1 The Sources of EM Energy

- ***Source*** – Sources include both intended and unintended sources that electromagnetically couple to and drive conductors (such that energy is conducted into areas that can energize and drive the electric machine to make a correct operation, or can cause problems with the correct operation of the victim devices).
- ***Physical Source Modeling*** – Sources may be characterized by their electrical size, the distance from materials with which they interact, their geometry, and the excitation applied to them.
- ***Source Excitation*** – Like fully specified circuit model sources, field sources must also be defined by their amplitude and impedance.

1.2.2 The Geometry of the Model and the Problem Space

The major concern of every CEM model is the geometry of the problem to be solved. A less complex representation must be created, which includes all the important details while avoiding unnecessary details. In addition to the fixed portions of the geometry, it is often necessary to include variables such as the range of positions in which a nearby wire – or any other conductor – could be placed. Together with the geometry of a problem, the properties of all materials used must also be included in the model. If the computational domain were of infinite extent, the simulation of free space would be involved. This can be achieved by using mesh truncation techniques or absorbing boundary conditions. These techniques require that extra free space is added around the model components.

1.2.3 Numerical Computation Methods

Substantial advancements have been made in enhancing the important numerical techniques – for example: the method of moments (MoM); the finite-difference time-domain (FDTD) method; the finite-element method (FEM); the proposed harmonic balance method (HBFEM) in this book; and the transmission line matrix (TLM) method. Many numerical methods were invented decades ago but, in all cases, additional novel ideas were required to make them applicable to today's real-world electromagnetic problems.

- **The quasi-static field** can be expressed by several different partial differential equations (PDEs). Although existing computational electromagnetic solvers provide preliminary insight, a multi-physics simulation system is needed to model coupled problems in their entity. **Multi-physics problems** are often related to more than two fields, such as thermal and **E** fields, or the **H** field, thermal dynamic field, and so forth. In the quasi-static field, the following methods are often used in FEM based EM computation:

- **Time-domain techniques** use a band-limited impulse to excite the simulation across a wide frequency range. The result obtained from a time-domain code is the model's response to this impulse. Where frequency-domain information is required, a Fourier transform is applied to the time-domain data.
- **Frequency-domain codes** solve for one frequency at a time. This is usually adequate for antenna work or electric machine simulation, and for examining specific issues. Frequency-domain codes are, in general, faster than their time-domain cousins. Therefore, several frequency-domain simulations can usually be run in the time it would take for a single time-domain simulation. However, in nonlinear EM field problems, there is a coalition between each frequency domain, particularly for solving harmonic problems in nonlinear time periodic problems. This can be called the **multi-frequency-domain or HBFEM**.

1.2.4 High-Performance Computation and Visualization (HPCV) in CEM

With the rapid growth of microelectronics and computer technologies, cluster-based high-performance parallel computers are becoming more and more powerful and cost-effective. This provides a new opportunity to apply computational electromagnetics technologies to challenging problems in EM computer modeling and simulation. Since the computational technique extends from numeric analysis to visualization analysis, the demand for innovative visualization techniques becomes higher and higher. Visualization is closely related to high-performance computation using visualization techniques to deal with the complex dynamic electromagnetic system problems.

Visualization techniques for computational electromagnetics in 2D and 3D promise to radically change the way data is analyzed. To minimize eddy current loss and other problems in nonlinear EM fields, the optimization algorithms have been considered in current computational electromagnetic (CEM) modeling approaches. In fact, the action of EM computer modeling and simulation involves several physical effects. Detailed knowledge of all these effects is a prerequisite for effective and efficient design. The first step in reducing the design time and allowing for aggressive design strategies is to use EM computer modeling techniques that will let designers try "what if" experiments in hours instead of months.

1.2.5 IEEE Standards 1597.1 and 1597.2 for Validation of CEM Computer Modeling and Simulations

IEEE P1597.1 and P1597.2 Standards, developed by the EMC community, were released in 2008 and 2010 respectively. IEEE Standard 1597.1-2008 is related to the

IEEE Standard for validation of computational electromagnetics computer modeling and simulations. IEEE Standard 1597.2-2010 was released for IEEE recommended practice for validation of computational electromagnetics computer modeling and simulations. The following highlighted descriptions are based on IEEE Standards 1597.1 and 1597.2 [15, 16].

The development of IEEE standards, and recommended practices for computational electromagnetics (CEM) computer modeling and simulation and code validation, has been a topic of much interest within the EMC community particularly since the mid-1980s [17]. This has been due to advances in computer hardware and software technologies,+ as well as the arrival of new CEM codes and applications. The areas of concern include, but are not limited to, high-frequency areas such as analyzing printed circuit boards (PCBs), radiated and conducted emissions/immunity, system-level electromagnetic compatibility (EMC), radar cross-section (RCS) of complex structures, and the simulation of various EM environment effects problems. In particular, there are concerns regarding the lack of well-defined methodologies to achieve code-to-code or even simulation-to-measurement validations with a consistent level of accuracy.

Since the mid-1960s, a number of CEM techniques have been developed, and numerical codes have been generated, to analyze various related electromagnetic problems, including electromagnetic compatibility (EMC). While each is based on classical electromagnetic theory and implements Maxwell's equations in one form or another, these techniques, and the manner in which they are used to analyze a given problem, can produce quite different results. A well-defined, mature, and robust methodology for validating computational electromagnetic techniques, with a consistent level of accuracy, is lacking. Indeed, this has eluded the EMC community for many years, and methods have been sought to address this deficiency. The EMC community has persisted regarding the validity, accuracy and applicability of existing numerical techniques to the general class of EMC problems of interest.

The IEEE P1597.1 Standard defines a method to validate computational electromagnetics (CEM) computer modeling and simulation (M&S) techniques, codes and models. It is applicable to a wide variety of electromagnetic (EM) applications, including (but not limited to) the fields of electromagnetic compatibility (EMC), radar cross-section (RCS), signal integrity (SI), and antennas. Validation of a particular solution data set can be achieved by comparing the data set obtained by measurements, alternate codes, canonical, or analytic methods.

IEEE P1597.2™, recommended practice for validation of computational electromagnetics computer modeling and simulation, has been developed to provide examples and problem sets for use in the validation of CEM computer modeling and simulation techniques, codes and models. It is applicable to a wide variety of electromagnetic applications. The recommended practice, in conjunction with this standard, shows how to validate a particular solution data set by comparing it to the data set obtained by measurements, alternate codes, canonical, or analytic methods. The key areas addressed include model accuracy, convergence, and techniques or code validity for a given set of canonical, benchmark, and standard validation models.

In fact, computer predictions have been compared to measurements to provide a first-order validation, but there is also much interest in how the techniques, when applied to a given problem or a class of problems, compare to each other and the fundamental theory upon which they are based. Hence, additional efforts are needed to establish a standardized method for validating these techniques, and to instill confidence in them. Therefore, the purpose of this first-of-its-kind standard is to define the specific process and steps that will be used to validate CEM techniques and to significantly reduce uncertainty (as it pertains to their implementation and application to practical EMC problem-solving tasks). The standardized process, based on the Feature Selective Validation (FSV) method, is used to validate various techniques against each other, as well as against measurement baselines (in order to determine the degree of agreement or convergence, and to identify the potential error sources that would lead to divergent trends).

In general, CEM techniques and codes, and the manner in which they are used to analyze a given problem, can produce quite different results. These results are affected by the way in which the underlying physic formalisms have been implemented within the codes, including the mathematical basis functions, numerical solution methods, numerical precision, and the use of building blocks (primitives) to generate computational models. Despite all CEM codes having their basis in Maxwell's equations in one form or another, their accuracy and convergence rate depends on how the physics equations are cast (e.g., integral or differential form, frequency or time domain), what numerical solver approach is used (full or partial wave, banded or partitioned matrix, non-matrix), inherent modeling limitations, approximations, and so forth. The physics formalism, available modeling primitives (canonical surface or volumetric objects, wires, patches, facets), analysis frequency, and time or mesh discretization further combine to affect accuracy, solution convergence and overall validity of the computer model.

The critical areas that must be addressed include model accuracy, convergence, and techniques or code validity for a given set of canonical, benchmark and standard validation models. For instance, uncertainties may arise when the predicted results using one type of CEM technique do not agree favorably or consistently with the results of other techniques or codes of comparable type, or even against measured data on benchmark models. Furthermore, it can be difficult to compare the results between certain techniques or codes, despite their common basis in Maxwell's equations. Exceptions can be cited, in particular, when comparing the results of "similar" codes grouped according to their physics, solution methods, and modeling element domains. Nevertheless, disparities even among codes in a certain "class" have been observed. Many examples can be cited where fairly significant deviations have been observed between analytical or computational techniques and empirical-based methods. Differences are not unexpected, but the degree of disparity in certain cases cannot be readily explained nor easily discounted. This has led to the often asked question: "Which result is accurate?"

1.3 HBFEM Used in Nonlinear EM Field Problems and Power Systems

Nonlinear phenomena in EM fields are caused by nonlinear materials used in electric machines. The nonlinear materials are normally field strength-dependent, which can cause harmonics. Therefore, when the time-periodic quasi-static EM field is applied to the nonlinear material, the electromagnetic properties of the material will be functions of the EM field, which is also time-dependent. On the other hand, harmonics can also be generated by power electronic devices and drives, which are largely used in power systems, renewable energy systems and microgrids. These power electronic devices and drives are used for power rectification, power conversion (e.g. DC/DC converter) and inversion (e.g. DC/AC inverter). In fact, HBFEM can be used to effectively solve these harmonic problems in nonlinear EM fields and power systems.

Since the harmonic balance FEM technique was introduced to analyze low-frequency electromagnetic (EM) field problems in the late 1980s [11, 12], various harmonic problems in nonlinear EM fields and power systems have been investigated and solved by using HBFEM [18–24]. Harmonic balance techniques were combined with the finite element method (FEM) to accurately solve the problems arising from time-periodic, steady-state nonlinear magnetic fields. The method can be used for weak and strong nonlinear time-periodic EM fields, as well as harmonic problems in renewable energy systems and microgrids with distributed energy resources.

The harmonic balance FEM (HBFEM) method uses a linear combination of sinusoids to build the solution, and represents waveforms using the sinusoid – coefficients combined with the finite element method. It can directly solve the steady-state response of the EM field in the multi-frequency domain. Thus, the method is often considerably more efficient and accurate in capturing coupled nonlinear effects than the traditional FEM time-domain approach when the field exhibits widely separated harmonics in the frequency spectrum domain (e.g. pulse width modulation (PWM)-based power electronic devices and drives). The HBFEM consists of approximating the time-periodic solution (magnetic potentials, currents, voltages, etc.) with a truncated Fourier series. Besides the frequency components of the excitation (e.g. applied voltages or current), the solution contains harmonics due to nonlinearity (magnetic saturation and nonlinear lumped electrical components), movement (e.g. rotation in electric machines), and power electronic devices and drives.

In order to solve time-periodic nonlinear magnetic field problems, a novel numerical computation method called HBFEM was developed, which is the combination of FEM and the harmonic balance method. The principle of a new approach of HBFEM is to drive a basic formulation of the harmonic balance finite element method (HBFEM). For simplicity of fundamental formulation, a time-periodic nonlinear magnetic field is assumed as two-dimensional in the (x, y) plane, and is quasi-stationary. Therefore, the vector potential A = (0, 0, A) satisfies in the region of interest surrounded with some boundary conditions. To calculate such a quasi-static magnetic field, the following equation (1-2) can be used:

$$\frac{\partial}{\partial x}\left(\nu\frac{\partial A}{\partial x}\right)+\frac{\partial}{\partial y}\left(\nu\frac{\partial A}{\partial y}\right)=-J_s+\sigma\frac{\partial A}{\partial t} \tag{1-2}$$

where ν and σ are magnetic reluctivity and conductivity.

Based on the harmonics balance theory, the governing equations of the quasi-static field containing harmonics can also be solved by using a FEM-based numerical approach. Assuming $\nabla\varphi = 0$ in the two-dimensional case, and using Galerkin's method to discretize, the governing equation for two dimensional problems can be written in an integral form that is given as:

$$G=\iint_S\left\{\frac{\partial N_i}{\partial x}\nu\frac{\partial A}{\partial x}+\frac{\partial N_i}{\partial y}\nu\frac{\partial A}{\partial y}\right\}dxdy-\iint_S\left\{J_s-\sigma\frac{\partial A}{\partial t}\right\}N_i dxdy=0 \tag{1-3}$$

where $N_i(x, y)$ is the interpolating function.

When the applied voltage waveform is a sinusoidal signal, the current can be considered as a non-sinusoidal waveform. The waveform may be distorted due to nonlinear load or power electronic devices. Therefore, the current excitation source will include harmonic components, and the resultant magnetic field will contain all harmonic components. The vector potential $\boldsymbol{A}$ and current density $\boldsymbol{J}$ are approximated as a summation of all harmonic solutions. According to the harmonic balance method, all variables (i.e. vector potentials, flux densities and applied current) are approximated as a summation of all harmonic solutions. Therefore, the time-periodic solution (harmonic problem) can be found when an alternating magnetizing current is applied. In fact, HBFEM has been successfully used to solve various nonlinear magnetic field problems, and the computation results have been verified by experimental results listed below (detailed results will be discussed in Chapter 3 and 4).

1.3.1 HBFEM for a Nonlinear Magnetic Field With Current Driven

The HBFEM differs from traditional finite element time-domain methods, transient analysis and other time harmonic methods. The harmonic balance method uses a linear combination of sinusoids to generate a solution, and represents waveforms using the coefficients of the sinusoids. It is combined with the finite element method to solve time-periodic, steady-state nonlinear electromagnetic field problems. The HBFEM directly solves the steady-state response of the electromagnetic field in the frequency domain, and so is often considerably more efficient than traditional time-domain methods when fields exhibit widely separated time constants and mildly nonlinear behavior. The electromagnetic field with harmonics satisfies Maxwell's equations. The magnetic core of the transformer and inductor, with nonlinear characteristics and hysteresis, is excited by a current source of current density *Js*.

When a nonlinear magnetic system is excited by a sinusoidal waveform, a number of harmonics will be generated in this nonlinear magnetic system. For the non-DC biased case, only odd harmonics can be generated in the magnetic field where the B-H curve, with and without hysteresis characteristics, is used.

1.3.2 HBFEM for Magnetic Field and Electric Circuit Coupled Problems

In most cases, pulse width modulation and zero-current switched resonant converters (including LLC converters, DC biased HVDC transformers and HV transformers) can be considered as a voltage source to the magnetic system, which is always coupled to the external circuits. In that case, the current in the input circuits will be unknown, and the saturation of the current waveform occurs because of the nonlinear characteristic of the magnetic core and power electronic device.

1.3.3 HBFEM for a Nonlinear Magnetic Field with Voltage Driven

When a high-frequency transformer of switching mode power supply is excited by a voltage source, such as pulse-width modulation (PWM) converters and zero-current switched (ZCS) resonant converters, the numerical analysis of the magnetic field should be carried out by taking account of the voltage source and the external circuits. If the excitation waveform is a square wave or triangular wave, it can be considered as a linear combination of harmonics. For a DC-biased transformer problem in the HVDC power system, the voltage source and the external circuits will be considered in the HBFEM simulation, which is a magnetic field- and electric circuit-coupled problem.

1.3.4 HBFEM for a Three-Phase Magnetic Tripler Transformer

A magnetic frequency tripler is a nonlinear magnetic system which is used for the production of triple-frequency output from a three-phase fundamental frequency source based on the nonlinear magnetic saturation characteristics. Although magnetic frequency triplers have been used extensively for certain applications, the design of these devices has, until the earlier 1990s, been largely empirical. The earlier, and some recent, papers [25–27] have discussed the analyses of magnetic frequency-tripling devices, based on an equivalent-circuit approach under various load conditions and the Preisach model.

However, the above methods are based on the equivalent circuit theory, magnetic nonlinear characteristics, hysteresis losses, eddy current losses, and magnetic flux distribution for each harmonic component cannot be calculated and presented. Therefore, the EM full wave solution can be obtained from an HBFEM-based numerical computation. HBFEM can provide magnetic flux distribution and eddy current losses at each harmonic [21].

1.3.5 HBFEM for a Three-Phase High-Speed Motor

The HBFEM taking account of external circuits and motion can be effectively used to solve the high-speed and hybrid induction motor problem [21]. A comparison is made between experimental and numerical results for the static model, and this has shown good consistency.

The high-speed hybrid induction motor consists of three-phase input windings and two-phase magnetic frequency tripler as an output, and an induction motor. The induction motor has two pairs of magnetic frequency triplers and four magnetic poles, with the air gap in the middle leg of the cores. The three-phase magnetizing windings are connected as a Scott connection, and four additional coils, connected with the capacitors for increasing output power, are put in the poles. When a 60 Hz commercial source is applied to the hybrid induction motor, the rotation speed (10 800 rpm) will be gained between the poles. The principle of the three-phase input windings and two-phase magnetic frequency tripler as an output is that two single-phase triplers (composed of three-legged cores) are connected in a Scott connection. Therefore, the input voltages shifted at 90 degrees are applied to two single-phase triplers.

Since the high-speed hybrid motor has a very complex configuration, an optimal design is requested in the design of electric machines. Optimal design ensures that the flux of the fundamental harmonic components does not pass through the poles and rotor, and only the three-times frequency flux passes through the poles and moves the rotor. In the HBFEM numerical analysis, the half model is used as an analysis area. The magnetizing windings are applied with a 60 Hz commercial three-phase voltage source. The harmonic components will be generated in the core when the core becomes saturated.

1.3.6 HBFEM for a DC-Biased 3D Asymmetrical Magnetic Structure Simulation

In high-frequency switching power supplies, the leakage inductance, skin and proximity effects, winding self-capacitance and inter-winding capacitance can cause some serious problems in high-frequency transformers. Detailed information about the distribution of eddy currents, flux density and harmonics distribution in the magnetic core and windings has to be known when designing a high-frequency transformer. Furthermore, the nonlinear nature and hysteresis of the core material can cause waveform distortion. These distortions cause further harmonics, which will increase power losses in both the winding and magnetic core, resulting in a loss of efficiency, as well as the possibility of parasitic resonance in the system.

A typical port core transformer has an axi-symmetrical structure and a B-H hysteresis curve. The transformer is excited by quasi-sinusoidal waveforms, which includes AC fundamentals, harmonics and DC components. When the magnetic core becomes saturated, the waveforms will be distorted, and harmonics will be generated in the magnetic field and circuits. The hysteresis loss in winding is also increased, due to the effect of high-frequency harmonics. This kind of problem can be easily solved by using HBFEM [22].

1.3.7 HBFEM for a DC-Biased Problem in HV Power Transformers

A typical DC transmission system consists of a DC transmission line connecting two AC systems. A converter at one end of the line converts AC power into DC power, while a similar converter at the other end reconverts the DC power into AC power. One converter acts as a rectifier, the other as an inverter. The basic purpose of the converter transformer on the rectifier side is to transform the AC network voltage to yield the DC voltage required by the converter. Three-phase transformers, connected in either wye-wye or wye-delta, are used.

The model has a voltage-driven source connected to the magnetic system, which is always coupled to the external circuits. The current in the input circuits will be unknown, but saturation of the current waveform occurs because of the nonlinear characteristic of the magnetic core. Considering a three-phase transformer, connected in wye-wye, a computer simulation model with a neutral *NN* (and external circuits for both primary and secondary windings) is obtained using the HBFEM technique. According to the Galerkin procedure, system matrix equations of HBFEM for the HVDC transformer can be obtained through Faraday's and Kirchhoff's laws for the transformer with external circuits [23, 24].

During geomagnetic disturbances, variations in the geomagnetic field induce quasi-DC voltages in the network, which drive geomagnetically induced currents (GIC) along transmission lines and through transformer windings to ground wherever there is a path for them to flow. The flow of these quasi-DC currents in transformer windings causes half-cycle saturation of transformer cores, which leads to increased transformer hotspot heating, harmonic generation, and reactive power absorption – each of which can affect system reliability. As part of the assessment of geomagnetic disturbances (GMDs) impacts on the Bulk-Power System, it is necessary to model the GIC produced by different levels of geomagnetic activity [28–30]. The HBFEM can be used for solving geomagnetically-induced currents (GIC) and harmonic problem directly, while commercially available GIC modeling software packages cannot solve the harmonic problem. The detailed theory and numerical model for GIC modeling will be explained later in Chapters 3 and 6.

References

[1] Ray, L., Hapeshis, L. (2011). *Power System Harmonic Fundamental Considerations: Tips and Tools for Reducing Harmonic Distortion in Electronic Drive Applications*. Schneider Electric, AT313, October.

[2] J.L. Hernández, Castro, M.A., Carpio, J. and Colmenar, A. (2009). Harmonics in Power Systems, International Conference on Renewable Energies and Power Quality, (ICREPQ'09) Valencia (Spain), 15th to 17th April, 2009.

[3] Churchill, L.D. (no date). *Electrical Harmonics: An Introduction and Overview of What That Means to You*. Available online at: http://itsyourenergy.com/Testimonials%20and%20Studies/Electrical-Harmonics-An-Introduction-and-Overview-of-What-That-Means-to-You.pdf

[4] Paice, D.A. (1996). *Power Electronic Converter Harmonics*. IEEE Press.

[5] IEEE Power and Energy Society Technical Council Task Force on Geomagnetic Disturbances (2013). Geomagnetic Disturbances – Their Impact on the Power Grid. *IEEE Power & Energy Magazine* **11**(4), 71–78.

[6] Radasky, W.A. (2011). *Overview of the Impact of Intense Geomagnetic Storms on the U.S. High Voltage Power Grid*. 2011 IEEE International Symposium on EMC, Aug, pp. 300–305.

[7] IEEE Std 519-1992 (1992). *IEEE Recommended Practices and Requirements for Harmonic Control in Electrical Power Systems*. IEEE: New York, NY.

[8] Ray, L., Hapeshis, L. (2011). *Power System Harmonic Fundamental Considerations: Tips and Tools for Reducing Harmonic Distortion in Electronic Drive Applications*. Schneider Electric AT313, October.

[9] Tihanyi, L. (1995). *Electromagnetic Compatibility in Power Electronics*. IEEE Press.

[10] Zhao, X., Lu, J., Li, L., Cheng, Z. and Lu, T. (2011). Analysis of the DC Biased Phenomenon by the Harmonic Balance Finite Element Method. *IEEE Transactions on Power Delivery* **26**(1), 475–484.

[11] Yamada, S. and Bessho, K. (1988). Harmonic field calculation by the combination of finite element analysis and harmonic balance method. *IEEE Transactions on Magnetics* **24**(6), 2588–2590.

[12] Yamada, S., Bessho, K. and Lu, J. (1989). HBFEM Applied to Nonlinear AC Magnetic Analysis. *IEEE Transactions on Magnetics* **25**(4), 2971–2973.

[13] Archambeault, B., Brench, C. and Ramahi, O.M. (2001). *EMI/EMC computational modelling Handbook*, Second Edition. Kluwer Academic Publishers.

[14] Zhou, P. and Lu, J. (2006). *EMC Computer Modelling*. China National Electric Power Industry Publisher.

[15] IEEE Standards 1597.1-2008 (2009). *IEEE Standard for Validation of Computational Electromagnetics Computer Modeling and Simulations*. IEEE Electromagnetic Compatibility Society, 18 May.

[16] IEEE Standards 1597.2-2010. (2011). *IEEE Recommended Practice for Validation of Computational Electromagnetics Computer Modeling and Simulations*. IEEE Electromagnetic Compatibility Society, February.

[17] Brüns, H-D, Schuster, C. and Singer, H. (2007). Numerical Electromagnetic Field Analysis for EMC Problems. *IEEE Transactions on Electromagnetic Compatibility* **49**(2), 253–262.

[18] Lu, J., Yamada, S. and Bessho, K. (1990). Development and Application of Harmonic Balance Finite Element Method in Electromagnetic Field. *International Journal of Applied Electromagnetics in Materials* **1**(2–4), 305–316.

[19] Lu, J., Yamada, S. and Bessho, K. (1990). Time-periodic Magnetic Field Analysis with Saturation and Hysteresis Characteristics by Harmonic Balance Finite Element Method. *IEEE Transactions on Magnetics* **26**(2), 995–998.

[20] Yamada, S., Biringer, P.P. and Bessho, K. (1991). Calculation of Nonlinear Eddy-current Problems by the Harmonic Balance Finite Element Method. *IEEE Transactions on Magnetics* **27**(5), 4122–4125.

[21] Lu, J., Yamada, S. and Bessho, K. (1991). Harmonic Balance Finite Element Method Taking Account of External Circuits and Motion. *IEEE Transactions on Magnetics* **27**(5), 4204–4207.

[22] Lu, J., Yamada, S. and Harrison, H.B. (1996). Application of HB-FEM in the Design of Switching Power Supplies. *IEEE Transactions on Power Electronics* **11**(2), 347–355.

[23] Zhao, X., Li, L., Cheng, Z. and Lu, J. (2010). *Research on Harmonic Balance Finite Element Method and DC Biased Problem in Transformers*. Proceedings of the CSEE, China, Vol. **30**, pp 103–108.

[24] Zhao, X., Lu, J., Li, L., Cheng, Z. and Lu, T. (2011). Analysis of the DC Biased Phenomenon by the Harmonic Balance Finite Element Method. *IEEE Transactions on Power Delivery* **26**(1), 475–484.

[25] Biringer, B.P. and Slemon, G.R. (1963). Harmonic analysis of the magnetic frequency tripler. *IEEE Transactions on Communication and Electronics* **82**, 327–332.

[26] Bendzsak, G.J. and Biringer, B.P. (1974). The influence of magnetic characteristics upon tripler performance. *IEEE Transactions on Magnetics* **10**(3), 961–964.

[27] Ishikawa, T. and Hou, Y. (2002). Analysis of a Magnetic Frequency Tripler Using the Preisach Model. *IEEE Transactions on Magnetics* **38**(2), 841–844.

[28] NERC (2013). *Computing Geomagnetically-Induced Current in the Bulk-Power System*. (Application Guide, Dec.).

[29] IEEE Power and Energy Society Technical Council Task Force on Geomagnetic Disturbances (2013). Geomagnetic Disturbances. *IEEE Power & Energy Magazine* July/August, pp 71–78.

[30] Samuelsson, O. (2013). *Geomagnetic disturbances and their impact on power systems*. Status report 2013. Division of Industrial Electrical Engineering and Automation, Lund University.

2

Nonlinear Electromagnetic Field and Its Harmonic Problems

2.1 Harmonic Problems in Power Systems and Power Supply Transformers

2.1.1 Nonlinear Electromagnetic Field

A nonlinear load, such as electric machines is, by definition, a device that naturally produces a non-sinusoidal current when energized by a sinusoidal voltage source. Nonlinear loads have high impedance during part of the voltage waveform, and when the voltage is at or near the peak, the impedance is suddenly reduced. Figure 2.1 illustrates characteristics of magnetic impedance associated with a B-H curve and permeability, and excitation current corresponding to a sinusoidal voltage excitation associated with a hysteresis B-H curve.

Current harmonics, as shown in Figure 2.1(b), are a problem because they cause increased losses in customer and utility power system components [1]. Single-phase nonlinear loads like magnetic component based devices, motors and transformers, and electronic devices and SMPS-based equipment (such as personal computers, electronic ballasts for fluorescent lights and other electronic equipment) generate odd harmonics (i.e. 3rd, 5th, 7th, 9th, etc.), as shown in Figure 2.2. When harmonic frequencies are prevalent, electrical power panels and transformers become mechanically resonant to the magnetic fields generated by higher frequency harmonics. When this happens, the

Harmonic Balance Finite Element Method: Applications in Nonlinear Electromagnetics and Power Systems, First Edition. Junwei Lu, Xiaojun Zhao and Sotoshi Yamada.

Companion website: www.wiley.com/go/lu/HBFEM

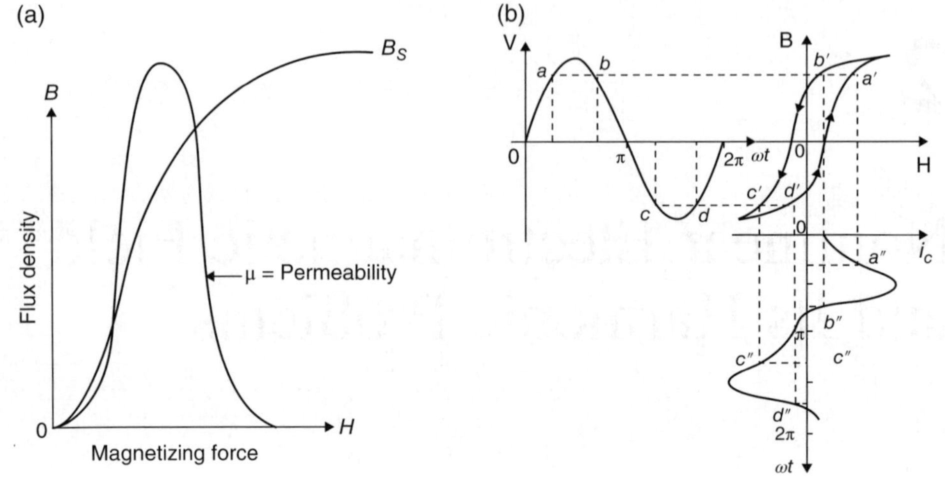

Figure 2.1 (a) Characteristics of magnetic impedance associated with a B-H curve and permeability. (b) Excitation current corresponding to a sinusoidal voltage excitation associated with a hysteresis B-H curve

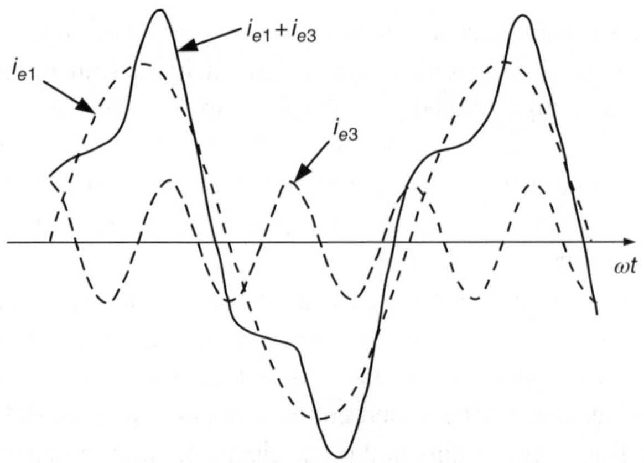

Figure 2.2 Excitation current corresponding to a sinusoidal voltage excitation

power panel or transformer vibrates and emits a buzzing sound for the different harmonic frequencies. Harmonic frequencies from the 3rd to the 25th are the most common range of frequencies measured in electrical distribution systems [2, 3].

Current distortion affects the power system and distribution equipment, and it may, directly or indirectly, cause the destruction of loads or loss of product. From a direct perspective, current distortion may cause transformers to overheat and fail, even though they are not fully loaded. Conductors and conduit systems can also overheat, leading to open circuits and down time.

Voltage distortion directly affects loads. Distorted voltage can cause motors to overheat and vibrate excessively. It can also cause damage to the motor shaft. Even nonlinear loads are prey to voltage distortion. Equipment ranging from computers to electronically ballasted fluorescent lights may be damaged by voltage distortion.

As the current distortion is conducted through the normal system wiring, it creates voltage distortion according to Ohm's Law. While current distortion travels only along the power path of the nonlinear load, voltage distortion affects all loads connected to that particular bus or phase. Each harmonic current in a facility's electrical distribution system will cause a voltage to exist at the same harmonic when the harmonic current flows into impedance, which results in voltage harmonics appearing at the load bus. For example, a 3rd harmonic current will produce a 3rd harmonic voltage, a 9th harmonic current will produce a 9th harmonic voltage, and so on. Another indirect problem introduced by current distortion is called resonance. Certain current harmonics may excite resonant frequencies in the system, and this resonance can cause extremely high harmonic voltages, possibly damaging sensitive electronic equipment.

2.1.2 Harmonics Problems Generated from Nonlinear Load and Power Electronics Devices

Electronic equipment (switching power supplies) draws current differently than nonelectronic equipment [4]. Instead of a load having a constant impedance drawing current in proportion to the sinusoidal voltage, electronic devices change their impedance by switching on and off near the peak of the voltage waveform. Switching loads on and off during part of the waveform results in short, abrupt, non-sinusoidal current pulses during a controlled portion of the incoming peak voltage waveform. These abrupt pulsating current pulses introduce unanticipated reflective currents (harmonics) back into the power distribution system. As mentioned earlier, these currents operate at frequencies other than the fundamental 50/60 Hz.

The harmonics problem in electrical circuits has been an issue since the establishment of the AC generators, where distorted voltage and current waveforms were observed in 20th century. For more than 100 years, harmonics have been reported to cause operational problems to power systems. From the above-mentioned harmonics in power systems, we find that harmonics are mainly generated from both the component level (nonlinear magnetic or dielectric components) and the circuit level (including power electronic circuits) [4, 5]. At the component level, the presence of harmonics in electromagnetic waves leads to a distorted signal for the E field or H field while, at the circuit level, the presence of harmonics in the voltage or current waveform leads to a distorted signal for voltage or current [6].

The harmonics generated from nonlinear electromagnetic fields (on nonlinear components) can cause voltage and current waveform distortion, due to the induced voltage and current carrying distorted waveforms and then propagating to circuits. Obviously, the source, or harmonics generator, is from the nonlinear component, and coupled with linear electric circuits. Figure 2.3 shows the nonlinear magnetic and nonlinear dielectric material, and the harmonics are generated from the B-H magnetic materials. The corresponding excitation current is non-sinusoidal, due to the nonlinear *B-H* relationship of the core, as shown in Figure 2.3(a). When only the fundamental component of the current is considered, however, the relationship between the phasors of voltage and current can be determined by a resistor (equivalent resistance of the core loss) in parallel with a lossless inductor (self-inductance of the coil), as illustrated in the diagram.

It is well known that matter can be classified by its electrical conductivity into conductors, semi-conductors, and insulators. Insulators, also called dielectrics, do not conduct electric current under the influence of an electric field. On the other hand, however, dielectrics can be polarized by an electric field. In a macroscopic sense, the same amount of positive and negative surface charge is induced on one surface side, and the opposite side perpendicular to the field, respectively. Microscopically, electrically charged particles constituting the materials (such as atomic nuclei, electrons, and ions) cannot be freely moved, but are displaced from their equilibrium position due to Coulombic force. This phenomenon is called dielectric polarization, as shown in Figure 2.3(b).

The amount of displacement is macroscopically characterized by so-called dielectric constants, and these are different in different materials. If the field is weak, the displacement is proportional to the intensity of the field. In this case, the dielectric constant can be considered as a proportional constant connecting the displacement and intensity of the field. However, if the field has a larger intensity, the situation can be different.

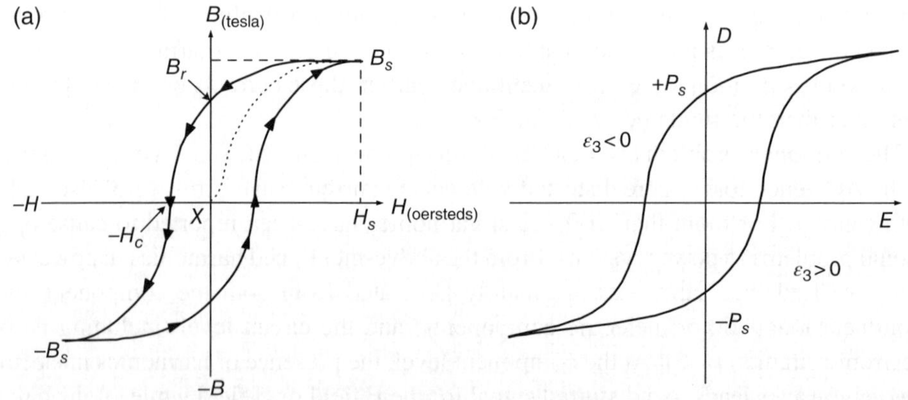

Figure 2.3 Nonlinear magnetic and nonlinear dielectric materials, (a) the B-H hysteresis loop of the magnetic material, and (b) the direction of the polarization (D-E) hysteresis loop of ferroelectric material

Actually, literalistic treatment of a dielectric constant as 'constant' is a kind of first-order or linear approximation of a nonlinear electric response. In reality, the dielectric constants are not constants, but vary depending on the external field. Table 2.1 presents the current distortion due to nonlinear load and power electronics devices, where the different waveform represents different load characteristic [1].

Some of the major effects include are listed below [1, 2]. These effects depend, of course, on the harmonic source, its location on the power system, and the network characteristics that promote propagation of harmonics.

Table 2.1 Current distortion due to nonlinear load and power electronics

Type of Load	Typical Waveform	Current Distortion	Weighting Factor
Single phase power supply	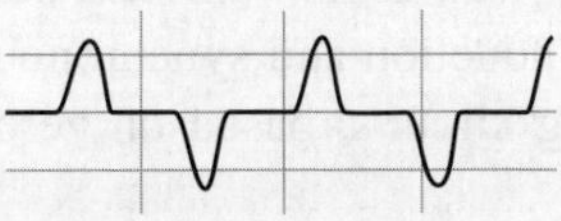	80% (high 3rd)	2.5
Semi-convertor	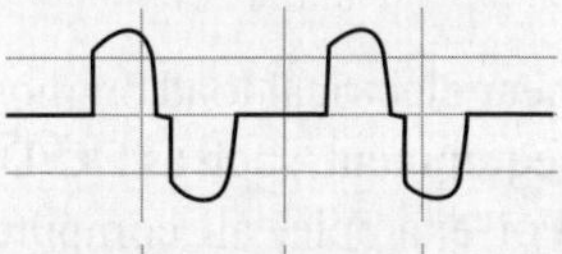	High 2nd, 3rd, 4th at partial load	2.5
Six pulse converter with capacitor smoothing, no inductance	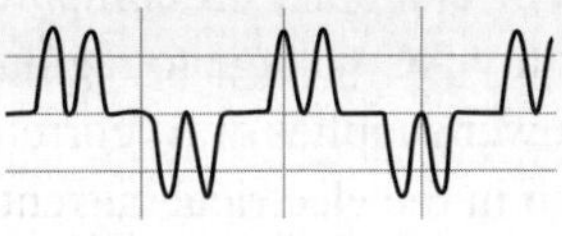	80%	2
Six pulse converter with capacitive smoothing, with series inductor >3%, or DC drive	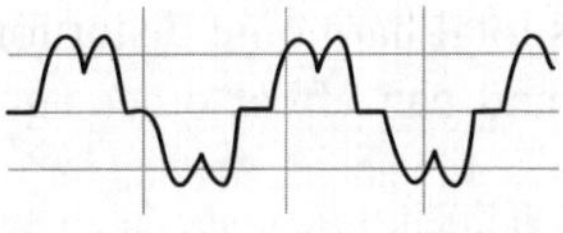	40%	1
Six pulse converter, with large inductor for current smoothing	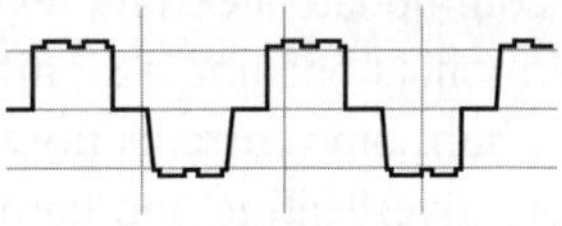	28%	0.8
12 pulse converter	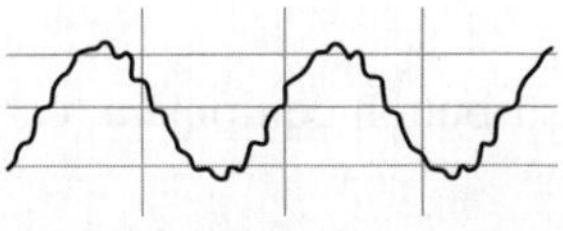	15%	0.5
AC voltage regulator	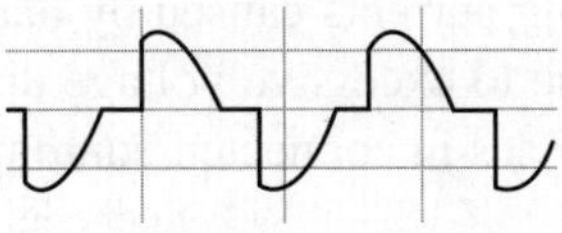	Varies with firing angle	0.7
Fluorescent lighting	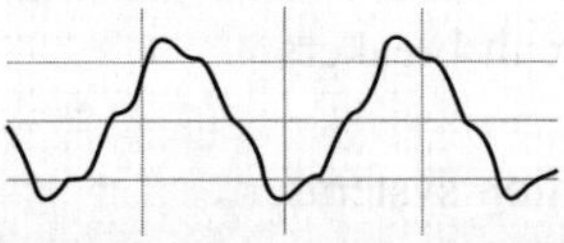	0.05	0.5

- Capacitor bank failure from dielectric breakdown or reactive power overload.
- Interference with ripple control and power line carrier systems, causing operational failure of systems which accomplish remote switching, load control, and metering.
- Excessive losses in – and heating of – induction and synchronous machines.
- Over-voltages and excessive currents on the system from resonance to harmonic voltages or currents on the network.
- Dielectric breakdown of insulated cables resulting from harmonic over-voltages on the system.
- Inductive interference with telecommunications systems.
- Errors in induction kWh meters.
- Signal interference and relay malfunction, particularly in solid-state and microprocessor-controlled systems.
- Interference with large motor controllers and power plant excitation systems (reported to cause motor problems, as well as non-uniform output).
- Mechanical oscillations of induction and synchronous machines.
- Unstable operation of firing circuits, based on zero voltage crossing detection or latching.

A large portion of the nonlinear electrical load on most electrical distribution systems comes from power electronics equipment, such as DC/DC converters or switching mode power supplies (SMPS) [7]. For example, all computer systems use switching mode power supplies that convert utility AC voltage to regulated low-voltage DC for internal electronics. These nonlinear power supplies draw current in high-amplitude short pulses that create significant distortion in the electrical current and voltage wave shape – *harmonic distortion*, measured as total harmonic distortion (THD). The distortion travels back into the power source, and can affect other equipment connected to the same source.

Most power systems can accommodate a certain level of harmonic current, but will experience problems when harmonics become a significant component of the overall load. As these higher-frequency harmonic currents flow through the power system, they can cause communication errors, overheating and hardware damage, such as:

- Overheating of electrical distribution equipment, cables, transformers, standby generators and so on.
- High voltages and circulating currents caused by harmonic resonance.
- Equipment malfunctions due to excessive voltage distortion.
- Increased internal energy losses in connected equipment, causing component failure and shortened lifespan.
- False tripping of branch circuit breakers.
- Metering errors.
- Fires in wiring and distribution systems.
- Generator failures.

- Crest factors and related problems.
- Lower system power factor, resulting in penalties on monthly utility bills.

2.1.3 Harmonics in the Time Domain and Frequency Domain

In electrical engineering, "time domain" is a term used to describe the analysis of electrical signals with respect to time. In the graph below, we have an example of an ideal undistorted alternating voltage or current signal. The values of the signal fluctuate between positive and negative amplitudes over time. As time progresses, the graphical representation clearly displays the variance in amplitude. One method of describing the non-sinusoidal waveform is called its Fourier Series. Jean Fourier was a French mathematician of the early 19th century who discovered a special characteristic of periodic waveforms.

2.1.3.1 A. Time Domain Model

In this model, it is assumed that the waveform under consideration consists of a fundamental frequency component and harmonic components with order of integral multiples of the fundamental frequency. It is also assumed that the frequency is known and constant during the estimation period. Consider a non-sinusoidal voltage given by a Fourier-type equation:

$$v(t) = \sum_{n=0}^{N} V_n \sin(n\omega_0 t + \phi_n) \tag{2-1}$$

where *v(t)* is the instantaneous voltage at time *t*, *n* is the order of the harmonic, V_n is the voltage amplitude of harmonic *n*, and *n* is the phase angle of harmonic *n*, 0 is the fundamental frequency, and *N* is the total number of harmonics.

The equation (2-1) can be rewritten as a summation of trigonometric function:

$$v(t) = \sum_{n=0}^{N} (V_{cn} \sin n\omega_0 t + V_{sn} \cos n\omega_0 t) \tag{2-2}$$

where

$$V_{cn} = V_n \cos\phi_n \tag{2-3}$$

and

$$V_{sn} = V_n \sin\phi_n \tag{2-4}$$

With the DC component case:

$$v(t) = V_0 + \sum_{n=1}^{N} V_n \sin(n\omega_0 t + \phi_n) \tag{2-5}$$

then equation (2-5) can be written as:

$$v(t) = V_0 + \sum_{n=0}^{N} (V_{cn} \sin n\omega_0 t + V_{sn} \cos n\omega_0 t) \tag{2-6}$$

Figure 2.4 shows sinusoidal waveform (a) and non-sinusoidal waveform (b) in the time domain. The distorted waveform includes a number of harmonics components. Periodic waveforms are those waveforms comprised of identical values that repeat in the same time interval, like those shown above. Fourier discovered that periodic waveforms can be represented by a series of sinusoids summed together. The frequency of these sinusoids is an integer multiple of the frequency represented by the fundamental periodic waveform. The waveform on the above, for example, is described entirely by one sinusoid – the fundamental – since it contains no harmonic distortion.

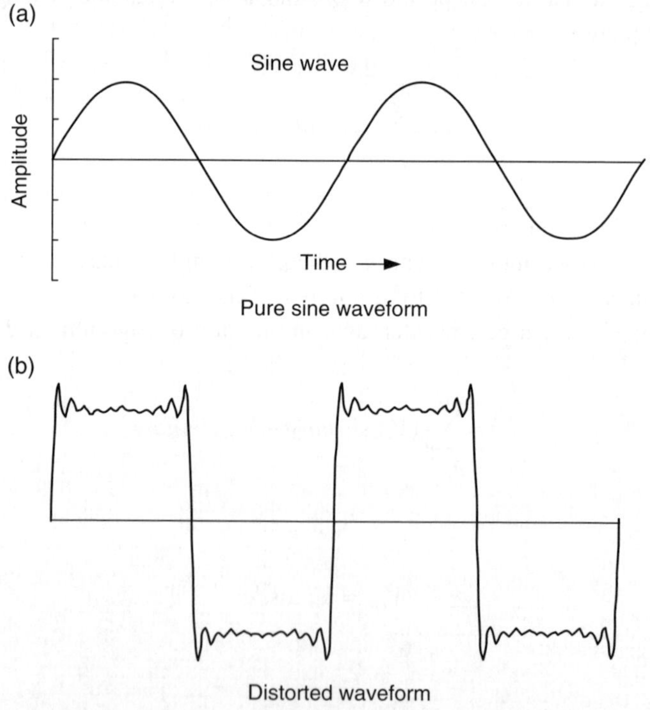

Figure 2.4 Harmonics in time domain presentation

2.1.3.2 B. Frequency Domain Model

All periodic waves can be generated with sine waves of various frequencies. The Fourier theorem breaks down a periodic wave into its component frequencies. Therefore, the frequency domain (FD) presentation allows us to see the amplitudes of the frequencies that make up a signal. Frequency domain presentation can be created by using Fourier series transforms, as shown in Figure 2.5.

The waveform can be defined as a summation of trigonometric function, and the complex Fourier series with k harmonics can be expressed as follows:

Trigonometry form:

$$f(t) = a_0 + \sum_{k=1}^{\infty} \{a_k \cos(k\omega_0 t) + b_k \sin(k\omega_0 t)\} \tag{2-7}$$

where:

$$a_0 = \frac{1}{T}\int_0^T f(t)dt \tag{2-8}$$

$$a_k = \frac{2}{T}\int_0^T f(t)\cdot\sin(n\omega t)dt \tag{2-9a}$$

$$b_k = \frac{2}{T}\int_0^T f(t)\cdot\cos(n\omega t)dt \tag{2-9b}$$

Exponential form:

$$f(t) = \sum_{k=-\infty}^{\infty} F_k e^{jk\omega_0 t} \tag{2-10}$$

where:

$$F_k = \frac{1}{T}\int_0^T f(t)\cdot e^{-j\omega_0 t}dt \tag{2-11}$$

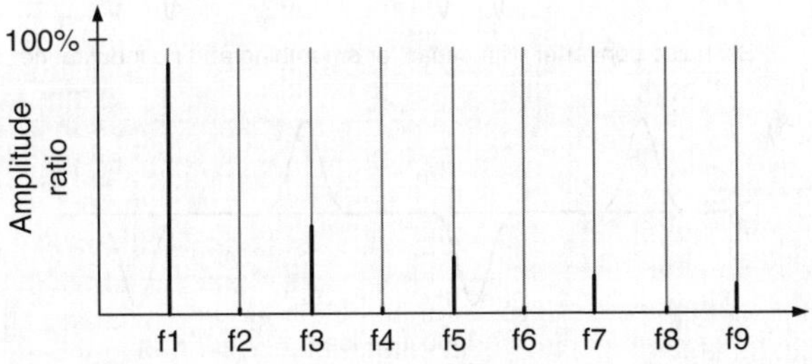

Figure 2.5 Frequency domain graphs – frequency spectrums

2.1.4 *Examples of Harmonic Producing Loads*

Harmonics are primarily caused by loads that draw current repetitively but in a non-sinusoidal manner. Harmonic loads include: ballast/fluorescent lighting and computer power supplies; uninterruptable power supplies (UPS); variable speed drives; charging circuits incorporating rectifiers; arc welders and three-phase machines. Examples of load current waveforms with harmonics are illustrated in Figure 2.6.

2.1.5 *Harmonics in DC/DC Converter of Isolation Transformer*

2.1.5.1 A. Isolation Transformer and Excitations

In switching mode power supplies, the isolation transformers are often used in DC-DC converter topologies like zero voltage or zero current switching resonant converters and push-pull current source converters. The HBFEM numerical modeling and computation results can provide more detailed information on harmonic distribution and power

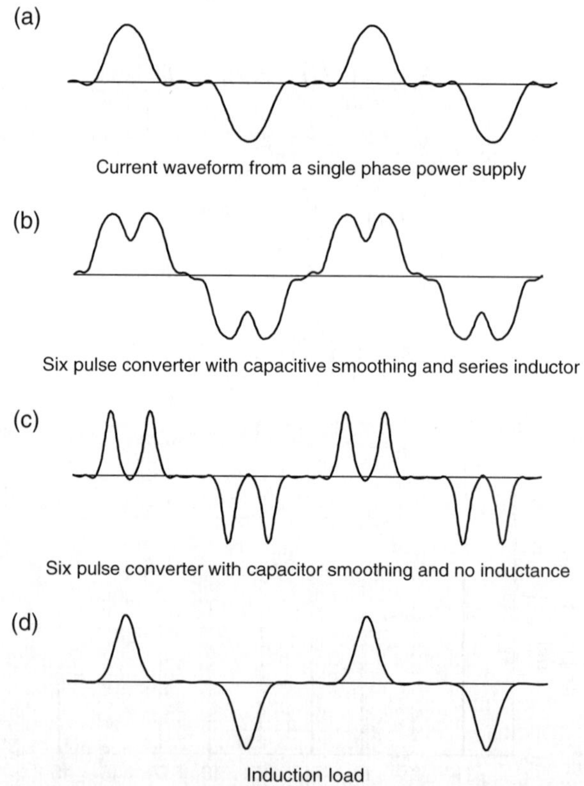

Figure 2.6 Examples of load current waveforms with harmonics

losses in transformer, and magnetic flux and eddy current distributions in core and windings respectively [8]. To understand harmonic distribution in the transformer the excitation voltage and magnetic flux waveforms and associated current waveforms should be investigated.

Sinusoidal Excitation:

Since v_1 (instantaneous value of the applied voltage) is sinusoidally varying, the flux must also be sinusoidal in nature varying with frequency *f*. Let:

$$\phi_m = \phi_{mp} \sin \omega t \tag{2-12}$$

where $\phi_{\rm mp}$ is the peak value of mutual flux. From Faraday's law, the voltage induced in the N-turn coil is:

$$e_1 = N_1 \frac{d\phi}{dt} = N_1 \omega \phi_{mp} \cos\omega t = E_{\max} \cos \omega t \tag{2-13}$$

The r.m.s. value of the induced voltage, E_1, is obtained by dividing the peak value in equation (2-13) by $\sqrt{2}$:

$$E_1 = \frac{E_{\max}}{\sqrt{2}} = \frac{N_1 \omega \phi_{mp}}{\sqrt{2}} 4.44 \phi_{mp} f N_1 \tag{2-14}$$

Square Wave Excitation:

With a square waveform excitation voltage, from Faraday's law, the voltage induced in the N-turn coil is:

$$v = e_1 = N_1 \frac{d\phi}{dt} \tag{2-15}$$

Time-dependent magnetic flux (triangular waveform) in the magnetic steady-state case can be expressed as:

$$\phi_m = \frac{1}{N_1} \int_0^t v dt = \frac{1}{N_1} Et = \frac{1}{2} \left(\frac{1}{N_1} \times E \frac{T}{2} \right) = \frac{ET}{4N_1} \tag{2-16}$$

Substituting $f = 1/T$, the magnetic flux is obtained as:

$$\phi_m = \frac{E}{4N_1 f} \tag{2-17}$$

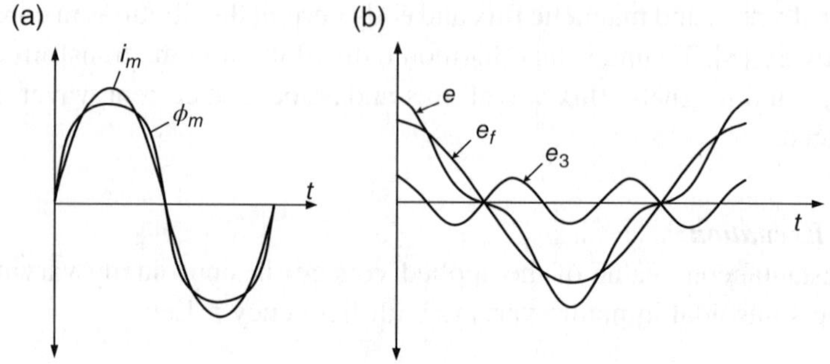

Figure 2.7 (a) Waveforms of flux; (b) voltage for sinusoidal magnetizing current in nonlinear magnetics

and the induced voltage, E, can be obtained as:

$$E = 4N_1 f \phi_m \tag{2-18}$$

Excitation Characteristics:

Figure 2.7 shows waveforms of flux and voltage for sinusoidal magnetizing current in nonlinear magnetics, where the distorted waveforms of flux and voltage contain component-only harmonics, as expressed in equations (2-19) and (2-20).

$$\phi_m = \phi_{mp_1}\sin\omega t + \phi_{mp_3}\sin 3\omega t + \phi_{mp_5}\sin 5\omega t + \phi_{mp_7}\sin 7\omega t + \cdots \tag{2-19}$$

and

$$e = \frac{d\phi}{dt} = \omega \begin{bmatrix} \phi_{mp1}\cos\omega t + 3\phi_{mp3}\cos 3\omega t + 5\phi_{mp5}\cos 5\omega t \\ + 7\phi_{mp7}\cos 7\omega t + \cdots \end{bmatrix} \tag{2-20}$$

Waveforms of flux and voltage for square voltage wave excitation in nonlinear magnetics, time-dependent magnetic flux (triangular waveform) in the magnetic steady state case will be illustrated as:

In the majority applications of switching mode power supplies, it is desirable to incorporate a transformer into the switching mode DC/DC converter, to obtain DC isolation between the converter input and output. In off-line power supply applications, isolation is usually required by regulatory agencies. This isolation could be obtained by simply connecting a 50 Hz or 60 Hz transformer at the power supply AC input terminals. However, since transformer size and weight vary inversely with frequency, incorporation of the transformer into the converter can make significant improvements; the transformer then operates at the converter switching frequency of tens or hundreds of kilohertz.

The size of modern ferrite power transformers is minimized at operating frequencies ranging from several hundred kilohertz to roughly one megahertz. These high frequencies lead to dramatic reductions in transformer size [9]. When a large step-up or step-down conversion ratio is required, the use of a transformer can allow better converter optimization. By proper choice of the transformer turns ratio, the voltage or current stresses imposed on the transistors and diodes can be minimized, leading to improved power efficiency and lower cost.

There are several ways of incorporating transformer isolation into any DC-DC converter, such as the full bridge, half-bridge, forward and, push-pull converters. Zero-voltage or current-switched resonant and LLC resonant converters are commonly used isolated versions of DC/DC converters. The flyback converter is an isolated version of the buck-boost converter. Isolated variants of the Cuk converter are also known. The full-bridge, forward, flyback and LLC converters, with isolated transformer, are briefly described in the following section.

2.1.5.2 B. Full-Bridge Buck-Derived Converter

The full-bridge transformer-isolated buck converter is sketched in Figure 2.9. Typical waveforms are illustrated in Figure 2.8. The transformer primary winding is driven symmetrically such that the net volt-seconds applied over two switching periods is equal to zero. During the first switching period, transistors Q_1 and Q_4 conduct.

2.1.5.3 C. Forward Converter

The forward converter is illustrated in Figure 2.10. This transformer-isolated converter is also based on the buck converter. It requires a single transistor, and therefore finds application at power levels lower than those encountered in the full bridge circuit.

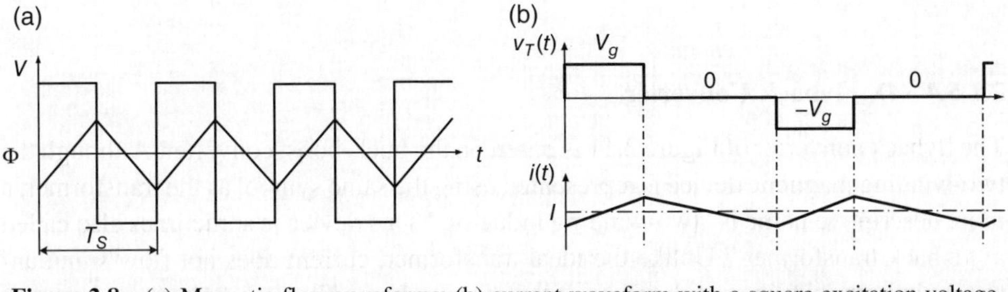

Figure 2.8 (a) Magnetic flux waveforms; (b) current waveform with a square excitation voltage

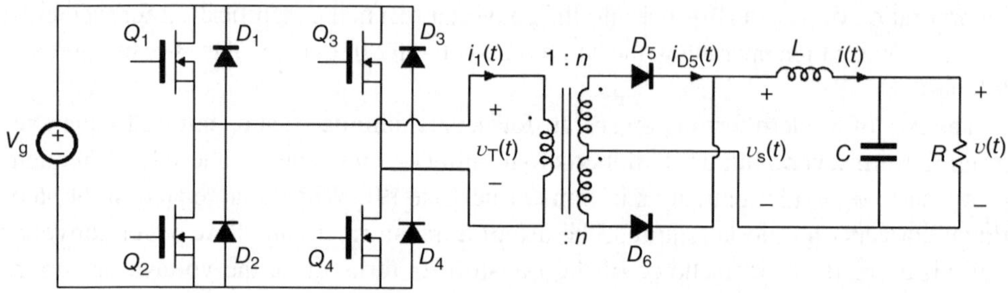

Figure 2.9 The full bridge transformer-isolated buck converter

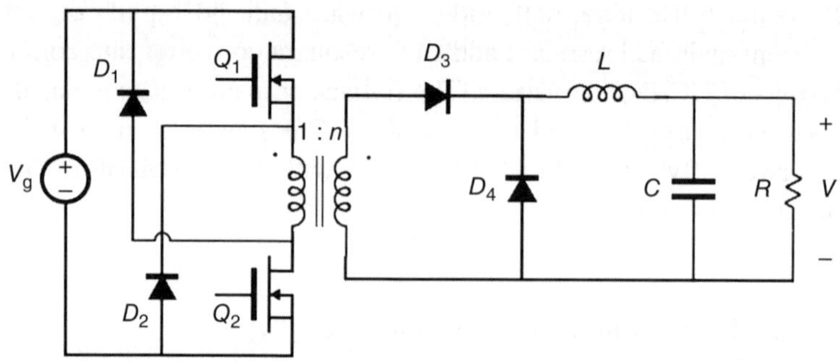

Figure 2.10 A two-transistor version of the forward converter

The maximum transistor duty cycle is limited in value; for the common choice $n1 = n2$, the duty cycle is limited to the range $D < 0.5$. The transformer is reset while transistor Q_1 is in the off state. While the transistor conducts, the input voltage V_g is applied across the transformer primary winding. This causes the transformer magnetizing current to increase. When transistor Q_1 turns off, the transformer magnetizing current forward biases diode D_1 and, hence, voltage $-V_g$ is applied to the second winding. This negative voltage causes the magnetizing current to decrease. When the magnetizing current reaches zero, diode D_1 turns off. Volt-second balance is maintained on the transformer windings, provided that the magnetizing current reaches zero before the end of the switching period.

2.1.5.4 D. Flyback Converter

The flyback converter of Figure 2.11 is based on the buck-boost converter. Although the two-winding magnetic device is represented using the same symbol as the transformer, a more descriptive name is "two-winding inductor." This device is sometimes also called a "flyback transformer." Unlike the ideal transformer, current does not flow simultaneously in both windings of the flyback transformer but, rather, the flyback transformer

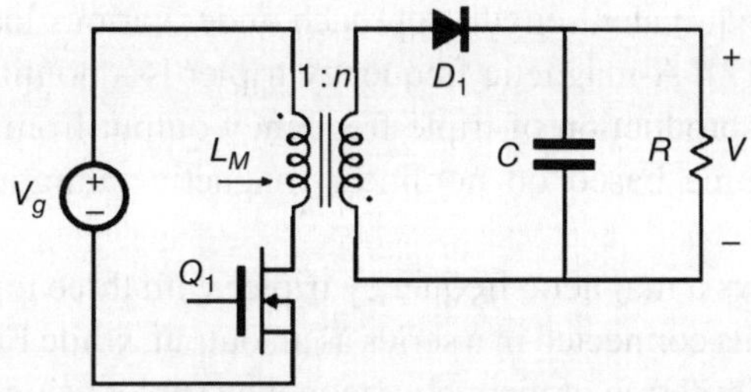

Figure 2.11 The flyback converter, a single-transistor isolated buck-boost converter

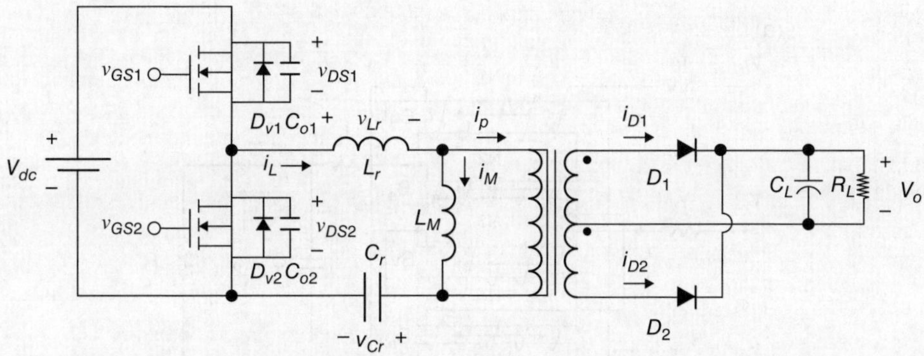

Figure 2.12 A half-bridge LLC converter

magnetizing inductance assumes the role of the inductor of the buck-boost converter. The magnetizing current is switched between the primary and secondary windings.

2.1.5.5 E. LLC Resonant Converter

A typical LLC converter is shown in Figure 2.12, where a load is attached; L_r and C_r are the main components affect the resonant frequency.

When the converter is at resonance state, the power of L_r and C_r cancels each other, while the voltage of L_m depends on the load voltage. If the load is not attached, the L_m will be involved in the resonance. The i_M always exists in an LLC transformer, and it requires a small L_m compared to the normal high-frequency transformers.

2.1.6 Magnetic Tripler

Magnetic frequency triplers have been used extensively for certain applications, and the design of these devices was, until the early 1990s, largely empirical. However, most published papers [10] only discussed the analyses of magnetic frequency-tripling devices, based on an equivalent-circuit approach under various load conditions and the Preisach model [11, 12]. A magnetic frequency tripler is a nonlinear magnetic system which is used for the production of triple-frequency output from a three-phase fundamental frequency source based on nonlinear magnetic saturation characteristics, as shown in Figure 2.13.

Figure 2.13(a) shows a magnetic frequency tripler with three input magnetizing coils and two secondary coils connected in a series as an output, while Figure 2.13(b) presents another winding configuration of magnetic frequency tripler using three secondary coils connected in a series. Since the magnetic frequency tripler usually works in the magnetic saturating state, the harmonic components will be generated in the magnetic core.

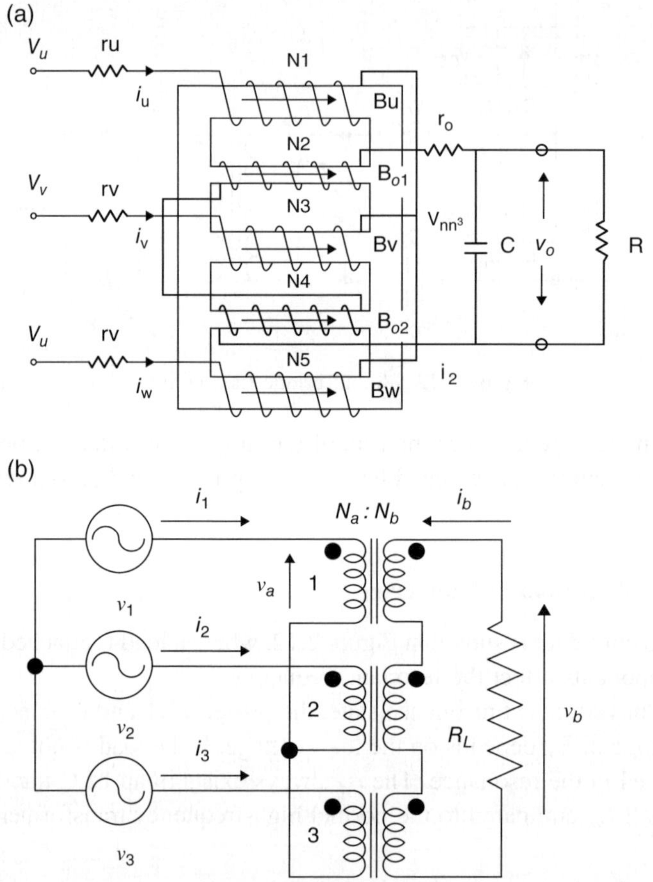

Figure 2.13 Magnetic triplers with three-phase input at 50Hz and single phase output at 150Hz. (a) magnetic frequency tripler with three input magnetizing coils and two secondary coils connected in a series as an output. (b) magnetic frequency tripler using three secondary coils connected in a series

2.1.7 Harmonics in Multi-Pulse Rectifier Transformer

Multi-pulse rectifier transformers have been used in various HVDC power supplies, including electrification systems, telecommunication base station UPS, electric power substation UPS, supercomputing or data centre UPS, and so forth [13, 14, 15].

2.1.7.1 6-Pulse Rectifier Transformer

Conventional three-phase rectifiers cause harmonics pollution at input current side, and therefore bring some hazard to AC electric network and the other electronic equipment, as shown in Figure 2.14. A three phase rectifier transformer (6-pulse transformer) with Y connection at secondary side of transformer is shown in Figure 2.14(a), and waveforms of 6-pulse output voltage for three phase rectifier transformer with delta connection at the secondary side is shown in Figure 2.14(b).

The current in phase A can be expressed as below, where the existing harmonics are 5, 7, 11, 13, … Waveforms of input voltage and current for the 6-pulse transformer are presented in Figure 2.15. This 6-pulse rectifier transformer can reduce the total harmonics distortion (THD) of the input current.

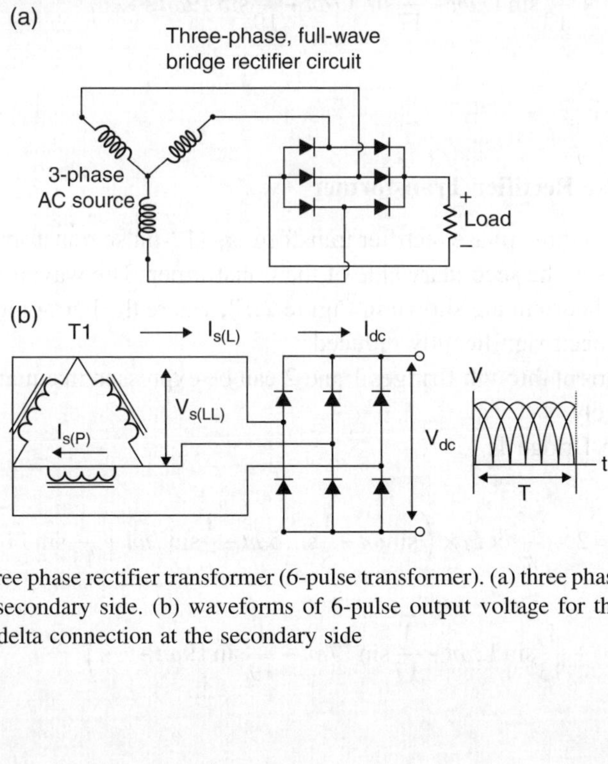

Figure 2.14 Three phase rectifier transformer (6-pulse transformer). (a) three phase transformer with Y connection at secondary side. (b) waveforms of 6-pulse output voltage for three phase rectifier transformer with delta connection at the secondary side

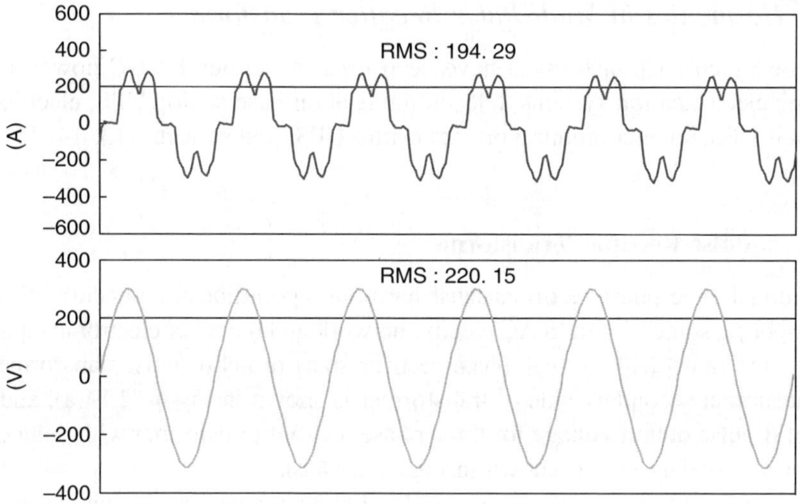

Figure 2.15 Waveforms of 6-pulse input voltage and current

$$i_A = 2\times\frac{\sqrt{3}}{\pi}\times I_d\times\left(\sin\omega t-\frac{1}{5}\sin 5\omega t-\frac{1}{7}\sin 7\omega t+\frac{1}{11}\sin 11\omega t\right.$$
$$\left.+\frac{1}{13}\sin 13\omega t-\frac{1}{17}\sin 17\omega t-\frac{1}{19}\sin 19\omega t+\cdots\right) \quad (2\text{-}21)$$

2.1.7.2 12-Pulse Rectifier Transformer

Figure 2.16 shows three phase rectifier transformers (12-pulse transformer) with Y and delta connections at the secondary side of the transformer. The waveforms of 12-pulse input voltage and current are shown in Figure 2.17, where the harmonics in the current waveform have been significantly reduced.

The output current through Bridges 1 and 2 can be expressed in equations (2-22) and (2-23), respectively.

The current in Bridge 1:

$$i_{IA} = 2\times\frac{\sqrt{3}}{\pi}\times I_d\times\left(\sin\omega t-\frac{1}{5}\sin 5\omega t-\frac{1}{7}\sin 7\omega t+\frac{1}{11}\sin 11\omega t\right.$$
$$\left.+\frac{1}{13}\sin 13\omega t-\frac{1}{17}\sin 17\omega t-\frac{1}{19}\sin 19\omega t+\cdots\right) \quad (2\text{-}22)$$

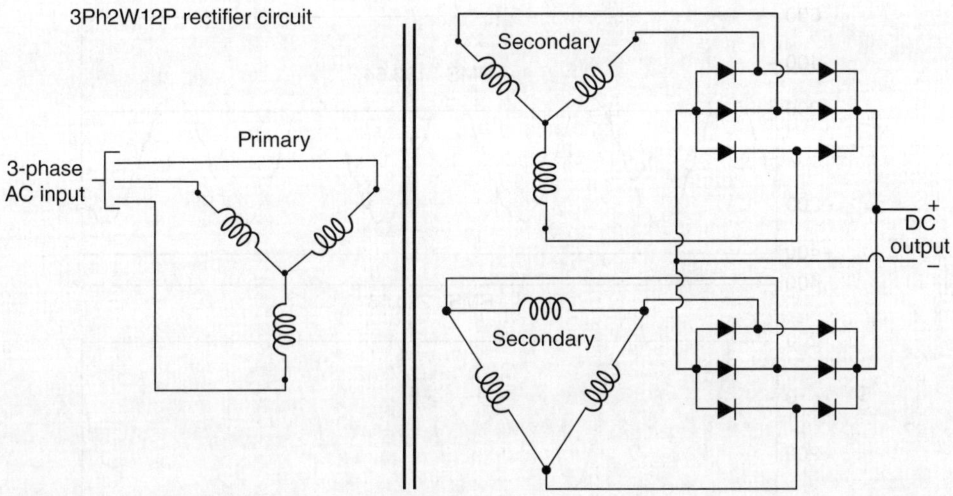

Figure 2.16 Three phase with 12-pulse rectifier transformer with Y and delta connections at the secondary side

The current (30° forward) in Bridge 2:

$$i_{IIA} = 2\times\frac{\sqrt{3}}{\pi}\times I_d\times\left(\sin\omega t+\frac{1}{5}\sin 5\omega t+\frac{1}{7}\sin 7\omega t+\frac{1}{11}\sin 11\omega t\right.$$
$$\left.+\frac{1}{13}\sin 13\omega t+\frac{1}{17}\sin 17\omega t+\frac{1}{19}\sin 19\omega t+\cdots\right) \tag{2-23}$$

Combining the current in Bridge 1 and Bridge 2, the following output current can be obtained as:

$$i_A = i_{IA} + i_{IIA} = 4\times\frac{\sqrt{3}}{\pi}\times I_d\times\left(\sin\omega t+\frac{1}{11}\sin 11\omega t+\frac{1}{13}\sin 13\omega t\cdots\right) \tag{2-24}$$

Harmonics 5, 7, 17, 19,… generated from Bridges 1 and 2 will cancel each other. Only harmonics 11, 13, 23, 25 remain.

A comparison of harmonics generated in 6-pulse and 12-pulse rectifier transformers is presented in Table 2.2, where the 12-pulse rectifier transformer has a significant harmonics reduction.

Obviously, the harmonics of the input current and flux, the effect and the principle of interphase reactor in rectifying circuit can be analyzed by HBFEM. The detailed HBFEM computation model and simulation results will be introduced in Chapters 4 and 6.

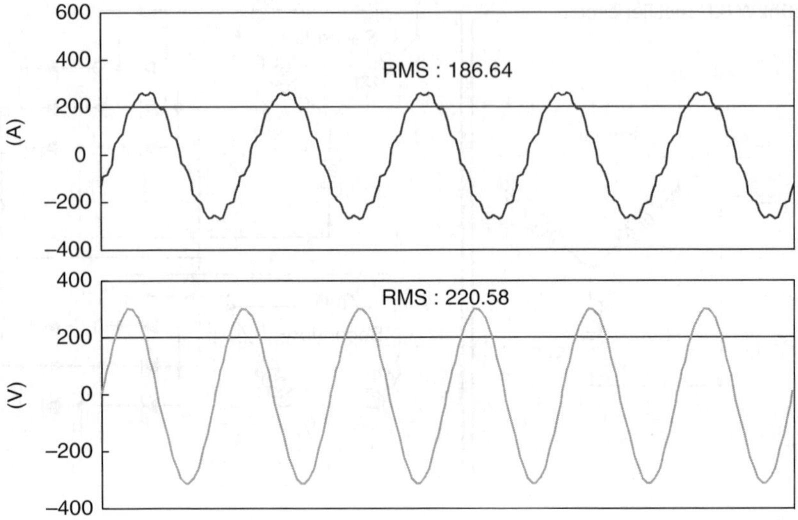

Figure 2.17 Waveforms of input voltage and current for 12-pulse rectifier transformer

Table 2.2 Harmonic components in 6-pulse and 12-pulse rectifier transformers

Harmonics	5th	7th	11th	13th	17th	19th	23rd
6-pulse rectifier transformer	20%	14%	9%	8%	6%	5%	4%
12-pulse rectifier transformer	0	0	9%	8%	0	0	4%

2.2 DC-Biased Transformer in High-Voltage DC Power Transmission System

2.2.1 Investigation and Suppression of DC Bias Phenomenon

The DC bias is that DC component appears in transformer-exciting currents, which is an abnormal running state. There are mainly four causes leading to DC bias:

(1) The operation of high-voltage direct current transmission systems (HVDC).
(2) The solar storm and the consequent geomagnetically induced current (GIC).
(3) AC/DC corridor-sharing high voltage power transmission systems.
(4) The unbalanced triggering angle of the converter valve.

(1), (3) and (4) are all related to the HVDC.

In order to increase power transmission capacity, provide stability for long-distance power delivery, and decrease transmission line losses, HVDC are utilized in China. Several ± 500 kV DC transmission lines have been built from west to east, each of them almost 1000 km long. However, the use of the HVDC also creates problems of its own. For example, in a monopolar HVDC system, the earth acts as the return path of the DC current, and a large direct current will flow in the earth, which will bring great potential differences in a large area. Thus, direct currents will flow through the transformers in the AC substations if their neutral points are grounded, and the transformers may be under DC bias.

As shown in Figure 2.18, in the normal state, the transformer works in a linear region of the magnetizing curve and the magnetic flux is sinusoidal. When the direct current flows in the windings, the transformer works in an abnormal state, where the direct current generates DC flux, which raises the waveform of the magnetic flux. Consequently, the transformer core is saturated significantly, and the exciting current produces the amount of harmonics under DC bias, due to the nonlinearity of the transformer core. Furthermore, the DC bias may lead to acute vibration, great noise and local overheating of the transformer, and may even make the protection fail. The generated harmonics in the exciting current will increase the reactive power loss of the power grid.

Since the current of the DC grounding electrode is very large, it is difficult to prevent a transformer located in the substations near the grounding electrode being affected. Therefore, it is vital to choose the appropriate grounding site, considering the current distribution in the soil. Some practical and effective methods are required to eliminate or suppress the DC bias phenomenon.

The direct current can be eliminated by connecting a condenser to the neutral point of the transformer, as shown in Figure 2.19. The characteristic of obstructing the direct current is utilized to cut off the invasive passage of bias current. Such obstructing equipment were applied for a set of 240 MVA transformers in 1996 in China.

The neutral point of transformer and the ground can also be linked by the linear or nonlinear resistance. The direct current flowing through the neutral point can be limited in a specific range by adjusting the magnitude of the resistance equipment. Although the

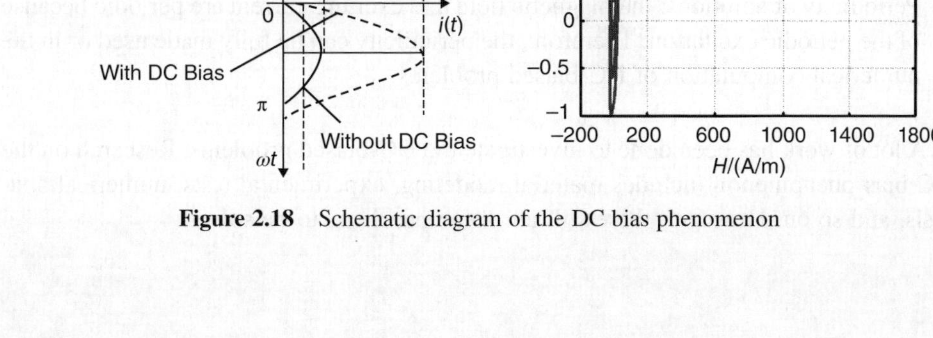

Figure 2.18 Schematic diagram of the DC bias phenomenon

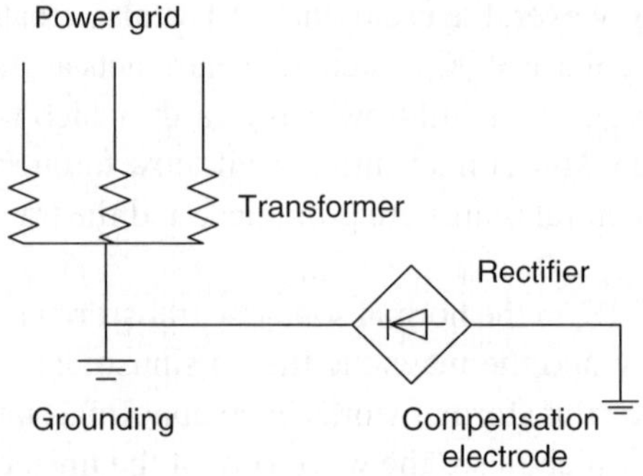

Figure 2.19 Inpouring reverse current for compensation

direct current is not eliminated thoroughly, the transformer will operate normally with slightly biased current.

The above two methods are effective in suppressing the influences of DC bias phenomena. However, possible side-effects as a result of the installed condenser or resistance should be considered, in order to protect the power grid from new hazards.

Another alternative method is to compensate the direct current by an additional power source. The device injects reverse current with the same strength as the biased current to the system. This method is comparably reliable and safe, and has been adopted in a substation in Jiangsu province, China.

2.2.2 Characteristics of DC Bias Phenomenon and Problems to be Solved

There are three main characteristics in the DC bias phenomenon:

1. Harmonic problems: due to invasion of the direct current, there are a large number of harmonics in exciting currents, including the DC component, odd and even harmonics. These harmonics may seriously affect the operation of power transformers.
2. Electromagnetic coupling: the winding of the power transformer is connected to the voltage source, while the direct current flows into the windings through the neutral point. In the numerical analysis of DC-biased problems in power transformers, the magnetic field coupled with electric circuits should be computed efficiently and accurately.
3. Periodicity of solutions: the magnetic field and exciting current are periodic because of the periodic excitation. Therefore, the periodicity can be fully made used of in the numerical computation of DC-biased problems.

A lot of work has been done to investigate the DC-biased problems. Research on the DC bias phenomenon includes material modeling, experimental tests, numerical analysis, and so on. However, there still are some problems to be solved:

1. Material property exhibits different characteristics under different excitations. Therefore, the magnetizing curve under DC-biased excitation will be different from that in sinusoidal excitation. It is obvious that the use of different magnetizing curves will affect the computed magnetic field, which should be investigated.
2. The AC and DC components of flux density in the magnetic core may vary nonlinearly with the direct current in the winding. A further investigation can help us to understand the DC bias phenomenon well, and to protect the power transformer from damage by DC-biased excitation.
3. Many periods are necessary to attain the steady state solution when the stepping method is used to solve eddy current problems. The time periodic method can also be used the compute the eddy current problem, making use of the symmetry of time domain solutions. However, a large amount of memory is required in the numerical computation. Therefore, the harmonic-balanced finite-element method (HBFEM) can be an optimal choice for periodic eddy problems.
4. The traditional finite element method in the frequency domain can be used when the problem is linear and the excitation is sinusoidal. However, it will not be effective for solving DC-biased problems in which the magnetic core is saturated significantly.
5. With the advent of power-electronic technology, the excitation conditions in transformers, motors, and so forth could be very atypical. An iron core under DC-biased magnetization will experience a distorted and asymmetrical hysteresis loop. Iron losses under DC-biased magnetization is also increased, compared with those under sinusoidal excitations. In such a case, the numerical computation should be carried out based on new magnetizing characteristics of the magnetic core.

The problems mentioned above will be investigated and can be solved by using HBFEM, which will be discussed in detail in Chapter 5 in this book. It is known that the power transformer is susceptible to DC bias in the operation of a high-voltage direct current (HVDC) transmission system. In-depth analysis of the mechanism of a DC bias phenomenon contributes to manufacture and maintenance of the transformer. The harmonic balance finite element method (HBFEM) is effective to solve the nonlinear time-periodic electromagnetic field. It is very important in aspects of both theory and application to study the HBFEM basic theory, computation method, and DC bias problem of a power transformer in frequency.

2.3 Geomagnetic Disturbance and Geomagnetic Induced Currents (GIC)

Solar activity includes coronal mass ejections, where plasma and magnetic fields erupt from the corona of the sun. The shock wave that precedes the coronal mass ejections also accelerates solar energetic particles, which are high-energy particles consisting of electrons and ions – mostly protons. The stream of solar energetic particles is called the solar wind. During low intensity, it is mostly deflected by the geomagnetic field of the Earth, but some solar energetic particles cause auroras in the polar regions.

In the case of a sufficiently large coronal mass ejection, however, the solar energetic particles travel faster and modify the magnetosphere, and they cause currents called electrojects in the ionosphere. This mostly occurs at the auroral circles, but may also happen at lower latitudes. The magnetic field associated with the electrojets perturbs the otherwise constant geomagnetic field, thus creating a geomagnetic disturbance (GMD) [17]. A similar overview was recently published by the IEEE power and energy society technical council task force on geomagnetic disturbances [18]. Olof Samuelsson's report and overview has provided more detailed information about GMD [19].

2.3.1 Geomagnetically Induced Currents in Power Systems

During geomagnetic disturbances, variations in the geomagnetic field induce quasi-DC voltages in the network, which drive geomagnetically-induced currents (GIC) along transmission lines and through transformer windings to the ground (wherever there is a path for them to flow). GIC are considered quasi-DC relative to the power system frequency, because of their low frequency (0.0001 Hz to 1 Hz); thus, from a power system modeling perspective, GIC can be considered as DC or quasi-DC. The flow of these quasi-DC currents in transformer windings causes half-cycle saturation of transformer cores, which leads to increased transformer hotspot heating, harmonic generation, and reactive power absorption – each of which can affect system reliability. As part of the assessment of the impacts of geomagnetic disturbances (GMDs) on the bulk-power system, it is necessary to model the GIC produced by different levels of geomagnetic activity.

The disturbance causes the Earth's magnetic field to change violently. A non-uniform geomagnetic storm profile, characterized by the rate of change of the magnetic field density vector, B, in units of nT/min, serves as an input to the model. The user has the ability to define the magnitude and direction of the storm. The magnetic field density data used as an input is data that has been recorded terrestrially by magnetometers placed in substations. Not only will the model produce the induced geoelectric field, it will also calculate the accumulated voltage between user-defined geographic coordinates, which are usually the start and end points of transmission lines. Therefore, the model approximates the interaction between the Sun and the Earth's magnetic fields by use of Faraday's law, as presented by Equation (2-25) [20].

$$E_{ge} = \int \frac{\partial B}{\partial t} \cdot dA = \frac{\partial \phi_{gm}}{\partial t} \tag{2-25}$$

The voltage induced in the transmission line can be expressed as:

$$V_{dc} = \oint \vec{E} \cdot d\vec{\ell} \tag{2-26}$$

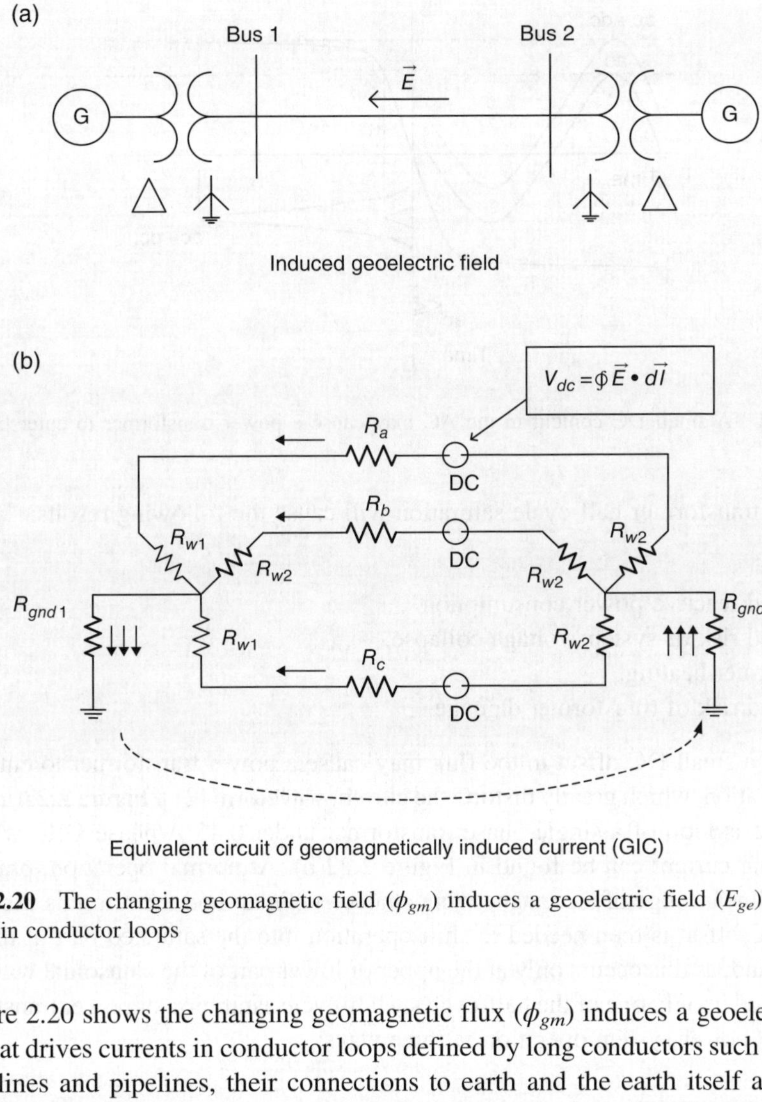

Figure 2.20 The changing geomagnetic field (ϕ_{gm}) induces a geoelectric field (E_{ge}) that drives currents in conductor loops

Figure 2.20 shows the changing geomagnetic flux (ϕ_{gm}) induces a geoelectric field (E_{ge}) that drives currents in conductor loops defined by long conductors such as electric power lines and pipelines, their connections to earth and the earth itself acting as a (return) conductor.

The current is referred to as geomagnetically induced current (GIC). Since the geomagnetic field variations are slow, a GIC is usually approximated by a DC current or quasi-DC. When modeling GIC in power systems, it is generally agreed that it is most correct to include DC voltage sources in the phase conductors of the power lines. Voltages and currents in electric power systems across the world are AC (alternating current) quantities, with a sinusoidal waveform and a frequency of 50 or 60 Hz. The levels of DC (direct current) current coming as GIC are very small compared to the amplitude of the AC currents, but may still have serious consequences. The reason is that even a small DC content in the AC may cause a power transformer to enter half-cycle saturation(see Figure 2.21).

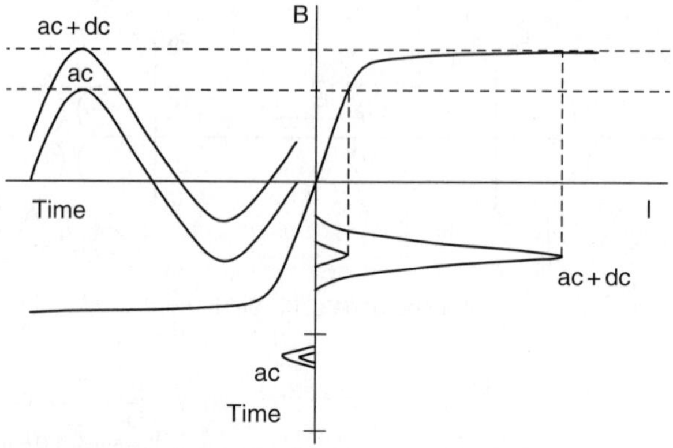

Figure 2.21 A small DC content in the AC may cause a power transformer to enter half-cycle saturation

In fact, transformer half-cycle saturation will cause the following results:

- Harmonics.
- Increased reactive power consumption.
- Increased risk of system voltage collapse.
- Transformer heating.
- Increased risk of transformer damage.

Adding a small DC offset in the flux may cause a power transformer to enter half-cycle saturation, which greatly distorts the current waveform [21]. Figure 2.22(a) shows the hysteresis loop of a single-phase transformer under 0.45 A/phase GIC, while the magnetizing current can be found in Figure 2.22(b). At normal operation, practically the entire linear range of the magnetizing characteristic around the origin is used. Only a small DC offset is then needed to shift operation into the saturated part of the characteristic and, as this occurs only at the upper or lower part of the sinusoidal waveform, it is distorted in only one of the half-cycles. Half-cycle saturation of power transformers may affect power system operation in three ways:

- Increased reactive losses stress the system;
- Waveform distortion causes relay disoperation;
- Heating may damage the transformer.

The peak value of the distorted sinusoidal waveform changes rapidly with GIC. Protective relays disconnect equipment to avoid overcurrent or overvoltage that may be damaging through overheating and arcing due to short-circuit faults. If such relays react on the peak value of the distorted waveform when peak value of the fundamental is more relevant, GIC will cause unwanted disconnection (tripping) of equipment. Loss of a line, a transformer or a reactive device such as an SVC (Static Var Compensator) typically weakens the system and increases the stress on other lines and transformers.

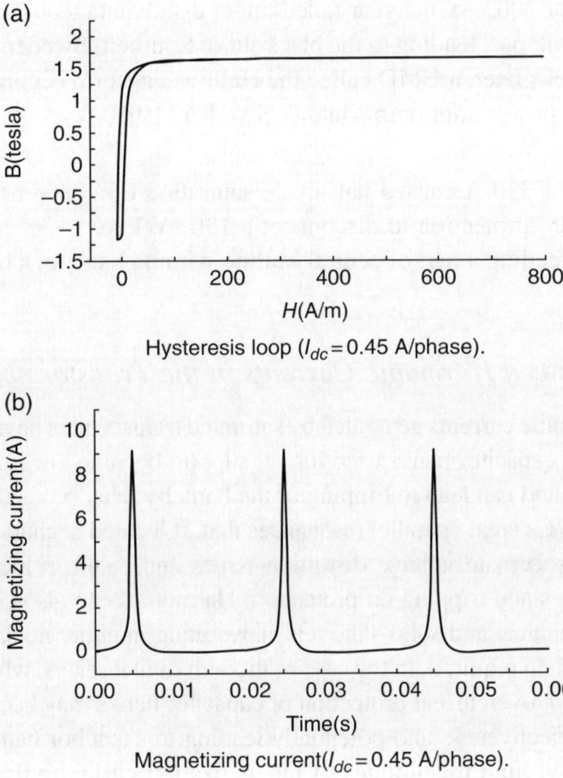

Figure 2.22 Magnetizing current and hysteresis loop under different direct magnetizing current (single phase transformer)

If the situation is already stressed, a voltage collapse may occur, leading to blackout in at least one part of the system. The rms value of the distorted current waveform increases with GIC, but less rapidly than the peak value. This has the effect of increased reactive losses in the transformer. To cover these losses, the transformer will draw more reactive power through the network. As with unwanted tripping, this adds stress to the system. Situations near voltage collapse are characterized by large flows of reactive power, which worsens the situation and brings the system closer to a blackout.

When the transformer core enters magnetic saturation, it cannot carry more magnetic flux. If the flux continues to increase, it therefore has to find new paths outside the core – typically in the tank, with oil in which the transformer is immersed. Since the tank and similar structures are not designed to carry AC magnetic flux, this will create eddy currents in the structures, leading to heating – typically at certain points, or so-called "hot spots". On the other hand, the magnetic saturation will generate many harmonics, which also cause eddy current losses. Sufficient heat will damage the transformer, which needs to be de-energized and taken out of service.

On 23 September 2003, switchgear failed under disadvantageous circumstances, and caused a voltage collapse, leading to the blackout of Southern Sweden and Eastern Denmark. Just five weeks later, a GMD called the Halloween storm occurred on 30 October 2003 and affected power supply in Malmö, Sweden [19]:

- Very high GIC of 330 A caused half-cycle saturation of a power transformer.
- Harmonics caused protection to disconnect a 130 kV line.
- As the line was feeding a part of central Malmö with no backup, a blackout occurred.

2.3.2 GIC-Induced Harmonic Currents in the Transformer

GIC-induced harmonic currents generated by saturated transformers have a significant systems impact. Shunt capacitor banks used for var support become low impedance paths for harmonic currents, and can lead to tripping of the bank by relay protection schemes. Harmonic filters for SVCs create parallel resonances that, if located at characteristic harmonic frequencies, can exacerbate voltage distortion issues and result in increased harmonics flow in these devices and tripping on protection. Harmonics can also cause the disoperation of electromechanical and solid-state relays, resulting in either nuisance operations or failure to operate when required. In the case of modern digital relays, where harmonic currents may be filtered, overcurrent protection of capacitor banks may become desensitized, thus reducing its effectiveness and potentially leading to capacitor bank damage.

During GIC flow, high magnitudes of magnetization current pulses and associated current harmonics produce increased, harmonic-rich stray flux. This results in much higher eddy and circulating current losses in the windings, as well as in the structural parts of the transformer. The resultant increase in load losses and temperatures of windings and structural parts must be assessed individually for each power transformer design, since the GIC-imposed thermal duty is outside standard service parameters.

A GMD event can last one to two days, and continually generates relatively low to moderate levels of GICs, with several intermittent periods of high GICs. For illustration purposes, the GICs have been scaled to produce 100 A per phase peak magnitude. GIC is a quasi-DC pattern, with hours-long periods of low to medium magnitudes relative to its GIC peaks. This sustained activity appears as a series of spikes (rather than constant DC) because of the observation time frame. The peaks of GIC activity are separated by an hour or more for this event and, in terms of the transformer's thermal time constant, each individual GIC peak can be considered an isolated event.

The technical report [19] presents recorded transformer data during a different GMD event on 10 May 1992. It includes neutral GIC, the real and reactive power loss, and the third and sixth harmonic currents in a 345/115 kV autotransformer. A 12-minute total duration portion of a GMD event rises to about a four-minute duration peak of GIC of 80 A per phase. During the peak, there is a significant jump in the transformer's reactive power demand, as well as harmonic current injection, as a result of the GIC flow in the transformer.

Quasi-DC current flow in transformer windings is characterized by a significant increase in excitation current (see Figure 2.21), as a result of half-cycle saturation. This excitation current is non-sinusoidal and is comprised of a large fundamental component, with even and odd harmonics. The immediate consequences of GICs and half-cycle saturation are the generation of even and odd harmonic currents, and an increase of reactive power absorbed by transformers. This can have a significant effect on var resources, var margins, generator performance, and protective relaying.

Unlike the traditional I^2X loss associated with load current flow through the transformer, var loss associated with half-cycle saturation is a shunt loss, where the transformer behaves as a large reactive load. Half-cycle var loss increases proportionally to GIC flow in transformer windings and, even for moderate amounts of GIC flow, this var loss increase may reduce system voltages to the point of encroaching on secure voltage limits. If var resources, such as capacitor banks, static var compensators (SVCs), and spinning reserves, are exhausted during a GMD event, a voltage collapse can occur on a single contingency. If var supplies are near the point of exhaustion, extreme operating actions (such as curtailing load) could be necessary to provide mitigation. Thus, the reactive power loss resulting from half-cycle saturation of transformers is one of the major concerns during GMD events.

Three-phase transformers, connected in either wye-wye or wye-delta, are used in power systems. During GMD events, the magnetostrictive strain is not truly sinusoidal in character, which leads to the introduction of the harmonics. With quasi-DC biased transformers, the saturation of magnetizing will also cause some harmonics. The harmonics in transformer noise may have a substantial effect on an observer, even though their level is 10 dB or more lower than that of the 100 Hz fundamental. In fact, the most striking point is the strength of the component at 100 Hz, or twice the normal operating frequency of the transformer.

Deviation from a "square-law" magnetostrictive characteristic would result in even harmonics (at 200, 400, 600 Hz, etc.), while the different values of magnetostrictive strain for increasing and decreasing flux densities – a pseudo-hysteresis effect – lead to the introduction of odd harmonics (at 300, 500, 700 Hz, etc.). If any part of the structure has a natural frequency at or near 100, 200, 300, 400 Hz, and so on, the result will be an amplification of noise at that particular frequency. However, the conventional GIC model cannot calculate GIC-induced harmonic currents [22]. The adoptions of the harmonic balance FEM GIC model can make it possible [23]. HBFEM is an accurate numerical method for computing GIC-induced harmonic currents. The detailed GIC computation model will be discussed in Chapter 6.

2.4 Harmonic Problems in Renewable Energy and Microgrid Systems

The integration of renewable energy sources (RES) to the grid would not be possible without the use of power electronic devices (e.g., DC and AC inverters). These devices are essential to interface these RES and distributed generators (DG) to the distribution system. Using power electronic devices (e.g., inverters) in order to connect the RES and DGs to the grid exhibits many advantages, such as faster voltage and frequency regulation, but it also displays one major disadvantage. The switching operation of the semiconductors included in the inverters causes voltage and current harmonic distortion via the distribution transformer to the grid.

This distortion increases for a number of other reasons, such as the behavior of nonlinear loads (diode rectifier bridges, personal computers, induction machines, etc.) and the switching operation of their converters. Specifically, harmonic distortion occurs due to loads supplied by converters, as well as to RES, interfaced to the grid through inverters. Harmonic distortion leads to poor power quality to the end user of the distribution system, as well as to increased value of line current.

For this reason, a lot of active filters and power conditioners have been proposed to alleviate the distribution system problems of current and voltage high-order harmonics, and power quality. Only a few of these proposals deal with improving power quality. Instead, the majority focus on harmonic cancellation and power quality improvement of critical loads against the high-order harmonic distortion of the grid that they are connected to. In this section, we will discuss the impact of harmonics on renewable energy and microgrid systems.

2.4.1 *Power Electronic Devices – Harmonic Current and Voltage Sources*

Power electronic devices, as used for renewable energy systems and microgrids, might be able to cause harmonics. The magnitude and order of harmonic currents injected by DC/AC inverters depends on the technology of the inverter and mode of operation [24]. Depending on the switching frequency, the harmonics produced may be significant, according to the capacitive coupling and the resonant frequency inside the PV installation. Moreover, every PV array is considered as an independent current source, with a DC current ripple independent of the converter ripple. These ripple currents are not synchronized with the converter, and produce subharmonics in the DC circuit, which increases the total harmonic distortion in the current waveform (THDI) [25].

The typical maximum harmonic order $h = 40$, defined in the power quality standards, corresponds to a maximum frequency of 2 kHz (with 50 Hz as the fundamental frequency) [26]. However, the typical switching frequency of DC/DC converters and DC/AC inverters, usually operated with the pulse width modulation (PWM) technique, is higher than 3 kHz. Hence, higher order harmonics, up to the 100th order, can be an important concern in large scale PV installations, where converters with voltage notching, high pulse numbers, or PWM controls result in induced noise interference, current distortion, and local GPR at PV arrays [27, 28].

Figure 2.23 shows a three-phase DC/AC inverter connected to a building transformer. The output voltage waveform and frequency spectrum generated by power electronics are illustrated in Figures 2.24 and 2.25 respectively.

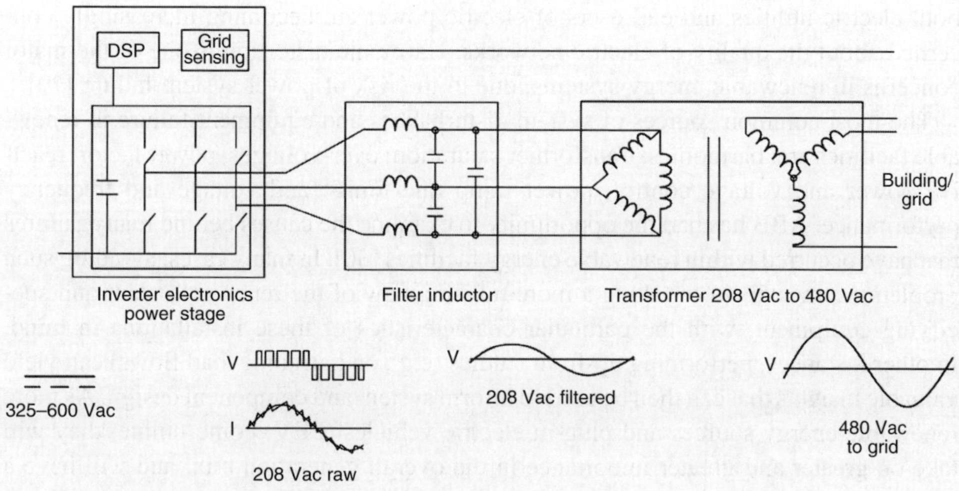

Figure 2.23 Three-phase DC/AC inverter connected to a building transformer

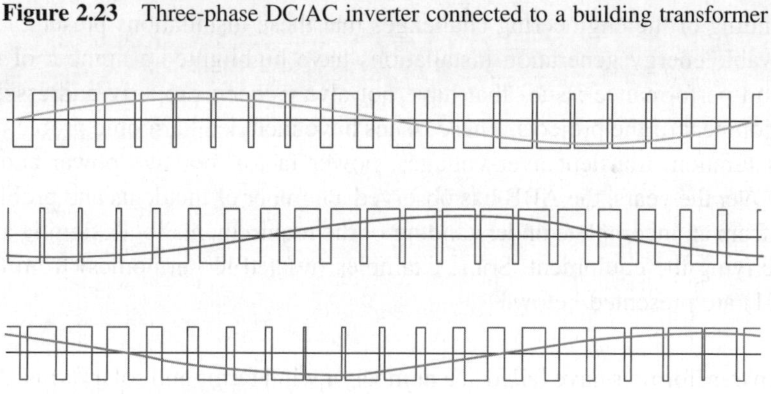

Figure 2.24 Output voltage waveform

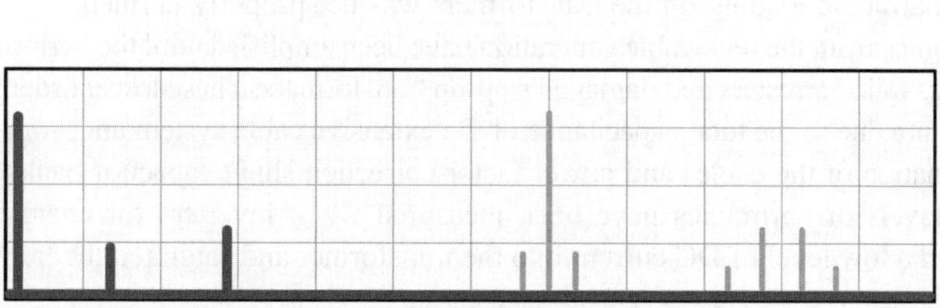

Figure 2.25 Frequency spectrum

2.4.2 Harmonic Distortion in Renewable Energy Systems

Renewable energy systems, such as wind farms and solar photovoltaic (PV) installations, are promising distributed generation (DG) sources to cover the increased demand for energy. With the incoming high penetration of distributed energy resources (DER), both electric utilities and end users of electric power are becoming increasingly concerned about the quality of electric networks. Harmonic distortion is one of the major concerns in renewable energy systems, due to the risk of power system failure [29].

The most common sources of system disturbances and equipment failure in renewable facilities are: harmonics; transformer saturation; over-voltages; power factor, reactive power and voltage control; power ramp rate limits; and voltage and frequency performance. ABB has had the opportunity to examine the causes behind many failures that have occurred within renewable energy facilities [30]. In many cases, avoiding such problems comes down to taking a more holistic view of the renewable plant and specifying equipment, with the particular characteristics of these installations in mind. In other instances, performing up-front studies (e.g., on harmonic load flows) can yield valuable insights that can then be used to inform system and component design. As more renewable energy sources and plug-in electric vehicles (PEV) come online, they will take on greater and greater importance in the overall generation mix, and will have a significant impact on conventional power systems. It is, therefore, vital to build our understanding of the engineering challenges that these installations present.

Renewable energy generation installations have highlighted a number of different design and performance issues that have not always been properly addressed during the development of the projects. These issues have included harmonic problems, transformer saturation, transient over-voltages, power factor, reactive power and voltage control. Over the years, the ABB has observed a number of incidents and problems that resulted from an incomplete understanding of the requirements for designing and properly specifying the equipment. Some examples (related to harmonics) from the ABB report [31] are presented below:

- Step-up transformers have failed at a number of wind farms and solar farms, due to the total loading of the transformers being more than specified. This was particularly true when harmonic loading on the transformers was not properly defined.
- Harmonics from the renewable generation have been amplified until the harmonic over-voltages failed arresters and damaged step-up transformers. These lower order resonant modes are due to the total capacitance of the extensive cable system and, sometimes, a combination of the cables and power factor correction shunt capacitor banks.
- High levels of harmonics have been measured when inverters for energy storage injected a low level of DC current into the transformer and saturated the transformer.

Harmonic loading, DC injection, voltage ripple, and voltage range of operation are key areas that influence the operation of the transformer. There is a trend in the industry now, widely seen in wind and in solar power, to provide specification information for the transformers used for the particular technology being provided. Harmonic problems mentioned in the above are more of a challenge with some technologies than others. For example, with wind, the full back-to-back conversion technology generates more harmonics than the doubly-fed induction generators with smaller converters and, in very weak systems, even harmonics from doubly-fed induction generators may cause problems.

The solution is to conduct harmonic studies for those technologies that have had harmonic problems and issues. Harmonic loading of transformers in PV generation systems and many wind technologies use converters to provide 50/60 Hz power to the collector system. These converters create harmonics of varying levels, which results in harmonic currents flowing through the step-up transformers from the converters to the collector system. Today, most wind turbine suppliers provide requirements for step-up transformers that include the harmonic loading and power factor capabilities required of the transformers. It is important to include the harmonics for the proper transformer design.

Harmonic over-voltages (HOVs) are typically over-voltages resulting from a harmonic resonance on the system. Resonance is a naturally occurring phenomenon on the power system, and occurs when the shunt capacitance of a system parallels the system inductance. These resonance conditions generally occur from the second to the fiftieth harmonic. It is when a resonance with minimal damping is excited by a harmonic source, or by transient events that create harmonics on the system, that high harmonic over-voltages can result.

In order to address the over-voltages and power quality issues caused by harmonics, a thorough harmonic analysis (harmonic load flow, frequency scan, and a detailed harmonic analysis using harmonic sources specified by measured data or turbine manufacturer provided data) needs to be performed. Existing shunt capacitor banks can be detuned to system resonance frequency in order to avoid any potential HOVs. If the system produces harmonics, the equipment in the system must be rated to withstand the effects of harmonics, along with the power quality at the point of interconnection. An important consideration is specifying the harmonic loading on the transformers, so they can withstand these harmonics and the overheating caused by the harmonics. Most utilities follow IEEE 519 guidelines for voltage and current distortions limits at the point of interconnection. The addition of tuned harmonic filters can reduce the harmonic impedance at the resonance frequency, thereby reducing the potential to cause HOVs.

Transformer saturation is not normally an issue with renewable generation sources. However, the inverter technology used in one instance resulted in injecting small levels of DC current into the transformer, and the resulting transformer saturation created significant harmonic currents. The results of the study show that several amperes of DC current injection on the low side could easily cause transformer core saturation and result in very high levels of even harmonics (2nd and 4th) on the high side. If the DC current injection is reduced, the harmonic amplification is reduced. Once the DC current injection was reduced to 25% of the measured values, there was no longer any saturation causing amplification of harmonics on the high side.

Most harmonics on a transmission system are caused by loads being served from the system. The one major exception is when a transformer or line/cable reactor is energized. When this is done, especially with random closing, a high level of unbalanced harmonic currents are injected into the system for a limited time. The odd harmonics are the characteristic harmonics on the power system. Balanced 3rd, 9th, 15th, etc. are considered zero sequence harmonics, and they require a ground path, since the harmonics on each phase are in phase. The other odd harmonics are positive or negative sequence. The zero sequence harmonics will see the zero sequence impedance, and the positive and negative sequence harmonics will see the positive and negative sequence impedance. Therefore, it is important to look at both the positive and zero sequence impedance when analyzing power system resonance and harmonic interactions.

In general, a substation with generator step-up transformers and auto-transformers (or three winding) with delta tertiary windings will provide a low impedance zero sequence path, and the zero sequence resonance will be lower than the positive sequence. If there are no low impedance ground sources, such as a switching station without transformers, then the zero impedance will be high, and the zero sequence resonance will be higher than the positive sequence resonance [26].

2.4.3 Harmonics in the Microgrid and EV Charging System

A microgrid is defined by the US Department of Energy (DOE) as "a group of interconnected loads and DERs within clearly defined electrical boundaries that act as a single controllable entity with respect to the grid. A microgrid can connect and disconnect from the grid to enable it to operate in both grid-connected and island-mode" [31]. Microgrids are "building blocks of smart grids", which consist of several basic technologies for operation. These include: distributed generation, distributed storage, interconnection switches, and control systems. Figure 2.26 illustrates IEEE microgrid standards 1547.4-2011, and the IEEE Guide for design, operation, and integration of distributed resource island systems with electric power systems [32]. IEEE 1547.4 covers key considerations for planning and operating microgrids. This includes:

- impacts of voltage, frequency, power quality, inclusion of a single point of common coupling (PCC) and multiple PCCs;
- protection schemes and modifications, monitoring, information exchange and control, understanding load requirements of the customer, knowing the characteristics of the DER;
- identifying steady state and transient conditions;
- understanding interactions between machines, reserve margins, load shedding, demand response;
- cold load pickup, additional equipment requirements, and additional functionality associated with inverters.

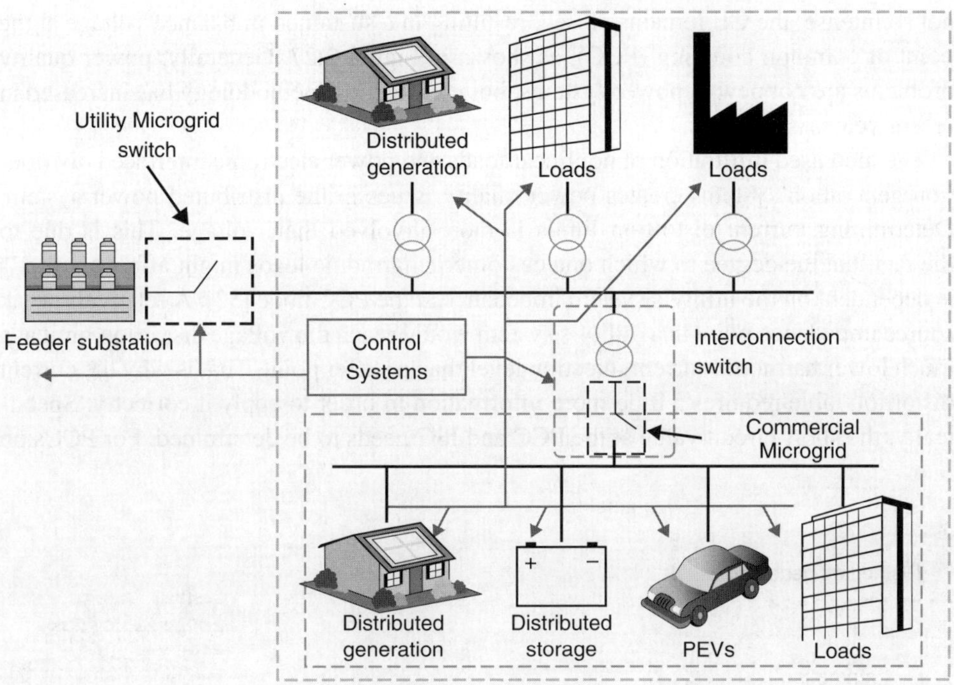

Figure 2.26 Microgrid coordinator/coordinated control system, conceptual commercial-level microgrid architecture

Distributed generation (DG) units are small sources of energy located at or near the point of use. DG technologies typically include photovoltaic (PV), wind, fuel cells, microturbines, and reciprocating internal combustion engines with generators. Distributed storage (DS) technologies are used in microgrid applications where the generation and loads of the microgrid cannot be exactly matched. The interconnection switch is the point of connection between the microgrid and the rest of the distribution system or smart grid. The control system of a microgrid is designed to safely operate the system in grid-parallel and stand-alone modes.

The concept of the microgrid is an effective way to integrate all kinds of distributed generators (DGs) as a utility-friendly customer. A typical AC microgrid usually consists of DGs such as wind generation, photovoltaic generation, fuel cell generation, energy storage systems (ESS) like batteries, super-capacitors, flywheels, local loads like lighting, air-conditioners and computers. Most of the DGs and ESS are DC form or have middle DC bus; thus, VSIs are usually adopted as the interfaces to the AC bus [33].

The impact of power quality hitches while linking the microgrid to the main grid is concerning, and it could become an important area to investigate. If an imbalance in voltage is alarming, the solid state circuit breaker (CB), connected between the microgrid and utility grid, will open to isolate the microgrid. When the voltage imbalance is not so intense, the CB remains closed, resulting in a sustained imbalance voltage at the point of common coupling (PCC), as shown in Figure 2.27. Generally, power quality problems are not new in power systems, but rectification methodology has increased in recent years.

The increased infiltration of nonlinear loads and power electronic interfaced distribution generation systems creates power quality issues in the distributed power system. Determining current distortion limits is more involved than voltage. This is due to the fact that the degree to which one customer's harmonic loads might affect another's is dependent on the utility's system impedance at the PCC node [32]. A relatively weak source impedance point in a utility's system would reach the voltage distortion limit at a much lower harmonic current injection level than a stiffer point. That is why the current distortion table requires a little more information in order to apply it correctly. Specifically, the short-circuit value at the PCC and ISC needs to be determined. For PCCs on

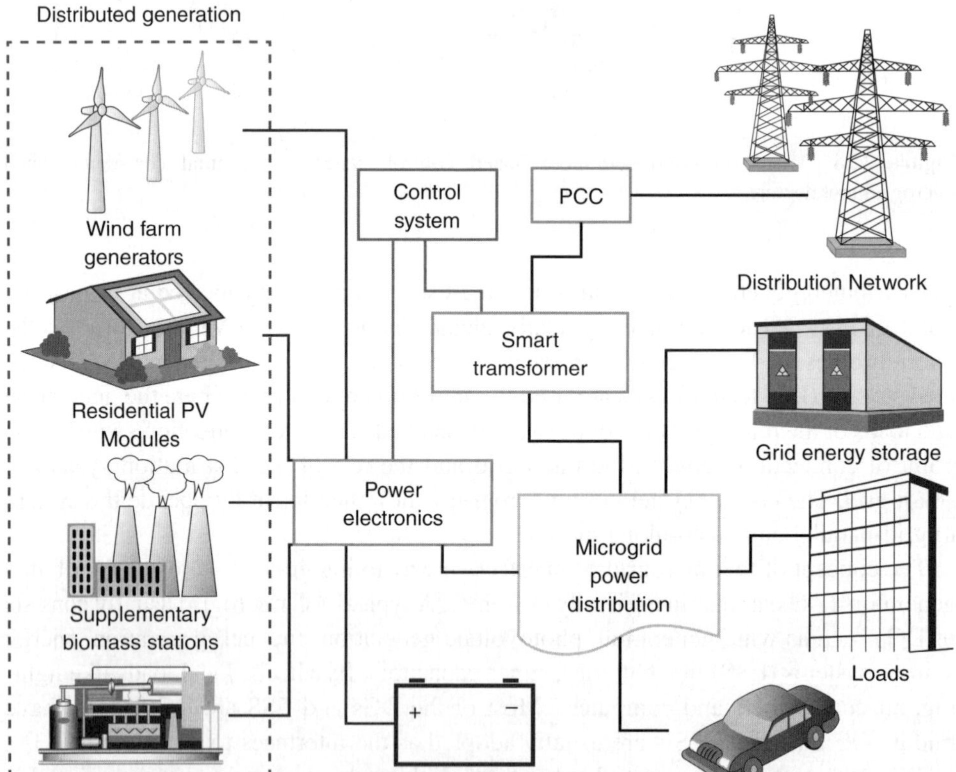

Figure 2.27 Basic concept of the microgrid connected through a PCC to the grid

the electric utility circuit, the utility usually provides this number. The second required value is the estimated or measured demand current of the customer's total load.

As discussed, the circuit node at which harmonic current and voltage limits are to be evaluated is that point on the electric utility system at which other customers could be served. This PCC is described graphically, as shown above. In the case of a microgrid with a grid connected, current harmonics generated by power electronics and drives will be injected into the main grid through the PCC. The voltage waveform at the PCC, and VSC current waveforms from the inverter, are illustrated in Figure 2.28.

Plug-in electric vehicles (PEVs), including all-electric vehicles and plug-in hybrid electric vehicles (PHEVs), provide a new opportunity to reduce oil consumption and play an important role in decarbonising road transport. Electricity suppliers and business investors will need to anticipate the long-term investments that will be needed to respond to this emerging trend, both in developing PEVs, the electricity network and associated PEV charging infrastructure. As the use of PEVs grows, PEVs could have a significant impact on the grid (or future smart grid) and load demand. This new type of electricity load will need careful management, in order to minimize the impact on peak electricity demand.

The integration of smart grid and microgrid technology with vehicle-to-grid (V2G) can enable PEVs to be used as distributed energy storage devices, feeding electricity stored in their battery tanks and sending it back into the grid when electric power is needed. Smart grids, and microgrids using PEVs, can help to reduce electricity system costs by providing a cost-effective means of providing regulation services, spinning reserves and peak-shaving capacity. However, there are a number of technical, environmental and economic barriers to such a development, including PEV charging infrastructure. Battery technology will be critical to the future of V2G supply [34].

The V2G facility of the microgrid and smart grid envisages the ability of electricity generating utilities to level the demand on their generating capacity. This is done by drawing energy from the batteries of electric vehicles (EVs) connected to the grid during the daylight hours of peak demand, and returning it to the vehicles throughout periods of low demand, during the night. Apart from the inconvenience, it would require charging stations to be capable of bi-directional power transfer, incorporating inverters with

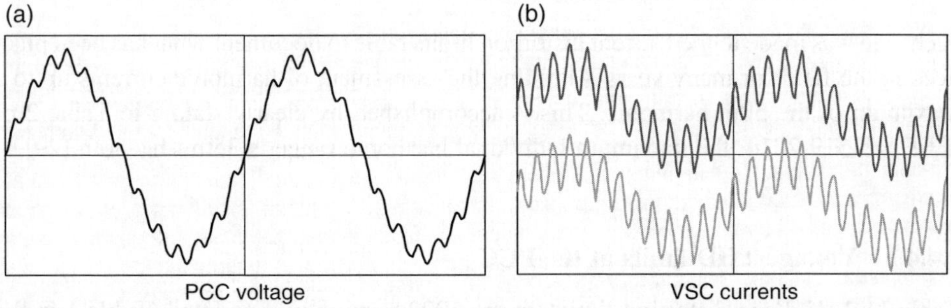

Figure 2.28 (a) Voltage waveform at the PCC; (b) VSC current waveforms from the inverter

precisely controlled voltage and frequency output to feed the energy back into the grid. Like PV generation systems and many wind technologies, V2G uses inverters to provide 50/60 Hz power to the power grid. These converters create harmonics of varying levels that result in harmonic currents flowing through the step-up transformers from the inverters to the collector system.

2.4.4 IEEE Standard 519-2014 [24]

IEEE Std 519-2014 is a newly published revision to the IEEE recommended practice and requirements for harmonic control in electric power systems [29], and it supersedes the IEEE Std 519-1992 revision. The overarching goal of the 2014 revision is the same as the 1992 version; to define the specific and separate responsibilities for each participant – utilities and users – to maintain the voltage THD within acceptable limits at the point of common coupling (PCC) between the utility and the user, and protect the user and utility equipment from the negative impact of harmonics. The separate individual responsibilities are:

- User – limit harmonic currents at the PCC to prescribed levels.
- Utility – limit voltage distortion at the PCC to prescribed levels by maintaining system impedance as necessary.

The 2014 version re-emphasizes and clarifies IEEE Std 519, as written, and is to be applied at the PCC – the point of common coupling between the utility and the user.

The size reduction of the document and the removal of conflicting material aids tremendously in clarifying:

- The standard is designed to be applied at the PCC
- The PCC is the point of common coupling between the utility and user

2.4.4.1 Current THD limits at the PCC

A change was made to the Current distortion limits table to document what has been practiced in the field for many years – limiting the assessment of harmonic currents up to a maximum of the 50th harmonic. This is accomplished by clearly stating in Table 2 of IEEE Std 519-2014; the maximum individual harmonic range is 35th ≤ h ≤ 50th [29].

2.4.4.2 Voltage THD limits at the PCC

Table 11-1, Voltage distortion limits, in the 1992 version was updated (Table 1 in the 2014 version) with the addition of a new voltage range and limits.

A new lower PCC voltage range of $V \leq 1.0\,kV$ was defined, with higher allowable harmonic voltage limits: Individual harmonic at 5% and Total harmonic distortion at 8%. These limits are higher than the next highest voltage range $1.0\,kV < V \leq 69\,kV$.

2.4.4.3 High-frequency Current Allowance in Low Current Distortion Systems

IEEE 519-2014 provides for an allowance of higher high-order harmonic current limits at a PCC that has low lower-order harmonics. The allowance is applied to Table 2, Current distortion limits, if a prescribed minimum performance level is met. For example, if a power system with $Isc/IL < 20$ has 5th and 7th harmonic currents at < 1%, then all other harmonic limits in Table 2 may be exceeded up to a factor of 1.4 and still be in compliance.

2.4.4.4 Harmonic Distortion Evaluations and Controlling Harmonics

Control only when harmonics create a problem. Types of problems:

- load harmonic currents are too large;
- path for harmonic currents is too long electrically (too much impedance) producing voltage distortion or communication-line interference;
- response of system magnifies one or more harmonics.

2.4.4.5 Reducing Load Harmonic Current

Sometimes transformer connections can be changed. For example:

- phase shift on some transformers supplying 6-pulse converters;
- delta windings block triplen currents;
- zig-zag transformers can supply triplens.

2.4.4.6 Harmonic Studies

Perform harmonic studies when:

- a problem occurs, to find a solution
- planning large capacitor bank installation on either utility or industrial systems
- planning installation of large nonlinear loads such as adjustable speed motor drives
- designing a harmonic filter or converting a capacitor to a harmonic filter

Since harmonic loading, DC injection, voltage ripple, and voltage range of operation are key areas that influence the operation of the transformer in renewable energy and microgrid systems, the HBFEM will be an effective method to analyze the harmonic problem in the transformer connected with voltage or current sources. The detailed HBFEM numerical model will be discussed in Chapter 6.

References

[1] Ray, L., Hapeshis, L. (2011). *Power System Harmonic Fundamental Considerations: Tips and Tools for Reducing Harmonic Distortion in Electronic Drive Applications*. Schneider Electric, AT313, October.

[2] IEEE (1992). *IEEE Recommended Practices and Requirements for Harmonic Control in Electrical Power Systems*. IEEE Std. 519–1992.

[3] IEEE (2004). *IEEE Guide for Applying Harmonic Limits on Power Systems – Unpublished Draft*. IEEE Std P519.1TM/D9a, January.

[4] Paice, D.A. (1996). *Power Electronic Converter Harmonics*. IEEE Press.

[5] Tihanyi, L. (1995). *Electromagnetic Compatibility in Power Electronics*. IEEE Press.

[6] Brüns, H-D, Schuster, C. and Singer, H. (2007). Numerical Electromagnetic Field Analysis for EMC Problems. *IEEE Transactions on Electromagnetic Compatibility* **49**(2), 253–262.

[7] Dugan, R.C. McGranaghan, M.F. and Gunther, E.W. (1992). *Electrical Power System Harmonics Design Guide*. Electrotek Concepts, Inc., Knoxville, TN.

[8] Lu, J., Yamada, S. and Harrison, H.B. (1996). Application of HB-FEM in the Design of Switching Power Supplies. *IEEE Transactions on Power Electronics* **11**(2), 347–355.

[9] Water, W. and Lu, J. (2013). Improved High-Frequency Planar Transformer for Line Level Control (LLC) Resonant Converters. *IEEE Magnetics Letters* **4**, Institute of Electrical and Electronics Engineers. DOI: 10.1109/LMAG.2013.2284767.

[10] Biringer, B.P. and Slemon, G.R. (1963). Harmonic analysis of the magnetic frequency tripler. *IEEE Transactions on Communication and Electronics* **82**, 327–332.

[11] Bendzsak, G.J. and Biringer, B.P. (1974). The influence of magnetic characteristics upon tripler performance. *IEEE Transactions on Magnetics* **10**(3), 961–964.

[12] Ishikawa, T. and Hou, Y. (2002). Analysis of a Magnetic Frequency Tripler Using the Preisach Model. *IEEE Transactions on Magnetics* **38**(2), 841–844.

[13] Venkatesh, P. and Dinesh, M.N. (2014). Harmonic Analysis of 6-Pulse and 12-Pulse Converter Models. *International Journal of Modern Engineering Research (IJMER)* **4**(9), 31–36.

[14] Nesan, R.T. and Jegadhish, J. (2015). Designing and Harmonic analysis of a new 24-Pulse Rectifier using diodes with Phase Shifting Transformer. *International Journal of Science, Engineering and Technology Research (IJSETR)* **4**(2), 273–276.

[15] ABB (2013). Technical guide No. 6: *Guide to harmonics with AC drives*.

[16] Zhang, J. Zhang, H. and Xiong, M. (2009). Application of a series capacitor based transformer neutral DC current blocking device. *Power System Technology* **33**(20), 147–151.

[17] NERC (2013). Application Guide: *Computing Geomagnetically-Induced Current in the Bulk-Power System*.

[18] IEEE Power and Energy Society Technical Council Task Force on Geomagnetic Disturbances (2013). Geomagnetic Disturbances – Their Impact on the Power Grid. *IEEE Power & Energy Magazine* **11**(4), 71–78.

[19] Samuelsson, O. (2013). *Geomagnetic disturbances and their impact on power systems*. Status report 2013, Division of Industrial Electrical Engineering and Automation, Lund University.

[20] Hutchins, T. (2012). *Geomagnetically induced currents and their effect on power systems*. PhD thesis, University of Illinois.

[21] Power IT Lab (no date). *GIC Impact for Transformer Harmonics and Reactive Power*. The University of Tennessee. http://powerit.utk.edu/GIC_impact.html

[22] Gilbert, J.L., Radasky, W.A. and Savage, E.B. (2012). *A Technique for Calculating the Currents Induced by Geomagnetic Storms on Large High Voltage Power Grids*. IEEE International Symposium on Electromagnetic Compatibility (EMC), Aug. 2012, pp. 323–328.

[23] Zhao, X., Lu, J., Li, L., Cheng, Z. and Lu, T. (2011) Analysis of the DC Biased Phenomenon by the Harmonic Balance Finite Element Method. *IEEE Transactions on Power Delivery* **26**(1), 475–485.

[24] IEEE Std. 519-2014 (2014: revision of IEEE Std. 519, 1992). *IEEE Recommended Practices and Requirements for Harmonic Control in Electric Power Systems*. Approved 27 March 2014.

[25] García-Gracia, M., El Halabi, N., Alonso, A. and Paz Comech, M. (2011). *Harmonic Distortion in Renewable Energy Systems: Capacitive Couplings*. Available online at www.intechopen.com.

[26] IEC (2010). IEC 61000-4-7 Ed. 2.1: *Updated Standard for Harmonics Measurements*, 7/2010

[27] Chayawatto, N. Kirtikara, K. Monyakul, V. Jivacate, C. and Chenvidhya, D. (2009). DC/AC switching converter modeling of a PV grid-connected system under islanding. *Phenomena, Renewable Energy* **34**(12), 2536–2544.

[28] Chicco, G. Schlabbach, J. and Spertino, F. (2009). Experimental assessment of the waveform distortion in grid-connected photovoltaic installations. *Solar Energy* **83**(1), 1026–1039.

[29] Altamaly, A.M. (no date). *Harmonic reduction techniques in renewable energy interface converter*. Available online at: www.intechopen.com.

[30] Vadlamani, V. and Martin, D. (2013). *Renewable energy design considerations*. Available online at: www.abb.com.

[31] Asmus, P. (2014). Why Microgrid are moving into the mainstream. IEEE Electrification Magazine **2**(1), 12–19.

[32] IEEE Standards Coordinating Committee (2011). 21, *IEEE Guide for Design, Operation, and Integration of Distributed Resource Island Systems with Electric Power Systems*. IEEE Std 1547.4™-2011, 20 July 2011.

[33] Bouloumpasis, I. Vovos, P. Georgakas, K. and Vovos, N.A. (2015). Current Harmonics Compensation in Microgrids Exploiting the Power Electronics Interfaces of Renewable Energy Sources. *Energies* **8**(4), 2295–2311.

[34] Lu, J. and Hossain, J. (2015). *Vehicle-to-Grid: Linking Electric Vehicles to the Smart Grid*. IET Digital Library.

3

Harmonic Balance Methods Used in Computational Electromagnetics

3.1 Harmonic Balance Methods Used in Nonlinear Circuit Problems

3.1.1 The Basic Concept of Harmonic Balance in a Nonlinear Circuit

The harmonic balance method is a powerful numerical technique for the analysis of high-frequency nonlinear circuits, and has been used to solve nonlinear microwave circuit problems since the 1960s. Over the last three decades, it has been reformulated into an accurate method for finding numerical solutions of a differential equation driven by sinusoids (without having to approximate the nonlinearities with polynomials). In fact, the harmonic balance method was firmly established in the 1970s and was widely used in solving nonlinear microwave circuit problems in the 1980s [1–3].

In electrical and electronic circuits, a signal whose domain is "time" is called a waveform (illustrated in Figure 3.1a), and one whose domain is "frequency" is called a spectrum (as shown in Figure 3.1b). All waveforms are assumed R-valued, whereas all spectra are assumed C-valued. The distorted waveforms in the system are usually symmetrical and time-periodic sinusoidal if $p(t) = p(t + nT)$ for all t. $P(T)$ denotes the set of all periodic functions with period T that can be uniformly approximated by the sum of an infinite number of time-periodic sinusoids [1]. Thus, the waveform can be defined as a summation of trigonometric function, as follows:

$$P(t) = \sum_{k=0}^{\infty} (P_{ks}\sin(k\omega t) + P_{kc}\cos(k\omega t)) \tag{3-1}$$

Harmonic Balance Finite Element Method: Applications in Nonlinear Electromagnetics and Power Systems, First Edition. Junwei Lu, Xiaojun Zhao and Sotoshi Yamada.

Companion website: www.wiley.com/go/lu/HBFEM

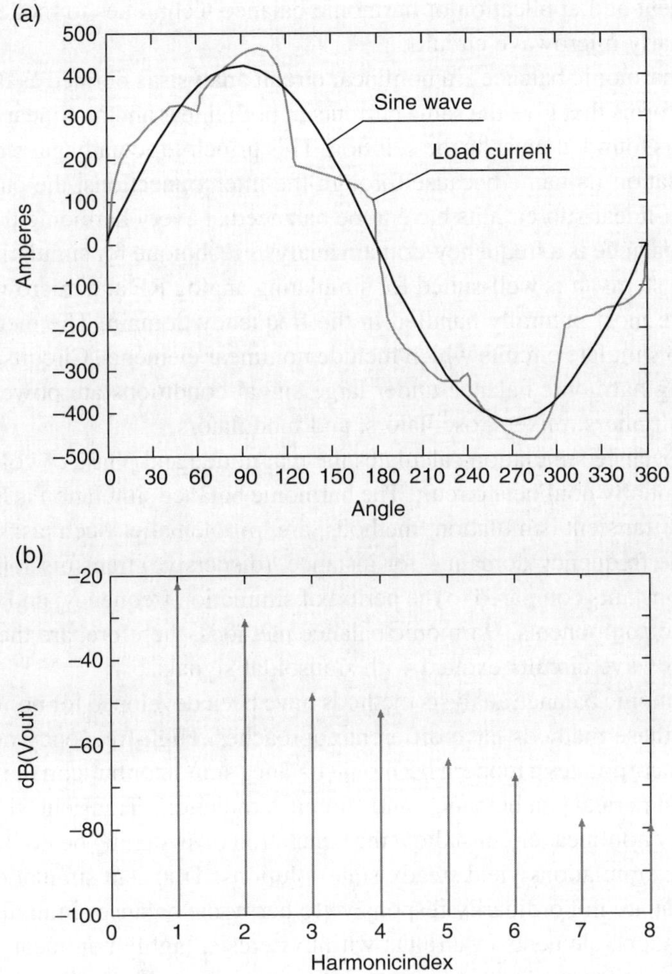

Figure 3.1 Distorted waveform and spectrum. (a) Distorted waveforms; (b) Spectrum

where $\omega = 2\pi/T$, $P_{ks}, P_{kc} \in R$ are harmonic coefficients, and the pair $P_k = [P_{ks}, P_{kc}] \in C$ are the Fourier coefficients of the Fourier series.

The harmonic balance method is a technique for the numerical solution of nonlinear analog circuits operating in a periodic, or quasi-periodic, steady-state regime. The method can be used to efficiently derive the continuous-wave response of numerous nonlinear microwave components, including amplifiers, mixers, and oscillators. Its efficiency is derived from imposing a predetermined steady-state form for the circuit response onto the nonlinear equations representing the network, and solving for the set of unknown coefficients in the response equation. Its attractiveness for nonlinear microwave applications results from its speed and ability to simply represent the dispersive, distributed elements that are common at high frequencies. The last decade has seen the development and application of harmonic balance techniques to model analog circuits, particularly microwave circuits.

The term "harmonic balance" in nonlinear circuit analysis is defined as the set of port voltage waveforms that give the same currents in both linear and nonlinear sub-circuits; when that set is found, it must be the solution. This principle actually gave the harmonic balance simulation its name because through the interconnections, the currents of the linear and non-linear sub-circuits have to be balanced at every harmonic frequency [1].

Harmonic balance is a frequency-domain analysis technique for simulating nonlinear circuits and systems. It is well-suited for simulating analog RF and microwave circuits, since these are most naturally handled in the frequency domain. The method is commonly used to simulate circuits which include nonlinear elements. Circuits that are best analyzed using harmonic balance under large signal conditions are power amplifiers, frequency multipliers, mixers, oscillators, and modulators.

Harmonic balance simulation calculates the magnitude and phase of voltages or currents in a potentially nonlinear circuit. The harmonic balance simulation is ideal for situations where transient simulation methods are problematic, such as: components modeled in the frequency domain – for instance, (dispersive) transmission lines; large circuit time constants compared to the period of simulation frequency; and circuits with lots of reactive components. Harmonic balance methods, therefore, are the best choice for most microwave circuits excited with sinusoidal signals.

Several harmonic balance analysis methods have been developed for nonlinear circuit analysis, and these methods have different approaches. High-frequency nonlinear circuit analysis incorporates a modern harmonic balance simulator built on the latest developments in numerical mathematics and circuit simulation. Transient simulations in high-frequency nonlinear circuits allow the simulation of switching behavior, while harmonic balance simulations yield steady-state solutions. Transient simulations can handle circuits that are not ordinarily responsive to harmonic balance simulations, such as frequency dividers, elements or circuits with hysteresis, highly nonlinear circuits, and digital circuits with memory. This capability broadens the applicability of high-frequency nonlinear circuits to more design types than harmonic balance alone.

In cases where both harmonic balance and transient analysis are applicable, the better choice is dependent on the excitation source, circuit type, available element models, and the desired measurements. Harmonic balance is a frequency domain solver, while transient analysis is in the time domain. For example, transient analysis is the better choice if the excitation sources are not periodic and have short rising or falling edges, the circuit shapes pulse using a large number of transistors, and the measurement is rise or fall time. Harmonic balance is the better choice if the excitation source has a discrete spectrum, the models are S-parameter files, and the desired measurement is power at a specific frequency.

The harmonic balance method is a powerful technique for the analysis of high-frequency nonlinear circuits such as mixers, power amplifiers, and oscillators. The method matured in the early 1990s, and quickly became recognized as the simulator

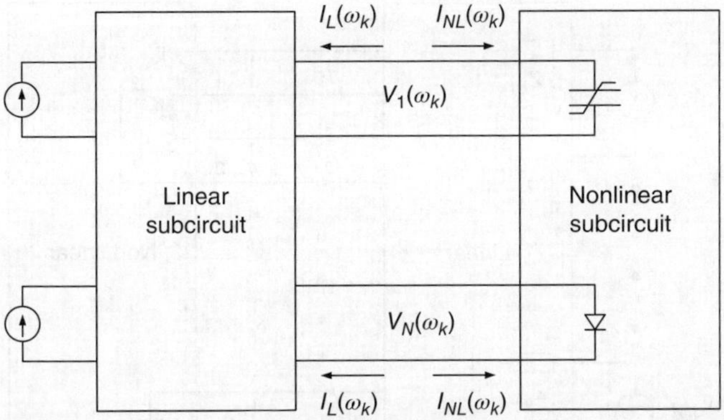

Figure 3.2 The concept of harmonic balance for a non-linear circuit

of choice for relatively small, high-frequency building blocks. More recently, with the adoption of new developments in the field of numerical mathematics, the range of applicability of harmonic balance has been extended to very large nonlinear circuits, and to circuits that process complicated signals composed of hundreds of spectral components.

Harmonic balance is a non-linear, frequency-domain, steady-state simulation. The voltage and current sources create discrete frequencies, resulting in a spectrum of discrete frequencies at every node in the circuit. Linear circuit components are solely modeled in the frequency domain. Non-linear components are modeled in the time domain and are Fourier-transformed before solving each step. As the non-linear elements are still modeled in the time domain, the circuit must first be separated into linear and non-linear parts. The internal impedances of the voltage sources are also put into the linear part. Figure 3.2 illustrates the concept of harmonic balance for a non-linear circuit.

3.1.2 The Theory of Harmonic Balance Used in a Nonlinear Circuit

As the non-linear elements are still modeled in the time domain, the circuit must first be separated into linear and non-linear parts (repetition – this is mentioned in the previous paragraph). The internal impedances Zi of the voltage sources are also put into the linear part. Figure 3.3 illustrates a conceptual circuit diagram for using harmonic balance in a non-linear circuit. The following symbols can be defined as:

- M = number of (independent) voltage sources
- N = number of connections between the linear and non-linear sub-circuit
- K = number of calculated harmonics
- L = number of nodes in the linear sub-circuit

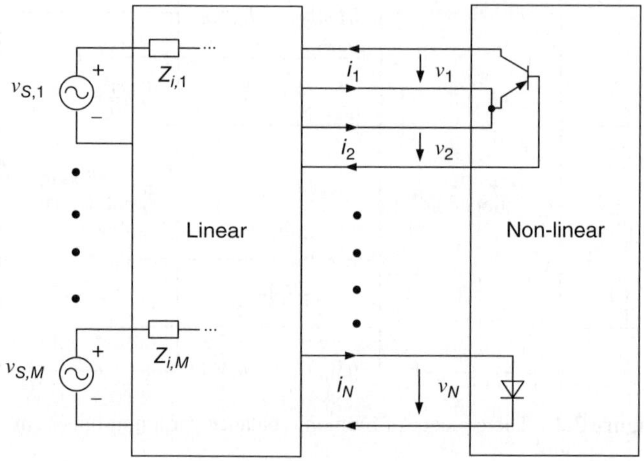

Figure 3.3 The circuit diagram for using harmonic balance in a non-linear circuit

The linear circuit is modeled by two transadmittance matrices. The first one (***Ys***) relates the source voltages $v_{s,1} \ldots v_{s,M}$ to the interconnection currents $i_1 \ldots i_N$. The second one (***Y***) relates the interconnection voltages $v_1 \ldots v_N$ to the interconnection currents $i_1 \ldots i_N$. Taking both, we can express the current flowing through the interconnections between the linear and non-linear sub-circuit:

$$I = Y_{s,NM} \cdot V_s + Y_{NN} \cdot V = I_s + Y \cdot V \tag{3-2}$$

Because V_s is known and constant, the first term can already be computed to give I_s. Taking the whole linear network as one block is called the "piecewise" harmonic balance technique.

The non-linear circuit is modeled by its current function:

$$i(t) = f_g(v_1, \cdots, v_P) \tag{3-3}$$

and by the charge of its capacitances:

$$q(t) = f_q(v_1, \cdots, v_Q) \tag{3-4}$$

These functions must be Fourier-transformed to give the frequency-domain vectors Q and I_G, respectively.

The non-linear equation system can be solved by the following equation, where the first term is the linear section and the second term is the nonlinear section:

$$F(V) = \{(I_S) + (Y) \cdot (V)\} + \{j \cdot \Omega \cdot Q + I_G\} = 0 \tag{3-5}$$

where matrix Ω contains the angular frequencies on the first main diagonal and zeros anywhere else, and 0 is the zero vector.

After each iteration step, the inverse Fourier transformation must be applied to the voltage vector V. Following this, the time domain voltages $v_{0,1}$, …, $v_{K,N}$ are put into (3-3) and (3-4), before a Fourier transformation gives the vectors Q and I_G for the next iteration step. After repeating this several times, a simulation result can be found. This result means the voltages v_1 … v_N are at the interconnections of the two sub-circuits. With these values, the voltages at all nodes can be calculated.

One significant difference between the harmonic balance and conventional simulation types (such as DC or AC simulation) is the structure of the matrices and vectors. A vector used in a conventional simulation contains one value for each node. In harmonic balance simulation, there are many harmonics and, thus, a vector contains $\boldsymbol{K}$ values for each node. This means that, within a matrix, there is a $\boldsymbol{K} \times \boldsymbol{K}$ diagonal sub-matrix for each node. Using this structure, all equations can be written in the usual way – that is, without paying attention to the special matrix and vector structure. In a computer program, however, a special matrix class is needed, in order to conserve memory for the off-diagonal zeros.

3.2 CEM for Harmonic Problem Solving in Frequency, Time and Harmonic Domains

3.2.1 Computational Electromagnetics (CEM) Techniques and Validation

A number of computational electromagnetics (CEM) techniques have been developed, and numerical codes have been generated to analyze various electromagnetics problems since the mid-1960s. While each is based on classical electromagnetic theory, and implements Maxwell's equations in one form or another, these techniques, and the manner in which they are used to analyze a given problem, can produce quite different results. The results are affected by the way in which the underlying physic formalisms have been implemented within the codes, including the mathematical basis functions, numerical solution methods, numerical precision, and the use of building blocks (primitives) to generate computational models.

Despite all CEM codes having their basis in Maxwell's equations of one form or another, their accuracy and convergence rate depends on how the physics equations are cast (e.g., integral or differential form, frequency or time domain), what numerical solver approach is used, inherent modeling limitations, approximations, and so forth. Although these techniques and codes have been applied to a myriad of electromagnetic problems, uncertainty still exists, and current validation practices have not always proved to be reliable. Computer predictions have been compared to measurements to provide a first-order validation, but there is also much interest in how the techniques, when applied to a given problem or a class of problems, compare to each other and the fundamental theory upon which they are based. Hence, additional efforts are needed to establish a standardized method for validating these techniques and to instill confidence in them [6, 7].

The physics formalism, available modeling primitives, analysis frequency, and time or mesh discretization further affect accuracy, solution convergence, and overall validity of the computer model. The critical areas that must be addressed include model accuracy, convergence, and techniques or code validity for a given set of canonical, standard validation, and benchmark models. For instance, uncertainties may arise when the predicted results using one type of CEM technique do not agree favorably or consistently with the results of other techniques or codes of comparable type, or even against measured data on benchmark models.

Furthermore, it can be difficult to compare the results between certain techniques or codes, despite their common basis in Maxwell's equations. Exceptions can be cited, in particular, when comparing the results of "similar" codes grouped according to their physics, solution methods, and modeling element domains. Nevertheless, disparities among codes in a certain "class" have been observed. Many examples can be cited where fairly significant deviations have been observed between analytical or computational techniques and empirical-based methods. Differences are not unexpected, but the degree of disparity in certain cases cannot be readily explained, nor easily discounted, which has led to the often asked-question, "*Which result is accurate?*" [6, 7]

3.2.2 Time Periodic Electromagnetic Problems Using the Finite Element Method (FEM)

The Finite Element Method (FEM), originating in the structural mechanics engineering discipline, solves PDEs for complex, nonlinear problems in magnetics and electrostatics using mesh elements. FEM techniques require the entire volume of the configuration to be meshed, as opposed to surface integral techniques, which only require the surfaces to be meshed. However, each mesh element may have completely different material properties from those of neighboring elements. In general, FEM techniques excel at modeling complex inhomogeneous configurations.

The method requires the discretization of the domain into a number of small homogeneous sub-regions or mesh cells and applying the given boundary condition, resulting in field solutions using a linear system of equations. The model contains information about the device geometry, material constants, excitations, and boundary constraints. The elements can be small where geometric details exist, and much larger elsewhere. In each finite element, a simple variation of the field quantity is assumed. The corners of the elements are called nodes. The FEM is to determine the field quantities at the nodes.

Most FEM methods are variational techniques that minimize or maximize an expression that is known to be stationary about the true solution. Generally, FEM techniques solve for the unknown field quantities by minimizing an energy quantity. This is applicable to a wide range of physical/engineering problems and frequencies, provided it can be expressed as a PDE.

The CEM problems can be solved by FEM in either the frequency domain or the time domain, depending on their accuracy and convergence rate of solution. For example, the time periodic nonlinear magnetic field can be described by the following equation:

$$\nabla \times \nu \nabla \times A + \sigma(\partial A/\partial t + \nabla \varphi) - J_s = 0 \tag{3-6}$$

where A is the vector magnetic potential, ν (=1/μ) is the reluctivity of magnetic material, σ is the conductivity, φ is the voltage, and current density J_s represents the current drive source.

Assuming $\nabla\varphi = 0$ in the two-dimensional case, and using Galerkin's method to discretize the governing equation and (3-6) for the two-dimensional problems, the weighted residual can be obtained as:

$$G = \iint_S \left\{ \frac{\partial N_i}{\partial x} \nu \frac{\partial A}{\partial x} + \frac{\partial N_i}{\partial y} \nu \frac{\partial A}{\partial y} \right\} dxdy - \iint_S \left\{ J_s - \sigma \frac{\partial A}{\partial t} \right\} N_i dxdy = 0 \tag{3-7}$$

where N_i is the interpolation function.

3.2.2.1 A. Time Domain Approach

Considering the orthogonal characteristic of trigonometric functions, substituting magnetic vector A, interpolation functions N_i, magnetic reluctivity ν and current density J_s into (3-7), the following FEM matrix equation can be obtained:

$$[S]\{A\} + [M]\frac{\partial}{\partial t}\{\dot{A}\} - [G]\{J\} = 0 \tag{3-8}$$

and the compact form can be written as:

$$[S]\{A\} + [M]\{\dot{A}\} - [K] = 0 \tag{3-9}$$

where [S] is the system coefficient matrix, $[\dot{A}]$ represents $\frac{\partial A}{\partial t}$, and $[M]$ can be obtained as:

$$[M] = \frac{\sigma \Delta^e}{12} \begin{bmatrix} 2 & 1 & 1 \\ 1 & 2 & 1 \\ 1 & 1 & 2 \end{bmatrix} \tag{3-10}$$

The equation can be solved using the Newton-Raphson method for nonlinear magnetic fields.

3.2.2.2 B. Frequency Domain Approach

For a frequency domain, the governing equation can be rewritten as:

$$\nabla \times \nu \nabla \times A + \sigma(j\omega A + \nabla \varphi) - J_s = 0 \tag{3-11}$$

In the two-dimensional case, and using Galerkin's method to discretize the governing equation and (3-11) for the two-dimensional problems, the weighted residual can be obtained as:

$$G = \iint_S \left\{ \frac{\partial N_i}{\partial x} \nu \frac{\partial A}{\partial x} + \frac{\partial N_i}{\partial y} \nu \frac{\partial A}{\partial y} \right\} dxdy - \iint_S \{J_s - \sigma j\omega A\} N_i dxdy = 0 \tag{3-12}$$

The system matrix can be obtained as follows:

$$[S]\{A\} + [M]\{A\} - [G]\{J\} = 0 \tag{3-13}$$

and the compact form can be written as:

$$([S] + [M])\{A\} - [K] = 0 \tag{3-14}$$

and $[S]$ is the system coefficient matrix, and $[M]$ matrix can be obtained as:

$$[M] = \frac{\omega \sigma \Delta^e}{12} \begin{bmatrix} 2 & 1 & 1 \\ 1 & 2 & 1 \\ 1 & 1 & 2 \end{bmatrix} \tag{3-15}$$

3.2.2.3 C. Multi-Frequency Domain Approach Using the Harmonic Balance Method

For a multi-frequency domain or harmonic domain problem, the governing equation for nonlinear magnetic field can be rewritten as:

$$G = \iint_S \left\{ \frac{\partial N_i}{\partial x} \nu \frac{\partial A_k}{\partial x} + \frac{\partial N_i}{\partial y} \nu \frac{\partial A_k}{\partial y} \right\} dxdy - \iint_S \left\{ J_0 - \sigma \frac{\partial A_k}{\partial t} \right\} N_i dxdy = 0 \tag{3-16}$$

Where magnetic vector potential A_k and current density J_0 consist of kth harmonics, σ is the conductivity of the conductors, the expression will have the form of:

$$A_k^i = \sum_{k=2n-1}^{\infty} \{A_{ks}^i \sin(k\omega t) + A_{kc}^i \cos(k\omega t)\} \tag{3-17}$$

$$J_0 = \sum_{k=2n-1}^{\infty} \{J_{ks} \sin(k\omega t) + J_{kc} \cos(k\omega t)\} \tag{3-18}$$

The nonlinear magnetic reluctivity ν corresponding to $B(t)$ can be expressed as:

$$\nu(t) = H(B(t))/B(t) = \nu_0 + \sum_{k=2n-2}^{\infty} \{\nu_{ks}\sin(k\omega t) + \nu_{kc}\cos(k\omega t)\} \tag{3-19}$$

where flux density B(t) is time-dependent. This can be expressed by the B-H curve including the hysteresis characteristic [8, 9]. ν (=1/μ) is the nonlinear magnetic reluctivity and the Fourier coefficients obtained from Equations (3-20) to (3-22), respectively.

$$\nu_0 = \frac{1}{T}\int_0^T \nu(t)dt \tag{3-20}$$

$$\nu_{ns} = \frac{2}{T}\int_0^T \nu(t)\cdot\sin(n\omega t)dt \tag{3-21}$$

$$\nu_{nc} = \frac{2}{T}\int_0^T \nu(t)\cdot cos(n\omega t)dt \tag{3-22}$$

If equations (3-17), (3-18) and (3-19) are substituted into (3-16), the HBFEM single element matrix equation can be derived as follows [8,9]:

$$\begin{aligned} G^e = &\frac{1}{4\Delta^e}\begin{bmatrix} (b_1b_1+c_1c_1)D & (b_1b_2+c_1c_2)D & (b_1b_3+c_1c_3)D \\ (b_2b_1+c_2c_1)D & (b_2b_2+c_2c_2)D & (b_2b_3+c_2c_3)D \\ (b_3b_1+c_3c_1)D & (b_3b_2+c_3c_2)D & (b_3b_3+c_3c_3)D \end{bmatrix}\begin{Bmatrix} A_1^e \\ A_2^e \\ A_3^e \end{Bmatrix} \\ &+\frac{\sigma\omega\Delta^e}{12}\begin{bmatrix} 2N & N & N \\ N & 2N & N \\ N & N & 2N \end{bmatrix}\begin{Bmatrix} A_1^e \\ A_2^e \\ A_3^e \end{Bmatrix} - \begin{Bmatrix} K_1^e \\ K_2^e \\ K_3^e \end{Bmatrix} = [S^e]\{A_i^e\} + [M^e]\{A_i^e\} - [K_i^e] \end{aligned} \tag{3-23}$$

The coefficients of the matrix D are determined by only the Fourier coefficients in Equations (3-19) to (3-22). The matrix D acts as a reluctivity and is called the *Reluctivity Matrix*. On the other hand, the matrix N is a constant concerned with harmonic orders, and is called the *Harmonic Matrix*. The magnetic reluctivity coefficient matrix, D, can be derived as:

$$D = \begin{bmatrix} d_{11} & d_{12} & d_{13} & d_{14} & \cdots \\ d_{21} & d_{22} & d_{23} & d_{24} & \cdots \\ d_{31} & d_{32} & d_{33} & d_{34} & \cdots \\ d_{41} & d_{42} & d_{43} & d_{44} & \cdots \\ \cdots & \cdots & \cdots & \cdots & \ddots \end{bmatrix} = \frac{1}{2} \begin{bmatrix} 2\nu_0 - \nu_{2c} & \nu_{2s} & \nu_{2c} - \nu_{4c} & -\nu_{2s} + \nu_{4s} & \cdots \\ & 2\nu_0 + \nu_{2c} & \nu_{2s} + \nu_{4s} & \nu_{2c} + \nu_{4c} & \cdots \\ & & 2\nu_0 - \nu_{6c} & \nu_{6s} & \cdots \\ & Symmetry & & 2\nu_0 + \nu_{6c} & \cdots \\ & & & & \ddots \end{bmatrix} \tag{3-24}$$

and the harmonic coefficient matrix, harmonics number N, can be derived as:

$$N = \begin{bmatrix} 0 & -1 & 0 & 0 & 0 & 0 & \cdots \\ 1 & 0 & 0 & 0 & 0 & 0 & \cdots \\ 0 & 0 & 0 & -3 & 0 & 0 & \cdots \\ 0 & 0 & 3 & 0 & 0 & 0 & \cdots \\ 0 & 0 & 0 & 0 & 0 & -5 & \cdots \\ 0 & 0 & 0 & 0 & 5 & 0 & \cdots \\ \cdots & \cdots & \cdots & \cdots & \cdots & \cdots & \ddots \end{bmatrix} \tag{3-25}$$

However, the harmonic matrix can include DC, as well as even and order harmonics, depending on the application problems.

If only the fundamental frequency and third harmonics are considered, $\{A\}$ and $\{k\}$ can be expressed as:

$$\begin{aligned} &\{A_i^e\} = \left(\{A_k^i\}^T \ \{A_k^i\}^T \ \{A_k^i\}^T \right) = \\ &\{ A_{1s}^1 \ A_{1c}^1 \ A_{3s}^1 \ A_{3c}^1, \ A_{1s}^2 \ A_{1c}^2 \ A_{3s}^2 \ A_{3c}^2, \ A_{1s}^3 \ A_{1c}^3 \ A_{3s}^3 \ A_{3c}^3 \}^T \end{aligned} \tag{3-26}$$

and:

$$\{K_i^e\} = \left(\{K_1\}^T \ \{K_2\}^T \ \{K_3\}^T \right) = \frac{\Delta^e}{3} \{J_{1s} \ J_{1c} \ J_{3s} \ J_{3c} \ \cdots, \ \cdots, \ \cdots\}^T \tag{3-27}$$

Finally, the HBFEM system matrix can be obtained as:

$$[S]\{A_k\} + [M]\{A_k\} - [G_k]\{J_k\} = 0 \tag{3-28}$$

and the compact form can be written as:

$$([S]+[M])\{A_k\}-[K]=0 \tag{3-29}$$

where $[M]$ is the harmonic matrix:

$$[M]=\frac{\omega\sigma\Delta^e}{12}\begin{bmatrix} 2N & N & N \\ N & 2N & N \\ N & N & 2N \end{bmatrix} \tag{3-30}$$

The time harmonic matrix N can be expressed as:

$$N=\begin{bmatrix} 0 & -1 & 0 & 0 \\ 1 & 0 & 0 & 0 \\ 0 & 0 & 0 & -3 \\ 0 & 0 & 3 & 0 \end{bmatrix} \tag{3-31}$$

where only the order harmonic components are considered.

3.2.3 *Comparison of Time-Periodic Steady-State Nonlinear EM Field Analysis Method*

The harmonic balance technique was first introduced to analyze low-frequency EM field problems in the late 1980s [8]. Harmonic balance techniques were combined with FEM to accurately solve the problems arising from time-periodic steady-state nonlinear magnetic fields. The harmonic balance FEM (HB-FEM) uses a linear combination of sinusoids to build the solution, and represents waveforms using the sinusoid; coefficients are combined with the finite element method. It can directly solve the steady-state response of the EM field in the multi-frequency domain. Thus, the method is often considerably more efficient and accurate in capturing coupled nonlinear effects than the traditional FEM time-domain approach (when the field exhibits widely separated harmonics in the frequency spectrum domain and mild nonlinear behavior). The method can be used for weak and strong nonlinear time-periodic EM fields, including DC-biased transformer problems.

The harmonic balance FEM consists of approximating the time-periodic solution (magnetic potentials, currents, voltages, etc.) by a truncated Fourier series. Besides the frequency components of the excitation (e.g. applied voltages), the solution contains harmonics due to nonlinearity (magnetic saturation and nonlinear lumped electrical components) and due to movement (e.g. rotation). The HBFEM leads to a single, very large system of algebraic equations. Depending on the problem at hand, it may be much more efficient than the time domain approach (time stepping). Indeed, the latter inevitably requires stepping through the transient phenomenon before reaching the quasi-steady-state.

The global HBFEM system of algebraic equations is derived in an original way. The Galerkin approach is applied to both the space and time discretization. The time harmonic basis functions are used for approximating the periodic time variation, as well as for weighing the time domain equations in the fundamental period. Magnetic saturation and nonlinear electrical circuit coupling are thus easily accounted for, by means of the Newton-Raphson method. Rotation in 2D FEM models of rotating machines, using the moving band technique, can also be considered. The HBFEM has been validated by applying it to several test cases (transformer feeding a rectifier bridge, various synchronous and asynchronous machines, etc.) [9–11]. The harmonic waveforms of the magnetic field, currents and voltages, and so on, are shown to converge well, compared with those obtained with time stepping, as the spectrum of the HBFEM analysis is extended.

The major differences between HBFEM and the traditional time-domain and transient analysis based FEM are shown in Table 3.1.

Table 3.1 Comparison of time-periodic steady-state nonlinear EM field analysis method

	Frequency domain method	Step-by-step method	Time-periodic method	Harmonic balance method
Computation domain	Single frequency domain	Time domain	Time domain	Multiple frequency domain
For nonlinear and harmonic problems	Yes, for weak nonlinear fields, but not for harmonic problems	Yes, for weak nonlinear fields and harmonic problems	Yes, for weak nonlinear fields and harmonic problems	Yes, for weak and strongly nonlinear fields and harmonic problems
Widely separated harmonics	Cannot compute	Difficult to compute	Difficult to compute	Easy to compute (e.g. PWM)
Computation time depending on:	Number of degrees of freedom	Time step and number of degrees of freedom	Time step and number of degrees of freedom	Harmonic number and number of degrees of freedom
Computation accuracy	Large error at high frequency harmonics	Truncation error	Truncation error	Number of harmonics considered
Calculation of harmonic components	Impossible	Calculate from computation results	Calculate from computation results	Calculate several harmonics simultaneously (the result itself)
Post-processing of harmonics	Impossible	Indirectly	Indirectly	Directly

3.3 The Basic Concept of Harmonic Balance in EM Fields

When linear EM field systems are excited by a sinusoid, the steady-state response is sinusoidal at the same frequency as the input. While nonlinear EM field systems can exhibit a variety of significant and bizarre behaviors, the systems of interest to designers generally have a periodic steady-state response to a sinusoidal input; the response period is usually equal to that of the input. Since the response is periodic, it is represented as a Fourier series – that is, as a linear combination of sinusoids, whose periods divide the period of the response. If the excited EM field system contains two or more sinusoids that are harmonically unrelated, the system responds in steady-state, with components having the sum and difference frequencies of the input sinusoids and their harmonics. If the response contains an infinite number of sinusoids, then usually all but a few are negligible.

3.3.1 Definition of Harmonic Balance

As mentioned in section 3.1.1, the waveform can be defined as a summation of trigonometric function, as follows:

$$P(t) = \sum_{k=0}^{\infty} (P_{ks}\sin(k\omega t) + P_{kc}\cos(k\omega t)) \tag{3-32}$$

where $\omega = 2\pi/T$, $P_{ks}, P_{kc} \in R$ are harmonic coefficients, and the pair $P_k = [P_{ks}, P_{kc}] \in C$ are the Fourier coefficients of the Fourier series.

The term "harmonic balance" in nonlinear circuit analysis is defined as the set of port voltage waveforms that give the same currents in both linear and nonlinear subcircuits; when that set is found, it must be the solution. This principle actually gave the harmonic balance simulation its name because, through the interconnections, the currents of the linear and non-linear subcircuits have to be balanced at every harmonic frequency [1–5].

3.3.2 Harmonic Balance in EM Fields

Harmonic balance can also be applied to EM field analysis, as the fields that contain the harmonics also satisfy Maxwell's equations. The harmonics generated in EM fields can be described in the following three ways [10–12]:

- When a linear EM object is excited by sources which contain the harmonics, it will exhibit a harmonic field.
- When a nonlinear EM object is excited by a sinusoidal signal, it will exhibit harmonic fields.
- When both linear and nonlinear EM objects are excited by the sources which contain the harmonics, the result is a complex harmonic field.

One of the most obvious properties of a nonlinear system is the generation of harmonics. For example, we use the following equations to describe the quasi-static EM fields. These can be defined as follows:

3.3.2.1 A. Nonlinear Magnetic Field

$$\nabla \times \nu \nabla \times A + \sigma(\partial A/\partial t + \nabla\varphi) - J_s = 0 \tag{3-33}$$

3.3.2.2 B. Nonlinear Electric Field

$$\nabla \cdot \{\sigma E + \varepsilon(\partial A/\partial t)\} = 0 \tag{3-34}$$

where the electric field E, magnetic vector potential A, scalar potential φ on the arbitrary node i in the descretized system, and the source current density J_s can be respectively expressed as:

$$A^i = A_0^i + \sum_{k=1}^{\infty}\{A_{ks}^i \sin(k\omega t) + A_{kc}^i \cos(k\omega t)\} \tag{3-35a}$$

$$\varphi^i = \varphi_0^i + \sum_{k=1}^{\infty}\{\varphi_{ks}^i \sin(k\omega t) + \varphi_{kc}^i \cos(k\omega t)\} \tag{3-35b}$$

$$E^i = E_0^i + \sum_{k=1}^{\infty}\{E_{ks}^i \sin(k\omega t) + E_{kc}^i \cos(k\omega t)\} \tag{3-35c}$$

$$J_s = J_0 + \sum_{k=1}^{\infty}\{J_{ks} \sin(k\omega t) + J_{kc} \cos(k\omega t)\} \tag{3-35d}$$

where the vector A_0, E_0, J_0 and scalar φ_0 are the DC components, respectively. In practical applications, harmonic k is not infinite. Only a finite number is required in the real system.

The above harmonic approximation can also be expressed by using complex Fourier series with k harmonics:

$$A^i = R_e\left\{A_0^i + \sum_{k=1}^{\infty} A_k^i e^{jk\omega t}\right\} \tag{3-36a}$$

$$\varphi^i = R_e\left\{\varphi_0^i + \sum_{k=1}^{\infty} \varphi_k^i e^{jk\omega t}\right\} \tag{3-36b}$$

$$E^i = R_e\left\{E_0^i + \sum_{k=1}^{\infty} E_k^i e^{jk\omega t}\right\} \tag{3-36c}$$

$$J^i = R_e\left\{J_0^i + \sum_{k=1}^{\infty} J_k^i e^{jk\omega t}\right\} \tag{3-36d}$$

3.3.3 Nonlinear Medium Description

Nonlinear phenomena in EM fields are caused by nonlinear materials, which are normally field strength-dependent. Therefore, when the time-periodic quasi-static EM field is applied to the nonlinear material, the electromagnetic properties of the material will be functions of the EM field. They will also be time-dependent. Simple nonlinear materials and their harmonic expressions are defined as follows:

3.3.3.1 A. Magnetic Medium:

The magnetic reluctivity ν corresponding to $B(t)$ can be expressed as:

$$\nu(B(t)) = H\{B(t)\}/B(t) = \nu_0 + \sum_{k=2n-2}^{\infty} \{\nu_{ks}\sin(k\omega t) + \nu_{kc}\cos(k\omega t)\} \tag{3-37}$$

where flux density $B(t)$ is time dependent. This can be expressed by the B-H curve including the hysteresis characteristic [11]. ν $(=1/\mu)$ is the nonlinear magnetic reluctivity and the Fourier coefficients obtained from Equations (38) to (40), respectively.

$$\nu_0 = \frac{1}{T}\int_0^T \nu(t)dt \tag{3-38}$$

$$\nu_{ks} = \frac{2}{T}\int_0^T \nu(t)\cdot\sin(k\omega t)dt \tag{3-39}$$

$$\nu_{kc} = \frac{2}{T}\int_0^T \nu(t)\cdot\cos(k\omega t)dt \tag{3-40}$$

3.3.3.2 B. Electric Medium:

The electrical conductivity σ related $\boldsymbol{E}$(t) can be expressed as:

$$\sigma(B(t)) = \sigma(E(t)) = \sigma_0 + \sum_{k=2n-2}^{\infty} \{\sigma_{ks}\sin(k\omega t) + \sigma_{kc}\cos(k\omega t)\} \tag{3-41}$$

where σ is the field strength-dependent conductivity, and the Fourier coefficients are obtained from:

$$\sigma_0 = \frac{1}{T}\int_0^T \sigma(t)dt \tag{3-42}$$

$$\sigma_{ks} = \frac{2}{T}\int_0^T \sigma(t)\cdot\sin(k\omega t)dt \tag{3-43}$$

$$\sigma_{kc} = \frac{2}{T}\int_0^T \sigma(t)\cdot\cos(k\omega t)dt \tag{3-44}$$

3.3.4 *Boundary Conditions*

Since the trigonometric functions are orthogonal functions, the harmonic potential P_k (degrees of freedom) on the boundary satisfies Dirichlet and Neumann boundary conditions. The frequency-domain representation, or spectrum on each boundary node, can then be expressed as follows:

3.3.4.1 A. Dirichlet Boundary Condition:

$$P_k = \{P_0, P_{1s}, P_{1c}, P_{2s}, P_{2c}, \cdots P_{ks}, P_{kc}\}^T \tag{3-45}$$

3.3.4.2 B. Neumann Boundary Condition:

$$\frac{\partial P_k}{\partial n} = \left\{\frac{\partial P_0}{\partial n}, \frac{\partial P_{1s}}{\partial n}, \frac{\partial P_{1c}}{\partial n}, \frac{\partial P_{2s}}{\partial n}, \frac{\partial P_{2c}}{\partial n}, \cdots \frac{\partial P_{kc}}{\partial n}, \frac{\partial P_{ks}}{\partial n}\right\}^T \tag{3-46}$$

where the potential P_k is the sum of harmonics on each boundary node i.

3.3.5 *The Theory of HB-FEM in Nonlinear Magnetic Fields*

3.3.5.1 A. Current Source-Driven Fields

If the excitation waveform is a sinusoidal signal, the current source can be considered as a sine wave (Figure 3.4(a) – primary coil is an excitation coil). The resulting magnetic field contains all harmonic components (order components only) when the magnetic field becomes saturated, as shown in Figure 3.4(b).

The magnetizing characteristic of the core can be expressed as a function of the flux density B – that is,

$$H(B) = aB + bB^3 + cB^5 + \cdots + jB^{(2n-1)} + \cdots \tag{3-47}$$

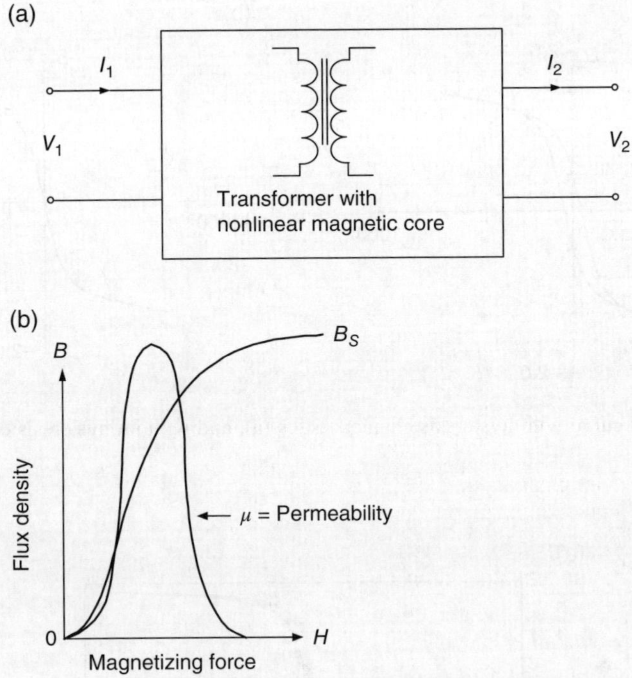

Figure 3.4 Magnetic core for a 2D transformer structure, and its typical B-H cure. (a) Transformer with nonlinear magnetic core. (b) B-H curve and permeability

where the hysteresis characteristic is neglected. The magnetic reluctivity ν can be written as:

$$\nu = \frac{1}{\mu} = H\left\{\frac{H(B)}{B}\right\} = a + bB^2 + cB^4 + \cdots + jB^{2n} + \cdots \tag{3-48}$$

When hysteresis characteristic is concerned as shown in Figure 3.5(a), the expression $H(B)$ can be obtained from a B-H curve data table, or a function of the flux density B with Hysteresis term [3], as follows:

$$H(B) \cong H_{Sat}(B) + H_{Hys}(B)\frac{dB}{dt} \tag{3-49}$$

However, in the DC-biased case [11,14], the B-H curve with hysteresis characteristics will be illustrated as Figure 3.6(b), and the expression of $H(B)$ can be obtained from a B-H curve data table.

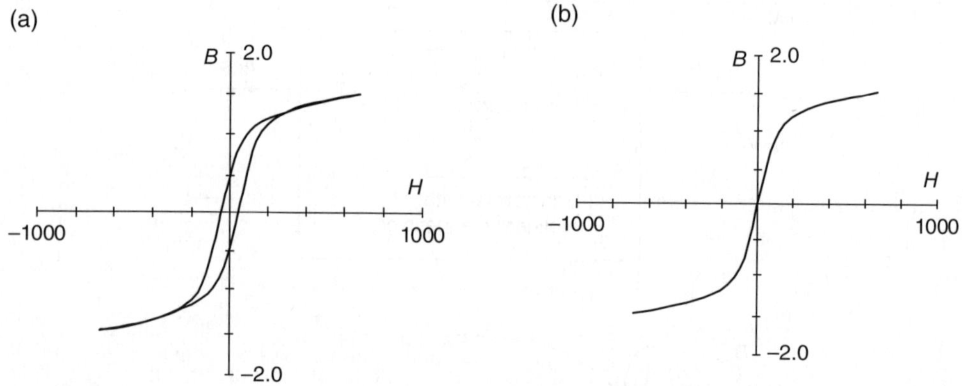

Figure 3.5 B-H curve with hysteresis characteristics (a), and without hysteresis characteristics (b)

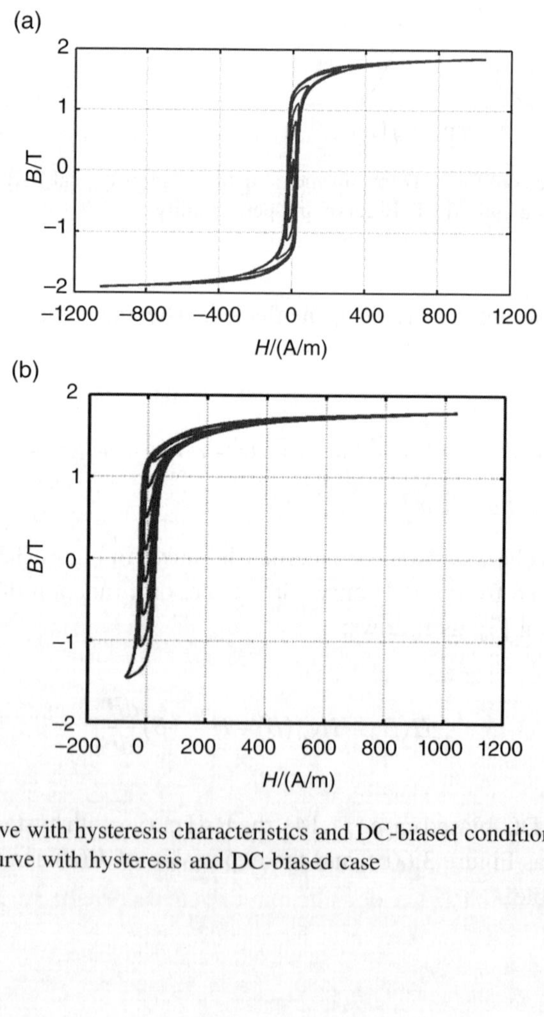

Figure 3.6 B-H curve with hysteresis characteristics and DC-biased condition. (a) H-B curve with hysteresis; (b) H-B curve with hysteresis and DC-biased case

3.3.5.2 B. HBFEM Formulation

In order to analyze a time-periodic nonlinear magnetic field, a combination of FEM and the harmonic balance method is developed. To explain the principle of a new approach, the basic formulation of the harmonic balance finite element method (HBFEM) can be derived. For simplicity of fundamental formulation, a time-periodic nonlinear magnetic field is assumed as 2-dimensional in the (x, y) plane, and quasi-stationary. Therefore, the vector potential $A=(0, 0, A)$ satisfies in the region of interest surrounded by some boundary conditions. To calculate such a quasi-static magnetic field, the following equation (3-50) can be used:

$$\frac{\partial}{\partial x}\left(\nu\frac{\partial A}{\partial x}\right)+\frac{\partial}{\partial y}\left(\nu\frac{\partial A}{\partial y}\right)-J_s+\sigma\frac{\partial A}{\partial t}=0 \tag{3-50}$$

where ν and σ are magnetic reluctivity and conductivity.

Based on harmonics balance theory, the governing equations containing harmonics can be also solved by using a FEM based numerical approach. Assuming $\nabla\varphi=0$ in the two-dimensional case, and using Galerkin's method to discretize the governing equation for two-dimensional problems, equation (3-50) can be written in an integral form that is given as:

$$G=\iint\limits_S\left\{\frac{\partial N_i}{\partial x}\nu\frac{\partial A}{\partial x}+\frac{\partial N_i}{\partial y}\nu\frac{\partial A}{\partial y}\right\}dxdy-\iint\limits_S\left\{J_s-\sigma\frac{\partial A}{\partial t}\right\}N_i dxdy=0 \tag{3-51}$$

where the interpolating function $N_i(x, y)$ is:

$$N_i=(a_i+b_ix+c_iy)/2\Delta \tag{3-52}$$

and where:

$$a_i=x_iy_k-x_ky_j \tag{3-53}$$

$$b_i=y_j-y_k \tag{3-54}$$

$$c_i=x_k-x_j \tag{3-55}$$

Δ is the cross-section of the element, $x_{i,j,k}$, $y_{i,j,k}$ are coordinates of the node.

The time-periodic solution (harmonic problem) will be investigated when an alternating magnetizing current is applied. According to the harmonic balance method, all variables (i.e. vector potentials, flux densities and applied current) are approximated as a summation of all harmonic solutions. For simplicity, the time-periodic solutions are given as the sum of the fundamental component and the third harmonic – that is:

$$A^i = \sum_{k=2n-1}^{\infty} \{A^i_{ks}\sin(k\omega t) + A^i_{kc}\cos(k\omega t)\} \tag{3-56a}$$

$$B_x = \sum_{k=2n-1}^{\infty} \{B_{xks}\sin(k\omega t) + B_{xkc}\cos(k\omega t)\} \tag{3-56b}$$

$$B_y = \sum_{k=2n-1}^{\infty} \{B_{yks}\sin(k\omega t) + B_{ykc}\cos(k\omega t)\} \tag{3-56c}$$

$$J_0 = \sum_{k=2n-1}^{\infty} \{J_{ks}\sin(k\omega t) + J_{kc}\cos(k\omega t)\} \tag{3-56d}$$

where ω is the fundamental angular frequency.

The magnetizing characteristics of a core can be expressed as a power series. The *B-H* curve is approximated as a third order power series – that is:

$$H(B) = aB + bB^3 \tag{3-57}$$

where the hysteresis characteristic is neglected. The magnetic reluctivity is written as:

$$\nu = H/B = a + bB^2 \tag{3-58}$$

where B is the magnitude of the flux density.

Substituting Equations (3-56) and (3-58) into Equation (3-51), carry out the integration in the ordinary FEM procedure. When the interpolating function is N_i and the integration is done in one triangular element, each term in Equation (3-51) is given as:

$$\begin{aligned}
&\iint_e \left\{\frac{\partial N_1}{\partial x}\left(\nu\frac{\partial A}{\partial x}\right)\right\}dxdy \\
&= \sum_{j=1,2,3}\frac{b_1 b_j}{4\Delta}\{(d_{11}A_{1s}{}^j + d_{12}A_{1c}{}^j + d_{13}A_{3s}{}^j + d_{14}A_{3c}{}^j)\sin\omega t \\
&+ (d_{21}A_{1s}{}^j + d_{22}A_{1c}{}^j + d_{23}A_{3s}{}^j + d_{24}A_{3c}{}^j)\cos\omega t \\
&+ (d_{31}A_{1s}{}^j + d_{32}A_{1c}{}^j + d_{33}A_{3s}{}^j + d_{34}A_{3c}{}^j)\sin 3\omega t \\
&+ (d_{41}A_{1s}{}^j + d_{42}A_{1c}{}^j + d_{43}A_{3s}{}^j + d_{44}A_{3c}{}^j)\cos 3\omega t\}
\end{aligned} \tag{3-59a}$$

$$\iint_e \left\{ \frac{\partial N_1}{\partial y} \left(\nu \frac{\partial A}{\partial y} \right) \right\} dxdy$$

$$\begin{aligned}
&= \sum_{j=1,2,3} \frac{c_1 c_j}{4\Delta} \{ \left(d_{11} A_{1s}{}^j + d_{12} A_{1c}{}^j + d_{13} A_{3s}{}^j + d_{14} A_{3c}{}^j \right) \sin\omega t \\
&+ \left(d_{21} A_{1s}{}^j + d_{22} A_{1c}{}^j + d_{23} A_{3s}{}^j + d_{24} A_{3c}{}^j \right) \cos\omega t \\
&+ \left(d_{31} A_{1s}{}^j + d_{32} A_{1c}{}^j + d_{33} A_{3s}{}^j + d_{34} A_{3c}{}^j \right) \sin 3\omega t \\
&+ \left(d_{41} A_{1s}{}^j + d_{42} A_{1c}{}^j + d_{43} A_{3s}{}^j + d_{44} A_{3c}{}^j \right) \cos 3\omega t \}
\end{aligned} \tag{3-59b}$$

$$\iint_e \left\{ \left(J_c - \sigma \frac{\partial A}{\partial t} \right) \right\} N_1 dxdy$$

$$\begin{aligned}
&= \left\{ \Delta J_{1s/3} + \sigma\omega \left(2A_{1c}{}^1 + A_{1c}{}^2 + A_{1c}{}^3 \right) \Delta/12 \right\} \sin\omega t \\
&+ \left\{ \Delta J_{1c/3} - \sigma\omega \left(2A_{1s}{}^1 + A_{1c}{}^2 + A_{1s}{}^3 \right) \Delta/12 \right\} \cos\omega t \\
&+ \left\{ \Delta J_{3s/3} + 3\sigma\omega \left(2A_{1s}{}^1 + A_{3c}{}^2 + A_{3c}{}^3 \right) \Delta/12 \right\} \sin 3\omega t \\
&+ \left\{ \Delta J_{3c/3} - 3\sigma\omega \left(2A_{3s}{}^1 + A_{3s}{}^2 + A_{3s}{}^3 \right) \Delta/12 \right\} \cos 3\omega t +
\end{aligned} \tag{3-59c}$$

and

$$\begin{aligned}
&\iint N_i N_j dxdy = 2\Delta/12 \quad if \quad i = j \\
&\iint N_i N_j dxdy = \Delta/12 \quad if \quad i \neq j
\end{aligned} \tag{3-59d}$$

When each coefficient at sine- and cosine-components is equated, the matrix equation can be obtained as (3-60):

$$\begin{aligned}
&\frac{1}{4\Delta} \{ (b_i b_1 + c_i c_1) D \quad (b_i b_2 + c_i c_2) D \quad (b_i b_3 + c_i c_3) D \} \{A\} \\
&+ \frac{\sigma\omega\Delta}{12} \{ 2N \quad N \quad N \} \{A\} - \{K_1\} = 0
\end{aligned} \tag{3-60}$$

where column vectors $\{A\}$ and $\{K_1\}$ are:

$$\{A\} = \{ A_{1s}^1 \ A_{1c}^1 \ A_{3s}^1 \ A_{3c}^1, \ A_{1s}^2 \ A_{1c}^2 \ A_{3s}^2 \ A_{3c}^2, \ A_{1s}^3 \ A_{1c}^3 \ A_{3s}^3 \ A_{3c}^3 \} \tag{3-61}$$

$$\{K_1\} = \frac{\Delta}{3} \{ J_{1s} \ J_{1c} \ J_{3s} \ J_{3c} \} \tag{3-62}$$

The matrices D and N are:

$$D=\begin{bmatrix} d_{11} & d_{12} & d_{13} & d_{14} \\ d_{21} & d_{22} & d_{23} & d_{24} \\ d_{31} & d_{32} & d_{33} & d_{34} \\ d_{41} & d_{42} & d_{43} & d_{44} \end{bmatrix}$$

$$=\begin{bmatrix} \alpha+\beta(B_0-B_{2c}/2) & \beta B_{2s}/2 & \beta(B_{2c}-B_{4c})/2 & \beta(B_{4s}-B_{2s})/2 \\ & \alpha+\beta(B_0+B_{2c}/2) & \beta(B_{2s}+B_{4s})/2 & \beta(B_{2c}+B_{4c})/2 \\ & & \alpha+\beta(B_0-B_{6c}/2) & \beta B_{6s}/2 \\ & symmetry & & \alpha+\beta(B_0+B_{6c}/2) \end{bmatrix} \tag{3-63}$$

where B can be obtained from the following expressions:

$$\begin{aligned}
B_0 &= \left(B_{x1s}^{\ 2}+B_{x1c}^{\ 2}+B_{x3s}^{\ 2}+B_{x3c}^{\ 2}\right)/2+\left(B_{y1s}^{\ 2}+B_{y1c}^{\ 2}+B_{y3s}^{\ 2}+B_{y3c}^{\ 2}\right)/2 \\
B_{2s} &= B_{x1s}B_{x1c}+B_{x1c}B_{x3s}-B_{x1s}B_{x3c}+B_{y1s}B_{y1c}+B_{y1c}B_{y3s}-B_{y1s}B_{y3c} \\
B_{2c} &= \left(B_{x1c}^{\ 2}-B_{x1s}^{\ 2}\right)/2+B_{x1s}B_{x3s}+B_{x1c}B_{x3c} \\
&\quad +\left(B_{y1c}^{\ 2}-B_{y1s}^{\ 2}\right)/2+B_{y1s}B_{y3s}+B_{y1c}B_{y3c} \\
B_{4s} &= B_{x1s}B_{x3c}+B_{x1c}B_{x3s}+B_{y1s}B_{y3c}+B_{y1c}B_{y3s} \\
B_{4c} &= B_{x1c}B_{x3c}-B_{x1s}B_{x3s}+B_{y1c}B_{y3c}-B_{y1s}B_{y3s} \\
B_{6s} &= B_{x3s}B_{x3c}+B_{y3s}B_{y3c} \\
B_{6c} &= \left(B_{x3c}^{\ 2}-B_{x3s}^{\ 2}\right)/2+\left(B_{y3c}^{\ 2}-B_{y3s}^{\ 2}\right)/2
\end{aligned} \tag{3-64}$$

The harmonic matrix can be expressed as:

$$N=\begin{bmatrix} 0 & -1 & 0 & 0 \\ 1 & 0 & 0 & 0 \\ 0 & 0 & 0 & -3 \\ 0 & 0 & 3 & 0 \end{bmatrix} \tag{3-65}$$

When three interpolations functions, N_i (i = 1, 2, 3) on the first order triangular element are applied, the combination of the matrix equations as Equation (3-60) is expressed as:

$$\frac{1}{4\Delta^e}\begin{bmatrix}(b_1b_1+c_1c_1)D & (b_1b_2+c_1c_2)D & (b_1b_3+c_1c_3)D\\ (b_2b_1+c_2c_1)D & (b_2b_2+c_2c_2)D & (b_2b_3+c_2c_3)D\\ (b_3b_1+c_3c_1)D & (b_3b_2+c_3c_2)D & (b_3b_3+c_3c_3)D\end{bmatrix}\begin{Bmatrix}A_1^e\\ A_2^e\\ A_3^e\end{Bmatrix}$$

$$+\frac{\sigma\omega\Delta^e}{12}\begin{bmatrix}2N & N & N\\ N & 2N & N\\ N & N & 2N\end{bmatrix}\begin{Bmatrix}A_1^e\\ A_2^e\\ A_3^e\end{Bmatrix}-\begin{Bmatrix}K_1^e\\ K_2^e\\ K_3^e\end{Bmatrix}=0 \tag{3-66}$$

Where:

$$\{K_i^e\}=(\{K_1\}^T\{K_2\}^T\{K_3\}^T)=\frac{\Delta^e}{3}\{J_{1s}\ \ J_{1c}\ \ J_{3s}\ \ J_{3c},\ \ \cdots,\ \ \cdots\}^T \tag{3-67}$$

The coefficients of the matrix D are determined by both parameters of the B-H curve and flux densities of each harmonic. Compared with the conventional FEM, it acts as a reluctivity and is called the *Reluctivity Matrix*. On the other hand, the matrix N is a constant concerned with harmonic orders, and is called the *Harmonic Matrix*.

The system equation for the entire region is obtained by the same procedure as the conventional FEM, and is also solved by the same iteration procedure for a nonlinear static field. The above method does not require the calculation concerned with time-variation, and can solve the time-periodic field distributions of AC machines with non-linear characteristics. However, the system matrix is four times bigger than the number of vertices when the fundamental and third harmonic components are taken into consideration.

3.3.6 The Generalized HBFEM

Considering the orthogonal characteristics of trigonometric functions, substituting magnetic vector A, interpolation functions N_i, magnetic reluctivity ν and current density J_s into Equation (3-56d), and using harmonic balance techniques, the following harmonic balance FEM matrix equation can be obtained (considering only odd harmonics for a single element):

$$G^e = \frac{1}{4\Delta^e}\begin{bmatrix}(b_1b_1+c_1c_1)D & (b_1b_2+c_1c_2)D & (b_1b_3+c_1c_3)D\\ (b_2b_1+c_2c_1)D & (b_2b_2+c_2c_2)D & (b_2b_3+c_2c_3)D\\ (b_3b_1+c_3c_1)D & (b_3b_2+c_3c_2)D & (b_3b_3+c_3c_3)D\end{bmatrix}\begin{Bmatrix}A_1^e\\ A_2^e\\ A_3^e\end{Bmatrix} + \frac{\sigma\omega\Delta^e}{12}\begin{bmatrix}2N & N & N\\ N & 2N & N\\ N & N & 2N\end{bmatrix}\begin{Bmatrix}A_1^e\\ A_2^e\\ A_3^e\end{Bmatrix} - \begin{Bmatrix}K_1^e\\ K_2^e\\ K_3^e\end{Bmatrix} \tag{3-68}$$

where the vector potential $\{A^e\}$ is called the frequency-domain representation, Δ is the area of the triangular elements and ω indicates the fundamental frequency. The constants b and c are obtained from the x and y coordinates, which are $b_{ie} = x_{je}y_{ke} - x_{ke}y_{je}$, $c_{ie} = y_{je} - y_{ke}$. The magnetic vector potential $\{A_i^e\}$ and current density $\{K_i^e\}$ are expressed as:

$$\{A^e\} = \{\, A_{1s}^1 \; A_{1c}^1 \; A_{3s}^1 \; A_{3c}^1, \; A_{1s}^2 \; A_{1c}^2 \; A_{3s}^2 \; A_{3c}^2, \; A_{1s}^3 \; A_{1c}^3 \; A_{3s}^3 \; A_{3c}^3 \,\} \tag{3-69}$$

$$\{K_i^e\} = \frac{\Delta^e}{3}\{J_{1s} \; J_{1c} \; J_{3s} \; J_{3c} \; \cdots, \cdots, \cdots\}^T \tag{3-70}$$

Using $\nu = \alpha + \beta B^2$, the matrix equation (3-63) can be made in a compact form, as:

$$D = \begin{bmatrix}\alpha+\beta(B_0-B_{2c}/2) & \beta B_{2s}/2 & \beta(B_{2c}-B_{4c})/2 & \beta(B_{4s}-B_{2s})/2\\ & \alpha+\beta(B_0+B_{2c}/2) & \beta(B_{2s}+B_{4s})/2 & \beta(B_{2c}+B_{4c})/2\\ & & \alpha+\beta(B_0-B_{6c}/2) & \beta B_{6s}/2\\ & symmetry & & \alpha+\beta(B_0+B_{6c}/2)\end{bmatrix} \tag{3-71}$$

The coefficients of matrix D are determined by only the Fourier coefficients in Equations (3-41) to (3-44). Matrix D acts as a reluctivity and is called the *Reluctivity Matrix*. On the other hand, the matrix N is a constant concerned with harmonic orders and is called the *Harmonic Matrix*. The magnetic reluctivity coefficient matrix, D, can be derived as:

$$D = \begin{bmatrix}d_{11} & d_{12} & d_{13} & d_{14} & \cdots\\ d_{21} & d_{22} & d_{23} & d_{24} & \cdots\\ d_{31} & d_{32} & d_{33} & d_{34} & \cdots\\ d_{41} & d_{42} & d_{43} & d_{44} & \cdots\\ \cdots & \cdots & \cdots & \cdots & \ddots\end{bmatrix} = \frac{1}{2}\begin{bmatrix}2\nu_0-\nu_{2c} & \nu_{2s} & \nu_{2c}-\nu_{4c} & -\nu_{2s}+\nu_{4s} & \cdots\\ & 2\nu_0+\nu_{2c} & \nu_{2s}+\nu_{4s} & \nu_{2c}+\nu_{4c} & \cdots\\ & & 2\nu_0-\nu_{6c} & \nu_{6s} & \cdots\\ & Symmetry & & 2\nu_0+\nu_{6c} & \cdots\\ & & & & \ddots\end{bmatrix} \tag{3-72}$$

and the harmonic coefficient matrix, harmonics number N, can be derived as:

$$N^e = \begin{bmatrix} 0 & -1 & 0 & 0 & 0 & 0 & \cdots \\ 1 & 0 & 0 & 0 & 0 & 0 & \cdots \\ 0 & 0 & 0 & -3 & 0 & 0 & \cdots \\ 0 & 0 & 3 & 0 & 0 & 0 & \cdots \\ 0 & 0 & 0 & 0 & 0 & -5 & \cdots \\ 0 & 0 & 0 & 0 & 5 & 0 & \cdots \\ \cdots & \cdots & \cdots & \cdots & \cdots & \cdots & \ddots \end{bmatrix} \tag{3-73}$$

Finally, the system matrix equation for current source excitation can then be written in a compact form:

$$[S]\{A\} + [M]\{A\} - \{K\} = 0 \tag{3-74}$$

where $[S]$ is the system matrix and $[M]$ is the harmonic related matrix. All harmonic components of magnetic vector potential A can be directly obtained by solving this system matrix equation.

The system equation for the entire region is obtained by the same procedure as the conventional FEM, and is solved by the iteration procedure for a nonlinear static field. Therefore, it is possible to obtain the approximate solutions by the sum of the finite harmonics, because the higher order components are reduced. In order to compare the size of the system matrix, we consider the magnetic field region where the number of unknown vector potential is n. It is assumed that the system matrix is the sparse matrix with the bandwidth k in the static field analysis. When the harmonic components, up to $(2m - 1)$ order, are considered in the HBFEM in the same problem, the size of the block matrices D and N are $2m^*2m$ respectively, and the order of the system matrix is $2m$ times bigger than the number of nodes in Figure 3.7. The width of the sparse matrix increases by $2m$ times.

3.4 HBFEM for Electromagnetic Field and Electric Circuit Coupled Problems

3.4.1 HBFEM in Voltage Source-Driven Magnetic Field

In most cases, HVDC transformer, voltage tripler, pulse width modulation (PWM) and zero-current switched resonant converters, including line level control (LLC) converters, as shown in Figure 3.8, can be considered as a voltage source to the magnetic system. They are always coupled to the external circuits. Therefore, the current in the input

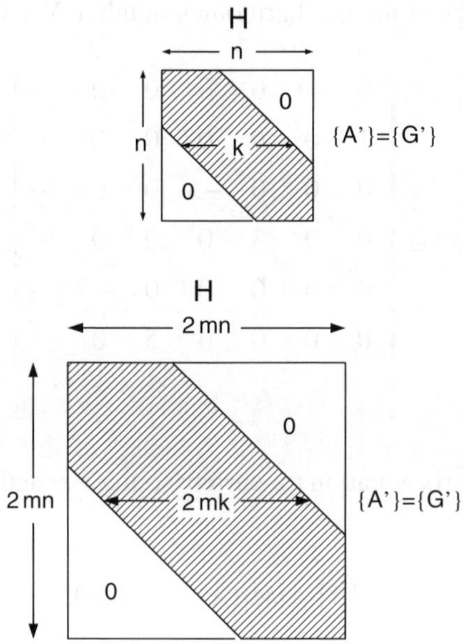

Figure 3.7 Size of the system matrix. (a) Static FEM; (b) HBFEM

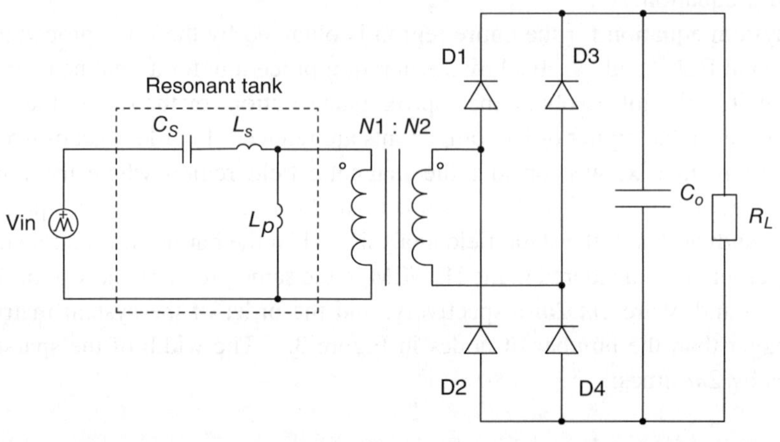

Figure 3.8 Simulation model of LLC converter with resonant tank with idea transformer

circuits will be unknown, and the saturation of the current waveform occurs because of the nonlinear characteristic of the magnetic core [10–12].

3.4.2 *Generalized Voltage Source-Driven Magnetic Field*

When electrical devices are excited by voltage source, such as LLC converter transformers and HVDC transformer under DC bias, the excitation current is unknown and

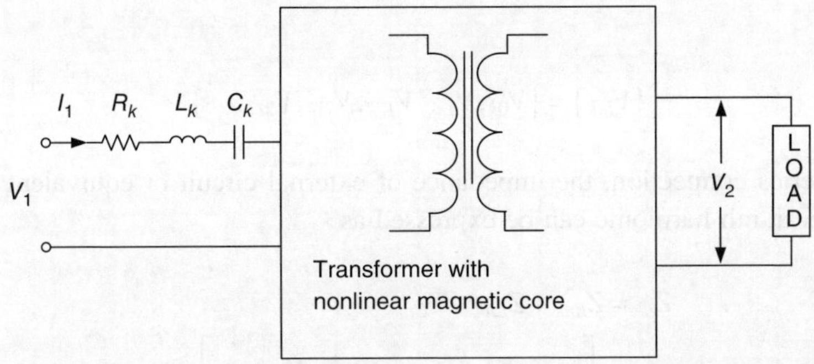

Figure 3.9 Coupling between the electric circuit and the magnetic field

Equation (3-68) is no longer applicable. In this case, the coupling between the electric circuit and the magnetic field should be taken into account [10]. The generalized electric circuit coupled with magnetic field can be illustrated as Figure 3.9.

According to Kirchhoff's law, the applied voltage on the external port can be defined as follows:

$$
\begin{aligned}
V_{ink} &= V_k + R_k I_k + L_k \frac{dI_k}{dt} + \frac{1}{C_k}\int I_k dt \\
&= \left[C_k^e\right]\{A\} + S_{ck}[Z_k]\{J_k\}
\end{aligned}
\tag{3-75}
$$

Where V_{ink} is the input voltage of circuit k, V_{k} is the corresponding induced electromotive force, S_{ck} and Z_{k} are the cross areas and impedance of winding k respectively, C_{k} and L_{k} are the capacitance and inductance respectively in circuit k. The induced electromotive force can be obtained based on Faraday's law:

$$
\begin{aligned}
V_k &= \frac{d\Psi}{dt} = \frac{d}{dt}\iint \boldsymbol{B}\cdot d\boldsymbol{S} \\
&= \frac{d}{dt}\iint \nabla\times\boldsymbol{A}\, d\boldsymbol{S} = \frac{d}{dt}\oint_f \boldsymbol{A} d\boldsymbol{l} \\
&= \frac{\omega N_k d_0}{3S_{ck}}[N\ \ N\ \ N]\begin{Bmatrix}\{A^1\}\\ \{A^2\}\\ \{A^3\}\end{Bmatrix}
\end{aligned}
\tag{3-76}
$$

where d_0 is the depth in the z-direction, N_k is the turn number of the k-th winding. V_k can be rewritten as a compact form as below:

$$
\{V_k\} = \left[C_k^e\right]\{A\} \tag{3-77}
$$

where

$$\{V_{ink}\} = \{V_{0k}, V_{1sk}, V_{1ck}, V_{2sk}, V_{2ck}, \cdots\}^T \tag{3-78}$$

In the series connection, the impedance of external circuit or equivalent circuit of transformer at nth harmonic can be expressed as

$$\begin{aligned} Z_{nk} &= Z_{Rnk} + Z_{Lnk} + Z_{Cnk} \\ &= \begin{bmatrix} R_k & -n\omega L_k + \dfrac{1}{n\omega C_k} \\ n\omega L_k - \dfrac{1}{n\omega C_k} & R_k \end{bmatrix} \end{aligned} \tag{3-79}$$

The expression of the impedance matrix is given by Equations (3-80) through (3-82), respectively,

$$[Z_{Rnk}] = \begin{bmatrix} R_k & 0 \\ 0 & R_k \end{bmatrix} \tag{3-80}$$

$$[Z_{Cnk}] = \frac{1}{n} \begin{bmatrix} 0 & \dfrac{1}{\omega C_k} \\ -\dfrac{1}{\omega C_k} & 0 \end{bmatrix} \tag{3-81}$$

$$[Z_{Lnk}] = n \begin{bmatrix} 0 & -\omega L_k \\ \omega L_k & 0 \end{bmatrix} \tag{3-82}$$

The generalized impedance matrix can be expressed as

$$[Z_k] = \begin{bmatrix} Z_{0k} & 0 & 0 & \cdots \\ 0 & Z_{1k} & 0 & \cdots \\ 0 & 0 & Z_{2k} & \cdots \\ \vdots & \vdots & \vdots & \ddots \end{bmatrix} \tag{3-83}$$

The input voltage can be defined as:

$$\{V_{ink}\} = \{V_k\} + S_{ck}[Z_k]\{J_k\} = [C_k]\{A\} + S_{ck}[Z_k]\{J_k\} \tag{3-84}$$

where matrix $[Z_k]$ is the circuit impedance including the resistance of windings and leakage inductance corresponding to the harmonics, and S_{ck} is the area of windings. The input voltages $\{V_{ink}\}$, including all harmonic components which have a known value, are expressed as follows:

$$\{V_{ink}\} = \{V_{0k}, V_{1sk}, V_{1ck}, V_{2sk}, V_{2ck}, \cdots\}^T \tag{3-85}$$

where $V_{in,0k}$ (=0) is a DC component.

The circuit-related matrix and FEM system matrix can be obtained as below respectively,

$$[C_k]\{A\} + S_{ck}[Z_k]\{J_k\} = \{V_{ink}\} \tag{3-86}$$

$$([S] + [N])\{A\} - [G_k]\{J_k\} = 0 \tag{3-87}$$

where $\{I_k\} = S_{ck}\{J_k\}$, and $[G_k]$ is the current density coefficient obtained from a single element of winding area, that is, $[G_e] = \Delta_e/3$.

Combining Equations (3-86) and (3-87), the global system matrix equations for multiple input and output are obtained. The harmonic balance FEM matrix equations for voltage source excitation can therefore be expressed as:

$$\begin{bmatrix} [H] & -[G_1] & -[G_2] & \cdots & -[G_k] & \cdots \\ [C_1] & S_{c1}[Z_1] & 0 & 0 & 0 & \cdots \\ [C_2] & 0 & S_{c2}[Z_2] & 0 & 0 & \cdots \\ \vdots & 0 & 0 & \ddots & 0 & \cdots \\ [C_k] & 0 & 0 & 0 & S_{ck}[Z_k] & \cdots \\ \vdots & \vdots & \vdots & \vdots & \vdots & \ddots \end{bmatrix} \begin{Bmatrix} \{A\} \\ \{J_1\} \\ \{J_2\} \\ \vdots \\ \{J_k\} \\ \vdots \end{Bmatrix} = \begin{Bmatrix} 0 \\ \{V_{in1}\} \\ \{V_{in2}\} \\ \vdots \\ \{V_{ink}\} \\ \vdots \end{Bmatrix} \tag{3-88}$$

where $\{A\}$ and $\{J_k\}$ are unknown and can be calculated by solving the system matrix equation, $[H]$ is the matrix obtained from $([S] + [N])$, and:

$$[C_k] = \frac{\omega d_0 \Delta}{3S_{ck}}[N \;\; N \;\; N] \tag{3-89}$$

is the geometric coefficient related to transformer windings.

From Equation (3-68), Equation (3-87) can be rewritten as:

$$G^e = \frac{1}{4\Delta^e}\begin{bmatrix}(b_1b_1+c_1c_1)D & (b_1b_2+c_1c_2)D & (b_1b_3+c_1c_3)D\\(b_2b_1+c_2c_1)D & (b_2b_2+c_2c_2)D & (b_2b_3+c_2c_3)D\\(b_3b_1+c_3c_1)D & (b_3b_2+c_3c_2)D & (b_3b_3+c_3c_3)D\end{bmatrix}\begin{Bmatrix}A_1^e\\A_2^e\\A_3^e\end{Bmatrix}$$
$$+\frac{\sigma\omega\Delta^e}{12}\begin{bmatrix}2N & N & N\\N & 2N & N\\N & N & 2N\end{bmatrix}\begin{Bmatrix}A_1^e\\A_2^e\\A_3^e\end{Bmatrix}-\begin{Bmatrix}K_1^e\\K_2^e\\K_3^e\end{Bmatrix}=[S^e]\{A^e\}+[N^e][A^e]-\{K^e\} \tag{3-90}$$

However, D and N have different details when the DC component is considered, where D is:

$$D=\begin{bmatrix}d_{11} & d_{12} & d_{13} & d_{14} & \cdots\\d_{21} & d_{22} & d_{23} & d_{24} & \cdots\\d_{31} & d_{32} & d_{33} & d_{34} & \cdots\\d_{41} & d_{42} & d_{43} & d_{44} & \cdots\\\cdots & \cdots & \cdots & \cdots & \ddots\end{bmatrix}=\frac{1}{2}\begin{bmatrix}2\nu_0-\nu_{2c} & \nu_{2s} & \nu_{2c}-\nu_{4c} & -\nu_{2s}+\nu_{4s} & \cdots\\ & 2\nu_0+\nu_{2c} & \nu_{2s}+\nu_{4s} & \nu_{2c}+\nu_{4c} & \cdots\\ & & 2\nu_0-\nu_{6c} & \nu_{6s} & \cdots\\ & Symmetry & & 2\nu_0+\nu_{6c} & \cdots\\ & & & & \ddots\end{bmatrix} \tag{3-91}$$

and the harmonic matrix, N, is:

$$N^e=\begin{bmatrix}0 & -1 & 0 & 0 & 0 & 0 & \cdots\\1 & 0 & 0 & 0 & 0 & 0 & \cdots\\0 & 0 & 0 & -3 & 0 & 0 & \cdots\\0 & 0 & 3 & 0 & 0 & 0 & \cdots\\0 & 0 & 0 & 0 & 0 & -5 & \cdots\\0 & 0 & 0 & 0 & 5 & 0 & \cdots\\\cdots & \cdots & \cdots & \cdots & \cdots & \cdots & \ddots\end{bmatrix} \tag{3-92}$$

The compact system matrix can be expressed as below:

$$\begin{bmatrix}[H] & [G]\\ [C_k] & S_{ck}[Z_k]\end{bmatrix}\begin{Bmatrix}A_k\\J_k\end{Bmatrix}=\begin{Bmatrix}0\\V_{ink}\end{Bmatrix} \tag{3-93}$$

where $[H]$ is the FEM system matrix, $[G]$ is the current density-related coefficient matrix, $[C]$ is the voltage-related coefficient matrix, and $[Z]$ is the impedance matrix.

3.5 HBFEM for a DC-Biased Problem in High-Voltage Power Transformers

3.5.1 DC-Biased Problem in HVDC Transformers

A typical HVDC transmission system is illustrated in Figure 3.10. It consists of a DC transmission line connecting two AC systems. A converter at one end of the line converts AC power into HVDC power, while a similar converter at the other end reconverts the DC power into AC power. Therefore, one converter acts as a rectifier, the other as an inverter. The basic purpose of the converter transformer on the rectifier side is to transform the AC network voltage to yield the AC voltage required by the converter [13]. Three-phase transformers, connected in either wye-wye or wye-delta, are used.

The magnetostrictive strain is not truly sinusoidal in character, which leads to the introduction of the harmonics. With a DC-biased transformer, magnetizing saturation will also cause some harmonics [14]. The B-H curve with hysteresis characteristics and a DC biased condition is illustrated in Figure 3.11, and the expression of $H(B)$ can be obtained from a B-H curve data table.

The harmonics in a transformer noise may have a substantial effect on an observer, even though their level is 10 dB or more lower than the 100 Hz fundamental. In fact, the most striking point is the strength of the component at 100 Hz, or twice the normal operating frequency of the transformer. Deviation from a "square-law" magnetostrictive characteristic would result in even harmonics (at 200, 400, 600 Hz, etc.), while the different values of the magnetostrictive strain for increasing and decreasing flux densities – a pseudo-hysteresis effect – lead to the introduction of odd harmonics (at 300, 500, 700 Hz, etc.). If any part of the structure has a natural frequency at or near 100, 200, 300, 400 Hz, and so on, the result will be an amplification of noise at that particular frequency.

3.5.2 HBFEM Model of HVDC Transformer

Consider a three-phase HVDC transformer model with a three-phase voltage-driven source connected to the magnetic system, which is always coupled to the external

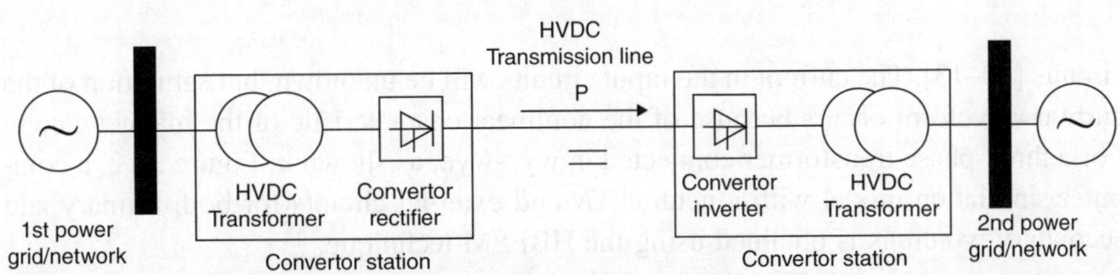

Figure 3.10 HVDC power transmission system

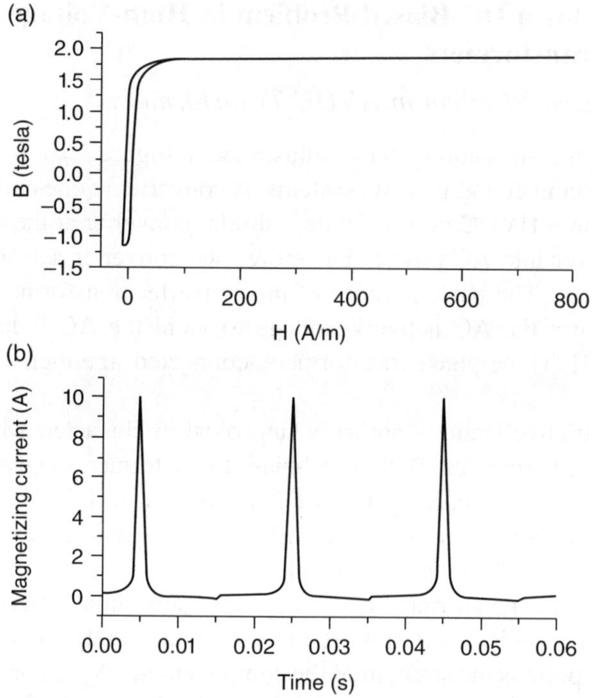

Figure 3.11 B-H curve with hysteresis characteristics and a DC biased condition. (a) DC-biased hysteresis loop; (b) Magnetizing current

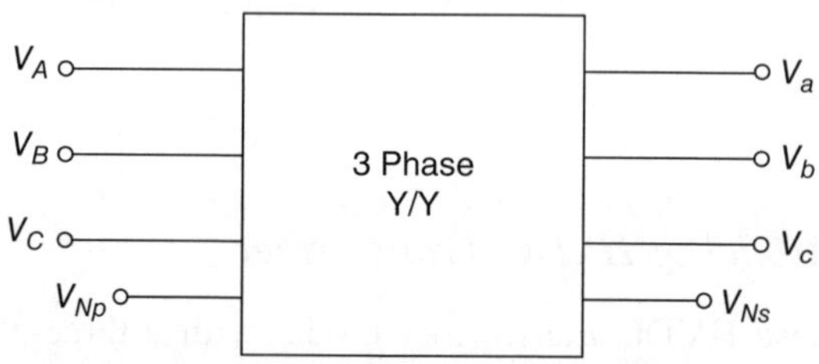

Figure 3.12 The block diagram of the three phase HVDC transformer including neutral points

circuits [10–13]. The current in the input circuits will be unknown, but saturation of the current waveform occurs because of the nonlinear characteristic of the magnetic core. For a three-phase transformer connected in wye-wye, as shown in Figure 3.12, a computer simulation model with a neutral *NN* and external circuits for both primary and secondary windings is obtained using the HBFEM technique.

According to the Galerkin procedure, system matrix equations of HBFEM for the HVDC three phase Y/Y connection transformer can be obtained through Faraday's and Kirchhoff's laws for the external circuit, as described below [15]:

$$\left.\begin{array}{l}
[H]\{A\}-[G_A]\{J_A\}-[G_B]\{J_B\}-[G_C]\{J_C\} \\
-[G_a]\{J_a\}-[G_b]\{J_b\}-[G_c]\{J_c\}=\{0\} \\
\\
[C_{A,in}]\{A\}+[I]\{V_{Np}\}+S_{A,in}[Z_{A,in}]\{J_A\}=[V_A\} \\
[C_{B,in}]\{A\}+[I]\{V_{Np}\}+S_{B,in}[Z_{B,in}]\{J_B\}=[V_B\} \\
[C_{C,in}]\{A\}+[I]\{V_{Np}\}+S_{C,in}[Z_{C,in}]\{J_C\}=[V_C\} \\
[C_{a,out}]\{A\}+[I]\{V_{Ns}\}-S_{a,out}[Z_{a,out}]\{J_a\}=[V_a\} \\
[C_{b,out}]\{A\}+[I]\{V_{Ns}\}-S_{b,out}[Z_{b,out}]\{J_b\}=[V_b\} \\
[C_{c,out}]\{A\}+[I]\{V_{Ns}\}-S_{c,out}[Z_{c,out}]\{J_c\}=[V_c\} \\
\\
S_{in,cA}[I]\{J_A\}+S_{in,cB}[I]\{J_B\}+S_{in,cC}[I]\{J_c\}=0 \\
S_{out,ca}[I]\{J_a\}+S_{out,cb}[I]\{J_b\}+S_{out,cc}[I]\{J_c\}=0
\end{array}\right\} \quad (3\text{-}94)$$

where $[Gk]$ is obtained from a single element – that is, $[G^e] = \Delta^e/3$. $[Z_{in}]$, $[Z_{out}]$ and S_{in}, S_{out} are external circuit impedances and cross-sectional areas of windings respectively, $[I]$ is unit matrix, V_{NN} is the voltage of primary or secondary neutral points when it is not grounded, $[C_{in}]$ and $[C_{out}]$ are geometric coefficients related to transformer windings, and current density J can be presented as:

$$\{J_k^e\} = \{J_0 \;\; J_{1s} \;\; J_{1c} \;\; J_{2s} \;\; J_{2c} \;\; J_{3s} \;\; J_{3c} \;\; \ldots\}^T \quad (3\text{-}95)$$

$[H]$ is the single-element matrix, and the detailed definitions can be expressed from:

$$\begin{aligned}
[H^e] = &\frac{1}{4\Delta^e}\begin{bmatrix} (b_1b_1+c_1c_1)D & (b_1b_2+c_1c_2)D & (b_1b_3+c_1c_3)D \\ (b_2b_1+c_2c_1)D & (b_2b_2+c_2c_2)D & (b_2b_3+c_2c_3)D \\ (b_3b_1+c_3c_1)D & (b_3b_2+c_3c_2)D & (b_3b_3+c_3c_3)D \end{bmatrix}\begin{Bmatrix} A_1^e \\ A_2^e \\ A_3^e \end{Bmatrix} \\
&+\frac{\sigma\omega\Delta^e}{12}\begin{bmatrix} 2N & N & N \\ N & 2N & N \\ N & N & 2N \end{bmatrix}\begin{Bmatrix} A_1^e \\ A_2^e \\ A_3^e \end{Bmatrix}
\end{aligned} \quad (3\text{-}96)$$

where matrix D and N are:

$$D=\frac{1}{2}\begin{bmatrix} 2\nu_0 & \nu_{1s} & \nu_{1c} & \nu_{2s} & \nu_{2c} & \nu_{3s} & \nu_{3c} & \cdots \\ 2\nu_{1s} & 2\nu_0-\nu_{2c} & \nu_{2s} & \nu_{1c}-\nu_{3c} & -\nu_{1s}+\nu_{3s} & \nu_{2c}-\nu_{4c} & -\nu_{2s}+\nu_{4s} & \cdots \\ 2\nu_{1c} & & 2\nu_0+\nu_{2c} & \nu_{1s}+\nu_{3s} & \nu_{1c}+\nu_{3c} & \nu_{2s}+\nu_{4s} & \nu_{2c}+\nu_{4c} & \cdots \\ 2\nu_{2s} & & & 2\nu_0-\nu_{4c} & \nu_{4s} & \nu_{1c}-\nu_{5c} & -\nu_{1s}+\nu_{5s} & \cdots \\ 2\nu_{2c} & & & & 2\nu_0+\nu_{4c} & \nu_{2s}+\nu_{4s} & \nu_{1c}+\nu_{5c} & \cdots \\ 2\nu_{3s} & & \text{Symmetry} & & & 2\nu_0-\nu_{6c} & \nu_{6s} & \cdots \\ 2\nu_{3c} & & & & & & 2\nu_0+\nu_{6c} & \cdots \\ \vdots & \vdots & \vdots & \vdots & \vdots & \vdots & \vdots & \ddots \end{bmatrix} \tag{3-97}$$

In the DC-biased case, the D matrix is different from a normal D matrix, due to the DC component ν_0 being involved in magnetizing, as shown in equation (3-98):

$$\nu(B(t))=H\{B(t)\}/B(t)=\nu_0+\sum_{k=2n-2}^{\infty}\{\nu_{ks}\sin(k\omega t)+\nu_{kc}\cos(k\omega t)\} \tag{3-98}$$

The N matrix is also different from the normal N matrix; an additional harmonic 0 is considered in the N matrix as presented below:

$$N^e=\begin{bmatrix} 0 & 0 & 0 & 0 & 0 & 0 & 0 & \cdots \\ 0 & 0 & -1 & 0 & 0 & 0 & 0 & \cdots \\ 0 & 1 & 0 & 0 & 0 & 0 & 0 & \cdots \\ 0 & 0 & 0 & 0 & -2 & 0 & 0 & \cdots \\ 0 & 0 & 0 & 2 & 0 & 0 & 0 & \cdots \\ 0 & 0 & 0 & 0 & 0 & 0 & -3 & \cdots \\ 0 & 0 & 0 & 0 & 0 & 3 & 0 & \cdots \\ \vdots & \vdots & \vdots & \vdots & \vdots & \vdots & \vdots & \ddots \end{bmatrix} \tag{3-99}$$

The system matrix equations of the HBFEM model in a compact form can be obtained as:

$$\begin{bmatrix} [H] & -[G_{in}] & -[G_{out}] & 0 & 0 \\ [C_{in}] & S_{in}[Z_{in}] & 0 & [I] & 0 \\ [C_{out}] & 0 & S_{out}[Z_{out}] & 0 & [I] \\ 0 & S_{in}[I] & 0 & 0 & 0 \\ 0 & & S_{out}[I] & 0 & 0 \end{bmatrix}\begin{Bmatrix} \{A_n\} \\ \{J_{k,in}\} \\ \{J_{k,out}\} \\ V_{NN,in} \\ V_{NN,out} \end{Bmatrix}=\begin{Bmatrix} 0 \\ \{V_{k,in}\} \\ \{V_{k,out}\} \\ 0 \\ 0 \end{Bmatrix} \tag{3-100}$$

where the input and output voltages can be defined as:

$$\{V_{k,in}\} = \{V_{k,Ain}, V_{k,Bin}, V_{k,Cin}\} \tag{3-101}$$

and

$$\{V_{k,out}\} = \{V_{k,Aout}, V_{k,Bout}, V_{k,Cout}\} \tag{3-102}$$

References

[1] Maas, S.A. (2003). *Nonlinear microwave and RF circuits*. Artech House.

[2] Nakhla, M.S. and Vlach, J. (1976). A piecewise harmonic balance technique for determination of periodic response of nonlinear systems. *IEEE Transactions on Circuits and Systems* **CAS-23**(2), 85–91.

[3] Kundert, K.S. and Sangiovanni-Vincentelli, A. (1986). Simulation of Nonlinear Circuits in the Frequency Domain, *IEEE Transactions on Computer-Aided Design* **CAD-5**(4), 521–535.

[4] Gilmore, R.J. and Steer, M.B. (1991). Nonlinear circuit analysis using the method of harmonic balance-A review of the art. Part I. Introductory concepts. *International Journal of Microwave and Millimeter-Wave Computer-Aided Engineering* **1**, 22–37.

[5] Brachtendorf, H.G., Welsch, G. and Laur, R. (1995). Fast simulation of the steady-state of circuits by the harmonic balance technique. *Proceedings, International Symposium on Circuits and Systems* **2**, 1388.

[6] IEEE Standards 1597.1-2008 (2009). *IEEE Standard for Validation of Computational Electromagnetics Computer Modeling and Simulations*. IEEE Electromagnetic Compatibility Society, 18 May.

[7] IEEE Standards 1597.2-2010. (2011). *IEEE Recommended Practice for Validation of Computational Electromagnetics Computer Modeling and Simulations*. IEEE Electromagnetic Compatibility Society, February.

[8] Yamada, S. and Bessho, K. (1988). Harmonic field calculation by the combination of finite element analysis and harmonic balance method. *IEEE Transactions on Magnetics* **24**(6), 2588–2590.

[9] Lu, J., Yamada, S. and Bessho, K. (1990). Time-periodic Magnetic Field Analysis with Saturation and Hysteresis Characteristics by Harmonic Balance Finite Element Method. *IEEE Transactions on Magnetics* **26**(2), 995–998.

[10] Lu, J., Yamada, S. and Bessho, K. (1990). Development and Application of Harmonic Balance Finite Element Method in Electromagnetic Field. *International Journal of Applied Electromagnetics in Materials* **1**(2–4), 305–316.

[11] Lu, J., Yamada, S. and Bessho, K. (1991). Harmonic Balance Finite Element Method Taking Account of External Circuits and Motion. *IEEE Transactions on Magnetics* **27**(5), 4204–4207.

[12] Lu, J., Yamada, S. and Harrison, H.B. (1996). Application of HB-FEM in the Design of Switching Power Supplies. *IEEE Transactions on Power Electronics* **11**(2), 347–355.

[13] Wildi, T. (2000). *Electrical Machines, Drives, and Power Systems*, Fourth Edition. Prentice Hall.

[14] Zhao, X., Lu, J., Li, L., Cheng, Z. and Lu, T. (2011). Analysis of the DC Biased Phenomenon by the Harmonic Balance Finite Element Method. *IEEE Transactions on Power Delivery* **26**(1), 475–484.

[15] Lu, J. (2009). *Harmonic Balance – Finite Element Method (HB-FEM) and Its Application in Low Frequency Electromagnetics*. Proceedings on Electromagnetic Theory and new Technology Conference, IEM and IEE China, April 2009.

4

HBFEM for Nonlinear Magnetic Field Problems

4.1 HBFEM for a Nonlinear Magnetic Field with Current-Driven Source

Numerical modeling techniques are becoming more popular for electromagnetic field analysis of high-frequency, high-density switching power supplies. Other numerical techniques for the solution of structural stress and thermal analysis are also finding favor with the design engineer. This paper discusses mainly the numerical modeling and analysis of electromagnetic fields in switching power supplies. The advantages of such power supplies are high efficiency, small size and lower weight [1–3].

These advantages do not come free of charge, because electromagnetic interference (EMI), leakage inductance, skin and proximity effects, winding self-capacitance and inter-winding capacitance can present serious problems [4]. Furthermore, the non-linear nature and hysteresis of the core material can cause waveform distortion. These waveform distortions cause harmonics, which will increase power losses in both the winding and magnetic core, and causes a loss of efficiency, as well as the possibility of parasitic resonance in the system, unless properly designed. Here, we use harmonic balance analysis, combined with the finite element method, to solve problems arising from non-linear, harmonic, eddy-current and power loss problems of transformers used in switching power supplies [5].

HBFEM differs from traditional finite element time-domain methods, transient analysis and other time harmonic methods [6]. The harmonic balance uses a linear

Harmonic Balance Finite Element Method: Applications in Nonlinear Electromagnetics and Power Systems, First Edition. Junwei Lu, Xiaojun Zhao and Sotoshi Yamada.

Companion website: www.wiley.com/go/lu/HBFEM

combination of sinusoids to generate a solution and represents waveforms using the coefficients of the sinusoids. It is combined with the finite element method to solve time-periodic, steady-state, nonlinear electromagnetic field problems. The HBFEM directly solves the steady-state response of the electromagnetic field in the frequency domain, and so is often considerably more efficient than traditional time-domain methods when fields exhibit widely separated time constants and mildly nonlinear behavior [7–8]. The electromagnetic field with harmonics satisfies Maxwell's equations. The magnetic core of a converter which has nonlinear characteristics and hysteresis, and is excited by a current source of current density J_s is considered. The detailed derivation has been introduced in Chapter 3.

In switching mode power supplies, some DC-DC converter topologies, such as zero voltage switching resonant converters and push-pull current source converters, can be considered as current-source to magnetic field systems, as shown in Figure 4.1. The HBFEM numerical modeling and computation results can provide more detailed information of harmonic distribution in magnetic core and eddy current distribution in windings which, compared with experimental results, are discussed in this section.

4.1.1 Numerical Model of Current Source to Magnetic Field

Assuming $\nabla\varphi = 0$ in the two-dimensional case, the governing equation (4-1), for two-dimensional problems can be written as:

$$G = \iint\limits_S \left\{ \frac{\partial N_i}{\partial x}\nu\frac{\partial A}{\partial x} + \frac{\partial N_i}{\partial y}\nu\frac{\partial A}{\partial y} \right\} dxdy - \iint\limits_S \left\{ J_s - \sigma\frac{\partial A}{\partial t} \right\} N_i dxdy = 0 \tag{4-1}$$

Considering the orthogonal characteristic of trigonometric functions, substituting magnetic vector A, interpolation functions N_i, magnetic reluctivity ν and current density J_s, and using Galerkin's method combined with harmonic balance techniques to discretize, the following harmonic balance matrix equation, considering only odd harmonics for a single element, can be obtained as:

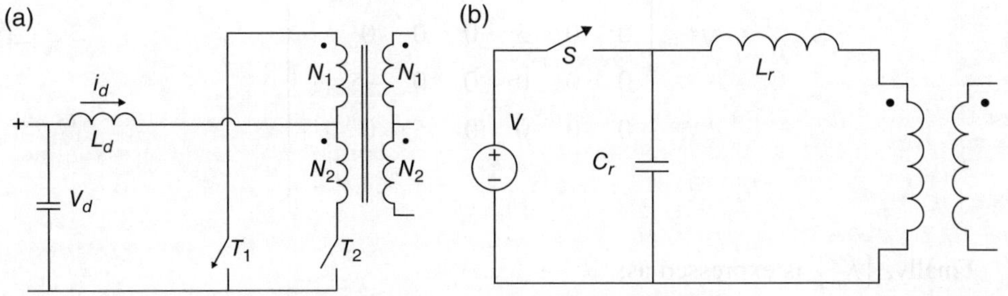

Figure 4.1 Current-source excitation to magnetic field. (a) Switch-mode push-pull converter; (b) Zero-voltage switched resonant converter

$$G^e = \frac{1}{4\Delta^e}\begin{bmatrix}(b_1b_1+c_1c_1)D & (b_1b_2+c_1c_2)D & (b_1b_3+c_1c_3)D \\ (b_2b_1+c_2c_1)D & (b_2b_2+c_2c_2)D & (b_2b_3+c_2c_3)D \\ (b_3b_1+c_3c_1)D & (b_3b_2+c_3c_2)D & (b_3b_3+c_3c_3)D\end{bmatrix}\begin{Bmatrix}A_1^e \\ A_2^e \\ A_3^e\end{Bmatrix}$$

$$+\frac{\sigma\omega\Delta^e}{12}\begin{bmatrix}2N & N & N \\ N & 2N & N \\ N & N & 2N\end{bmatrix}\begin{Bmatrix}A_1^e \\ A_2^e \\ A_3^e\end{Bmatrix}-\begin{Bmatrix}K_1^e \\ K_2^e \\ K_3^e\end{Bmatrix}=[S^e]\{A^e\}+[N^e][A^e]-\{K^e\} \qquad (4\text{-}2)$$

where the vector potential $\{A^e\}$ is called the frequency-domain representation, Δ is the area of the triangular elements, and ω indicates the fundamental frequency. b and c are obtained from x and y coordinates ($b_{ie} = x_{je}y_{ke} - x_{ke}y_{je}$, $c_{ie} = y, - y_{ke}$). The D and N matrix blocks represent nonlinear magnetic material related to harmonics and harmonic frequencies, respectively:

$$D=\begin{bmatrix}d_{11} & d_{12} & d_{13} & d_{14} & \cdots \\ d_{21} & d_{22} & d_{23} & d_{24} & \cdots \\ d_{31} & d_{32} & d_{33} & d_{34} & \cdots \\ d_{41} & d_{42} & d_{43} & d_{44} & \cdots \\ \cdots & \cdots & \cdots & \cdots & \ddots\end{bmatrix}=\frac{1}{2}\begin{bmatrix}2\nu_0-\nu_{2c} & \nu_{2s} & \nu_{2c}-\nu_{4c} & -\nu_{2s}+\nu_{4s} & \cdots \\ & 2\nu_0+\nu_{2c} & \nu_{2s}+\nu_{4s} & \nu_{2c}+\nu_{4c} & \cdots \\ & & 2\nu_0-\nu_{6c} & \nu_{6s} & \cdots \\ & Symmetry & & 2\nu_0+\nu_{6c} & \cdots \\ & & & & \ddots\end{bmatrix} \qquad (4\text{-}3)$$

and

$$N^e=\begin{bmatrix}0 & -1 & 0 & 0 & 0 & 0 & \cdots \\ 1 & 0 & 0 & 0 & 0 & 0 & \cdots \\ 0 & 0 & 0 & -3 & 0 & 0 & \cdots \\ 0 & 0 & 3 & 0 & 0 & 0 & \cdots \\ 0 & 0 & 0 & 0 & 0 & -5 & \cdots \\ 0 & 0 & 0 & 0 & 5 & 0 & \cdots \\ \cdots & \cdots & \cdots & \cdots & \cdots & \cdots & \ddots\end{bmatrix} \qquad (4\text{-}4)$$

Finally, $\{K_i^e\}$ is expressed as:

$$\{K_i^e\}=\frac{\Delta^e}{3}\{J_{1s}\ J_{1c}\ J_{3s}\ J_{3c}\ \cdots,\ \cdots,\ \cdots\}^T \qquad (4\text{-}5)$$

The system matrix equation for current source excitation can be then written as:

$$[S]\{A\} + [N]\{A\} - \{K\} = 0 \tag{4-6}$$

All the harmonic components of A can be obtained by solving this matrix equation.

4.1.2 *Example of Current-Source Excitation to Nonlinear Magnetic Field*

The magnetic components of power supplies are usually excited by time-periodic quasi-static signals and, therefore, the magnetic field can be considered as a quasi-static field. If the excitation is a sinusoidal wave, due to the magnetic saturation, harmonics occur in the magnetic field and current. These harmonics will increase power losses and EMI. This example shows the numerical analysis of nonlinear magnetic fields for power supplies.

In zero-voltage switched (ZVS) resonant converters and switching mode push-pull converters inserting an inductor at the input side, the excitation can be considered as a current-source to the magnetic field. The configuration of the magnetic system is shown in Figure 4.2, where the hysteresis characteristic is considered as shown in Figure 4.2(b). If the magnetic core is excited by a sinusoidal wave, a harmonic flux will occur in the magnetic field as shown in Figure 4.3, which is calculated by HBFEM.

To validate the numerical results, the analysis model with an air gap and two slots at the central leg, as shown in Figure 4.4, has been used. This structure significantly affects the magnetic flux distribution and density in the central leg.

The numerical results in Figure 4.5 demonstrate that the magnetic density at three different positions has good agreement with experimental results in the case of current source excitation. The computed results also present the magnetic flux distribution for fundamental and third harmonic components, as shown in Figure 4.6.

4.2 Harmonic Analysis of Switching Mode Transformer Using Voltage-Driven Source

In most cases, pulse width modulation and zero-current switched resonant converters can be considered as a voltage-source to the magnetic system which is always coupled to the external circuits, as shown in Figure 4.7. The current in the input circuits will be unknown, but saturation of the current waveform occurs because of the nonlinear characteristic of the magnetic core. This section discusses the HBFEM model and computed results, compared with experimental results.

4.2.1 *Numerical Model of Voltage Source to Magnetic System*

When a transformer of power supply is excited by a voltage source, such as pulse-width modulation (PWM) converters and zero-current switched (ZCS) resonant converters,

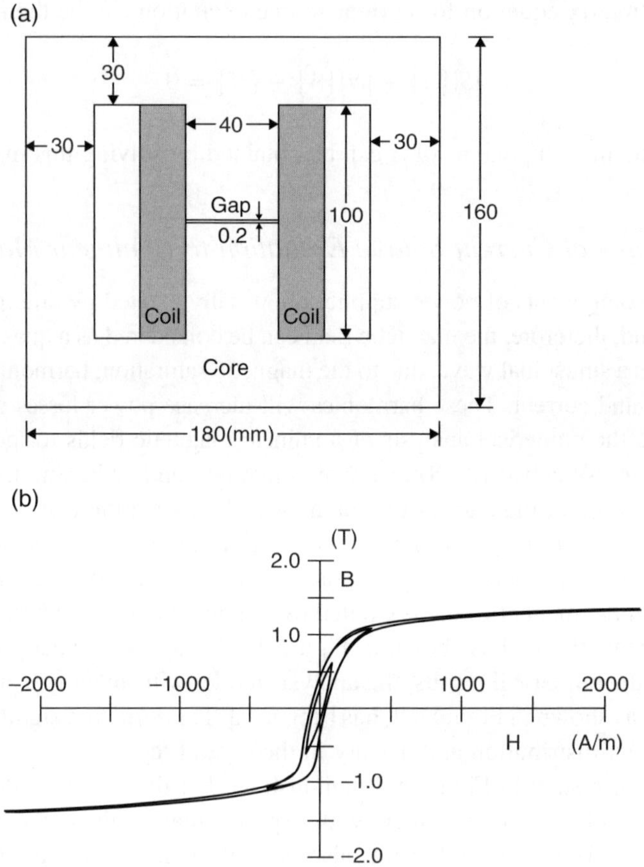

Figure 4.2 Magnetic system with current-source excitation. (a) Magnetic configuration; (b) Hysteresis characteristic

the numerical analysis of the magnetic field should be carried out by taking account of the voltage source and the external circuits. If the excitation waveform is a square wave or a triangular wave, it can be considered as a linear combination of harmonics. Figure 4.8 shows a generalized model of voltage-source to the magnetic system, where the input current is unknown.

According to Faraday's and Kirchhoff's laws, the relationship between the voltage and the magnetic field for a single element can be obtained from:

$$V_{ink} = V_k + R_k I_k + L_k \frac{di_k}{dt} + \frac{1}{C_k}\int i_k dt \tag{4-7}$$

where V_{ink} is the input voltage of circuit k, V_k is the corresponding induced electromotive force, S_{ck} and Z_k are the cross-areas and impedance of winding k, respectively. C_k and L_k are the capacitance and inductance, respectively, in circuit k.

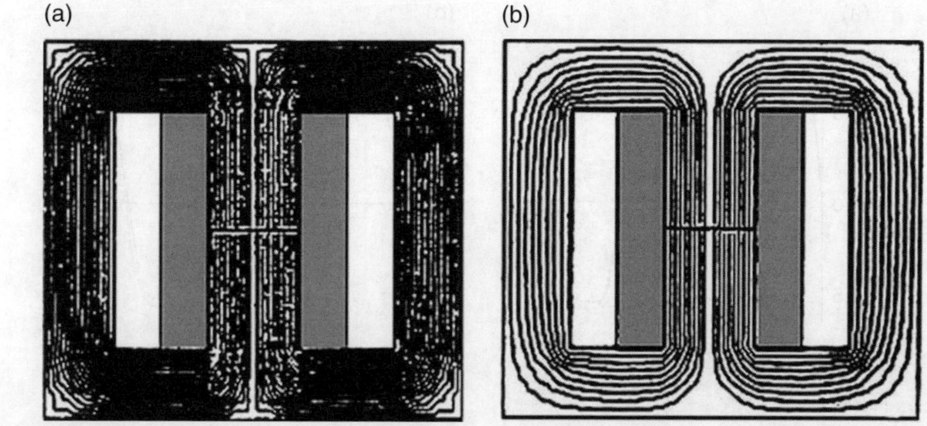

Figure 4.3 The magnetic flux distribution for fundamental and third harmonic components. (a) Fundamental component; (b) Third harmonic component

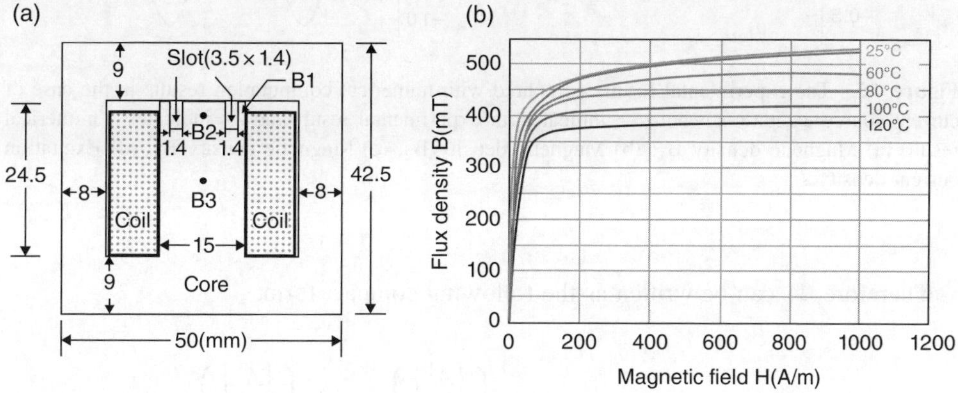

Figure 4.4 The analysis model with an air gap and two slots at the central leg, and B-H curve

The induced electromotive force can be obtained based on Faraday's law:

$$
\begin{aligned}
V_k &= \frac{d\Psi}{dt} = \frac{d}{dt}\iint \boldsymbol{B}\cdot d\boldsymbol{S} \\
&= \frac{d}{dt}\iint \nabla \times \boldsymbol{A} d\boldsymbol{S} = \frac{d}{dt}\oint_f \boldsymbol{A} d\boldsymbol{l} \\
&= \frac{\omega N_k d_0}{3S_{ck}}[N \;\; N \;\; N]\begin{Bmatrix} \{A^1\} \\ \{A^2\} \\ \{A^3\} \end{Bmatrix}
\end{aligned}
\tag{4-8}
$$

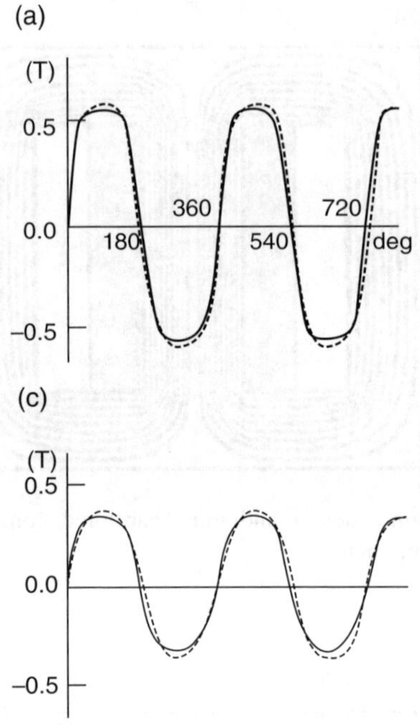

Figure 4.5 The experimental results compared with numerical computation results in the case of current source excitation, where … indicates an experimental result and — indicates a numerical result. (a) Magnetic density B_1; (b) Magnetic density B_2; (c) Magnetic density B_3; (d) Excitation current density J

Therefore, V_k can be written as the following compact form:

$$\{V_k\} = \frac{\omega d_0 \Delta^e}{3S_{ck}}[N \;\; N \;\; N]\begin{Bmatrix} \{A^1\} \\ \{A^2\} \\ \{A^3\} \end{Bmatrix} = [C_{ck}]\begin{Bmatrix} \{A^1\} \\ \{A^2\} \\ \{A^3\} \end{Bmatrix} \tag{4-9}$$

where matrix $[C_{ck}]$ is the coefficients related to the voltage:

$$[C_{ck}] = \frac{\omega d_0 \Delta^e}{3S_{ck}}[N \;\; N \;\; N] \tag{4-10}$$

and S_{ck} and d_0 are the area of windings and the depth in the z-direction respectively. Then the input voltage can be defined:

$$\{V_{ink}\} = \{V_k\} + S_{ck}[Z_k]\{J_k\} = [C_k]\{A\} + S_{ck}[Z_k]\{J_k\} \tag{4-11}$$

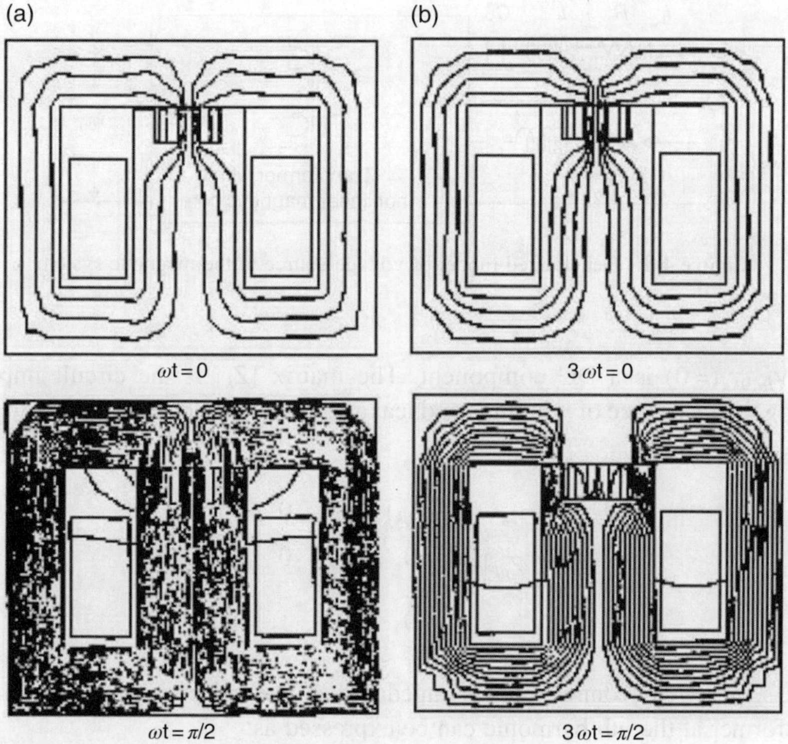

Figure 4.6 The magnetic flux distribution for fundamental and third harmonic components. (a) Fundamental harmonic component; (b) Third harmonic component

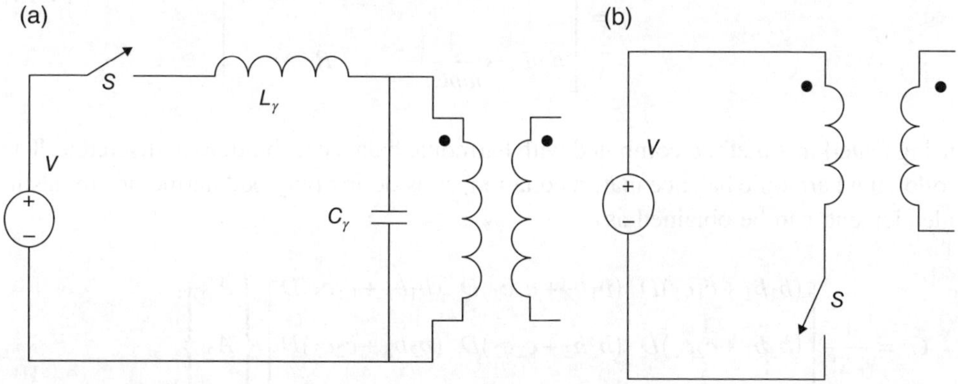

Figure 4.7 Voltage-source to the magnetic system used for switch mode transformers

where S_{ck} is the area of windings. The input voltage $\{V_{ink}\}$, including all harmonic components which have a known value, is expressed as follows:

$$\{V_{in,k}\} = \{V_{in,0k}, V_{in,1sk}, V_{in,1ck}, V_{in,2sk}, V_{in,2ck}, \cdots\}^T \tag{4-12}$$

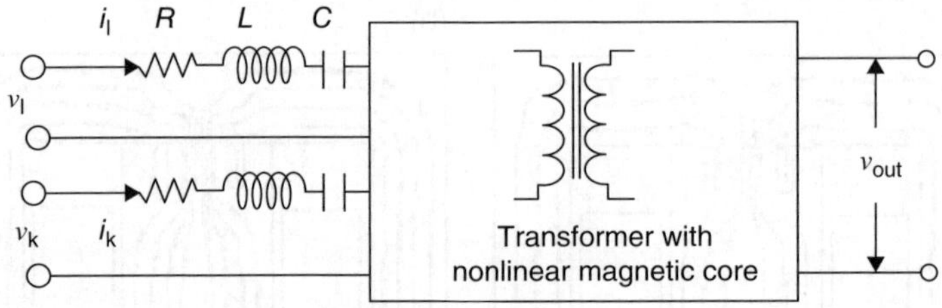

Figure 4.8 Generalized model of voltage-source to the magnetic system

where $V_{in,0k}$ $(=0)$ is a DC component. The matrix $[Z_k]$ is the circuit impedance, including the resistance of windings and leakage inductance corresponding to the harmonics, and

$$[Z_{nk}] = \begin{bmatrix} [Z_{0k}] & 0 & 0 \\ 0 & [Z_{1k}] & 0 \\ 0 & 0 & \ddots \end{bmatrix} \tag{4-13}$$

In the series circuit connection, the impedance of external circuit or equivalent circuit of transformer at the nth harmonic can be expressed as:

$$\begin{aligned} [Z_{nk}] &= [Z_{Rnk}] + [Z_{Lnk}] + [Z_{Cnk}] \\ &= \begin{bmatrix} R_k & -n\omega L_k + \dfrac{1}{n\omega C_k} \\ n\omega L_k - \dfrac{1}{n\omega C_k} & R_k \end{bmatrix} \end{aligned} \tag{4-14}$$

using Galerkin's method combined with harmonic balance techniques to discretize. The following harmonic balance matrix equation, considering only odd harmonics for a single element, can be obtained as

$$\begin{aligned} G^e = &\frac{1}{4\Delta^e}\begin{bmatrix} (b_1b_1+c_1c_1)D & (b_1b_2+c_1c_2)D & (b_1b_3+c_1c_3)D \\ (b_2b_1+c_2c_1)D & (b_2b_2+c_2c_2)D & (b_2b_3+c_2c_3)D \\ (b_3b_1+c_3c_1)D & (b_3b_2+c_3c_2)D & (b_3b_3+c_3c_3)D \end{bmatrix}\begin{Bmatrix} A_1^e \\ A_2^e \\ A_3^e \end{Bmatrix} \\ &+ \frac{\sigma\omega\Delta^e}{12}\begin{bmatrix} 2N & N & N \\ N & 2N & N \\ N & N & 2N \end{bmatrix}\begin{Bmatrix} A_1^e \\ A_2^e \\ A_3^e \end{Bmatrix} - \begin{Bmatrix} K_1^e \\ K_2^e \\ K_3^e \end{Bmatrix} = [S^e]\{A^e\} + [N^e][A^e] - \{K^e\} \end{aligned} \tag{4-15}$$

where the D and N matrix at no DC-biased case can be expressed as:

$$D=\begin{bmatrix} d_{11} & d_{12} & d_{13} & d_{14} & \cdots \\ d_{21} & d_{22} & d_{23} & d_{24} & \cdots \\ d_{31} & d_{32} & d_{33} & d_{34} & \cdots \\ d_{41} & d_{42} & d_{43} & d_{44} & \cdots \\ \cdots & \cdots & \cdots & \cdots & \ddots \end{bmatrix}=\frac{1}{2}\begin{bmatrix} 2\nu_0-\nu_{2c} & \nu_{2s} & \nu_{2c}-\nu_{4c} & -\nu_{2s}+\nu_{4s} & \cdots \\ & 2\nu_0+\nu_{2c} & \nu_{2s}+\nu_{4s} & \nu_{2c}+\nu_{4c} & \cdots \\ & & 2\nu_0-\nu_{6c} & \nu_{6s} & \cdots \\ & \mathit{Symmetry} & & 2\nu_0+\nu_{6c} & \cdots \\ & & & & \ddots \end{bmatrix} \tag{4-16}$$

and harmonic matrix N is:

$$N^e=\begin{bmatrix} 0 & -1 & 0 & 0 & 0 & 0 & \cdots \\ 1 & 0 & 0 & 0 & 0 & 0 & \cdots \\ 0 & 0 & 0 & -3 & 0 & 0 & \cdots \\ 0 & 0 & 3 & 0 & 0 & 0 & \cdots \\ 0 & 0 & 0 & 0 & 0 & -5 & \cdots \\ 0 & 0 & 0 & 0 & 5 & 0 & \cdots \\ \cdots & \cdots & \cdots & \cdots & \cdots & \cdots & \end{bmatrix} \tag{4-17}$$

The system matrix equation related to current therefore can be rewritten as

$$[S]\{A\} + [N]\{A\} - [G_k]\{J_k\} = 0 \tag{4-18}$$

where $[G_k]$ is obtained from a single element, that is:

$$[G^e]=\frac{\Delta^e}{3} \tag{4-19}$$

Combining Equations (4-11) and (4-18), the global system matrix equations for multiple input and output are obtained.

$$\begin{bmatrix} [H] & -[G_1] & -[G_2] & \cdots & -[G_k] & \cdots \\ [C_{c1}] & S_{c1}[Z_1] & 0 & 0 & 0 & \cdots \\ [C_{c2}] & 0 & S_{c2}[Z_2] & 0 & 0 & \cdots \\ \vdots & 0 & 0 & \ddots & 0 & \cdots \\ [C_{ck}] & 0 & 0 & 0 & S_{ck}[Z_k] & \cdots \\ \vdots & \vdots & \vdots & \vdots & \vdots & \ddots \end{bmatrix}\begin{Bmatrix} \{A\} \\ \{J_1\} \\ \{J_2\} \\ \vdots \\ \{J_k\} \\ \vdots \end{Bmatrix}=\begin{Bmatrix} 0 \\ \{V_{in1}\} \\ \{V_{in2}\} \\ \vdots \\ \{V_{ink}\} \\ \vdots \end{Bmatrix} \tag{4-20}$$

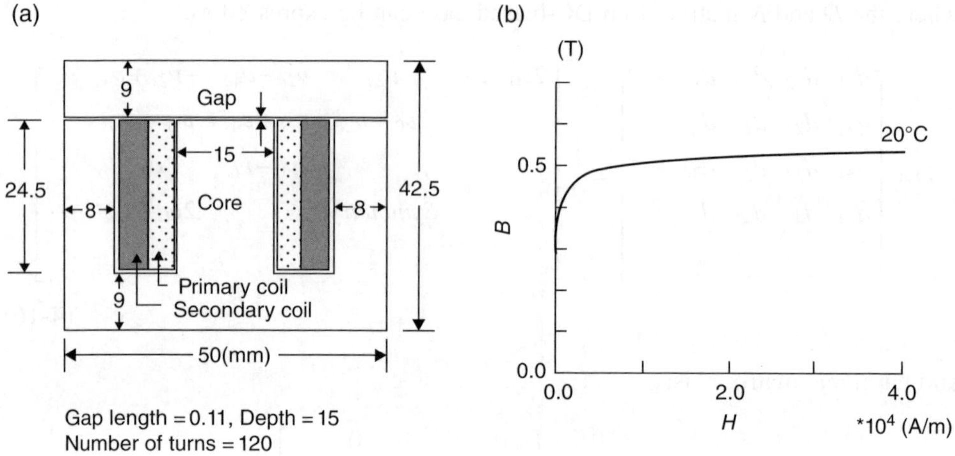

Figure 4.9 Magnetic core for a 2-D transformer structure and its B-H cure. (a) Magnetic core (b) B-H curve

where $[H]$ (= $[S] + [N]$) is FEM system matrix, [G] is current density-related coefficient matrix, $[C_{ck}]$ is voltage-related coefficient matrix, and $[Z]$ is impedance matrix. $\{A\}$ and $\{J_k\}$ are unknown, and can be calculated by solving the system matrix equation.

The harmonic balance FEM matrix equation with voltage driven source for a compact system matrix can be expressed as below:

$$\begin{bmatrix} [H] & [G] \\ [C_{ck}] & S_{ck}[Z_k] \end{bmatrix} \begin{Bmatrix} A_k \\ J_k \end{Bmatrix} = \begin{Bmatrix} 0 \\ V_{ink} \end{Bmatrix} \tag{4-21}$$

4.2.2 Example of Voltage-Source Excitation to Nonlinear Magnetic Field

As an example, a high-frequency transformer used in switching power supply is used to analyze the harmonic distribution. If the excitation waveform is a square wave, the source can be considered as a linear combination of harmonics. The nonlinear magnetic material of core will also generate some additional harmonics. Figure 4.9(a) shows the 2-D numerical model of voltage excitation to a nonlinear magnetic field. The magnetizing characteristic and B-H curve used in the simulation are illustrated in Figure 4.9(b). The magnetic flux distribution for each harmonic component can be calculated through HBFEM [5, 7, 9].

The distribution of magnetic flux for triangular wave excitation is shown in Figure 4.10. The harmonic components of current and magnetic flux can help to accurately analyze and design the magnetic core and winding structures, and further determine the power loss and leakage flux in the magnetic system.

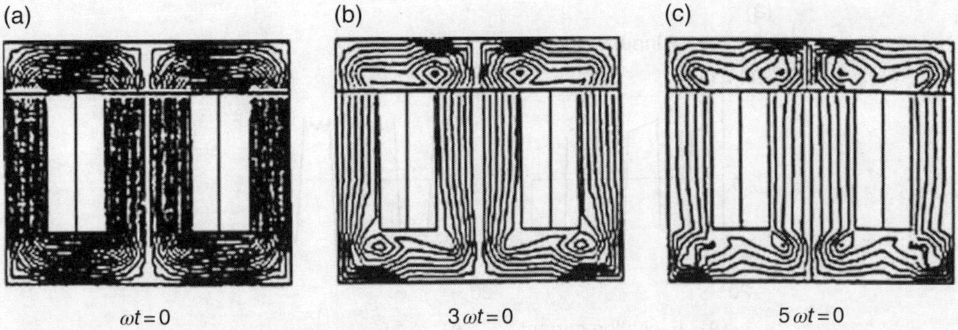

Figure 4.10 Magnetic flux distribution for their harmonic components at phase of zero degree. (a) fundamental component; (b) third harmonic component; (c) fifth harmonic component

Figure 4.11 also demonstrates the numerical analysis results with square and triangular waveform excitations. The numerical model of voltage excitation with a nonlinear magnetic field is more accurate than conventional magnetic field-only analysis. The harmonics can be also found from the resultant current waveform.

4.3 Three-Phase Magnetic Frequency Tripler Analysis

4.3.1 Magnetic Frequency Tripler

A magnetic frequency tripler is a nonlinear magnetic system which is used for the production of triple-frequency output from a three-phase fundamental frequency source based on the nonlinear magnetic saturation characteristics, as shown in Figure 4.12. Although magnetic frequency triplers have been used extensively for certain applications, the design of these devices was, until the early 1990s, largely empirical. Earlier and some recent papers [10–12] have discussed the analyses of magnetic frequency-tripling devices based on an equivalent-circuit approach under various load conditions and the Preisach model.

However, the above methods, based on the equivalent circuit theory, magnetic nonlinear characteristics, hysteresis losses and eddy current losses, and magnetic flux distribution for each harmonic component, cannot be calculated and presented. Therefore, the EM full wave solution can be obtained from HBFEM based numerical computation. HBFEM can provide magnetic flux distribution and eddy current losses at each harmonic.

4.3.2 Nonlinear Magnetic Material and its Saturation Characteristics

As discussed in Chapter 3, when a nonlinear magnetic system is excited by a sinusoidal waveform, a number of harmonics will be generated in this nonlinear magnetic system.

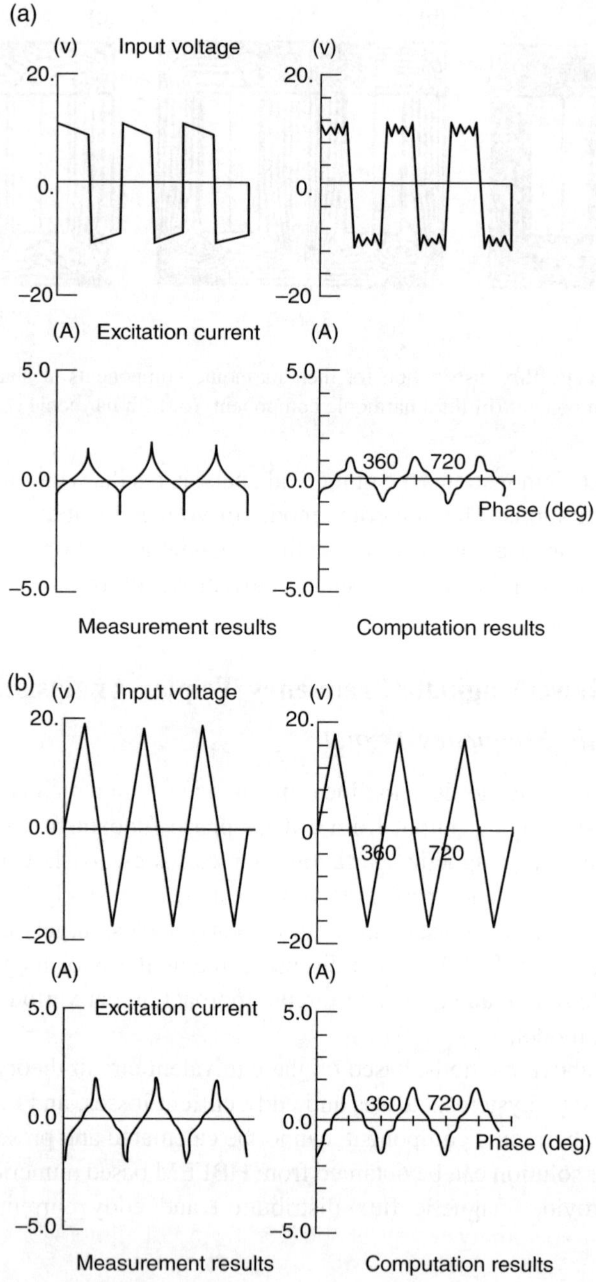

Figure 4.11 Comparison between computation and measurement. (a) Input voltage source; (b) Current caused by voltage source

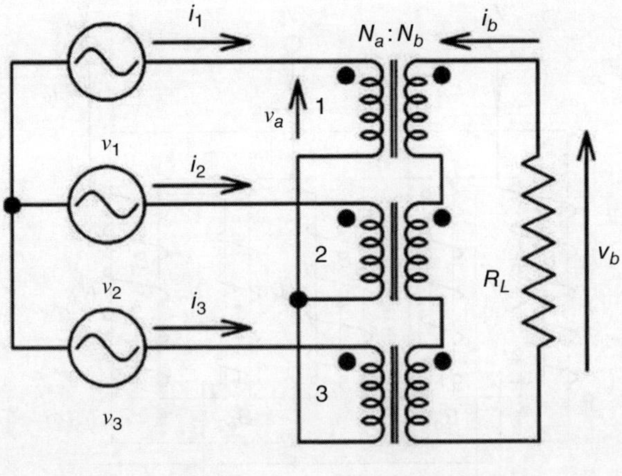

Figure 4.12 A circuit diagram of a magnetic frequency tripler

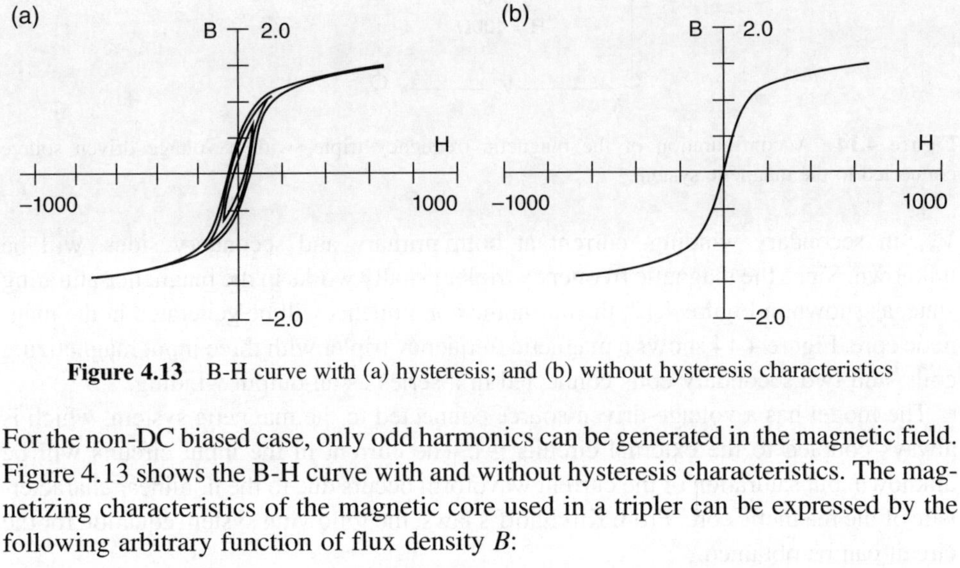

Figure 4.13 B-H curve with (a) hysteresis; and (b) without hysteresis characteristics

For the non-DC biased case, only odd harmonics can be generated in the magnetic field. Figure 4.13 shows the B-H curve with and without hysteresis characteristics. The magnetizing characteristics of the magnetic core used in a tripler can be expressed by the following arbitrary function of flux density B:

$$H(B) = H_0(B) + H_e \frac{dB}{dt} = \left(\alpha B + \beta B^{2n-1}\right) + \left(\frac{1}{f_s}\frac{dB}{dt}\right)\frac{dB}{dt} \tag{4-22}$$

where the B-H curve with hysteresis is obtained from Equation (4-22). Without hysteresis characteristics, it can be obtained from the first two terms of Equation (4-22).

4.3.3 Voltage Source-Driven Connected to the Magnetic Field

The computation model of magnetic frequency tripler can be built as a voltage source-driven CEM model, where input voltages V_{in} in primary are given. The output voltages

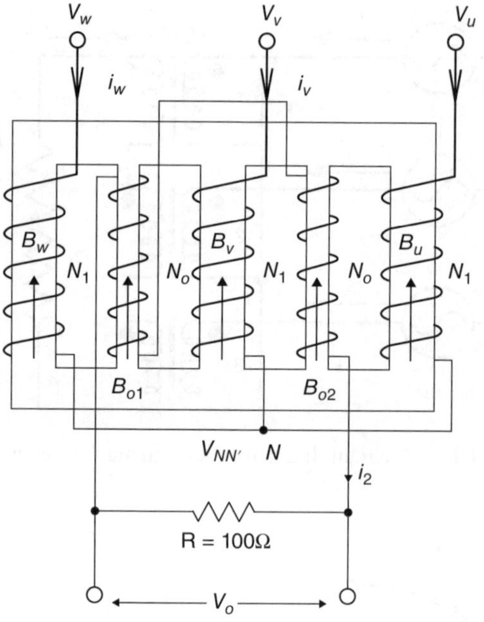

Figure 4.14 A configuration of the magnetic frequency tripler with a voltage driven source connected to the magnetic system

V_{out} in secondary winding, current at both primary and secondary sides, will be unknown. Since the magnetic frequency tripler usually works in the magnetic saturating state, as shown in Figure 4.13, the harmonic components will be generated in the magnetic core. Figure 4.14 shows a magnetic frequency tripler with three input magnetizing coils, and two secondary coils connected in a series as an output winding.

The model has a voltage-driven source connected to the magnetic system, which is always coupled to the external circuits [9]. The current in the input circuits will be unknown, but saturation of the current waveform occurs due to the nonlinear characteristic of the magnetic core. From Kirchhoff's laws, the following system equation for the circuit can be obtained:

$$\left.\begin{aligned}
&V_{aN'} + V_{NN'} + i_u r_u = V_u \\
&V_{bN'} + V_{NN'} + i_v r_v = V_v \\
&V_{cN'} + V_{NN'} + i_w r_w = V_w \\
&i_u + i_v + i_w = 0 \\
&V_2 + i_2(r_0 + R_2) = V_o
\end{aligned}\right\} \tag{4-23}$$

Considering a three-phase transformer, connected in wye, a computer simulation model with a neutral *NN* and external circuits for both primary and secondary windings is obtained using the HBFEM technique. According to the Galerkin procedure, system matrix equations of HBFEM for the tripler transformer can be obtained through Faraday's and Kirchhoff's laws for the external circuit. The matrix equation of input voltage can be defined as:

$$\left.\begin{aligned}
&[C_{Uin}]\{A\}+[I]\{V_{NNin}\}+S_{cu}[Z_{Uin}]\{J_{Uin}\}=[V_{Uin}]\\
&[C_{Vin}]\{A\}+[I]\{V_{NNin}\}+S_{cv}[Z_{Vin}]\{J_{Vin}\}=[V_{Vin}]\\
&[C_{Win}]\{A\}+[I]\{V_{NNin}\}+S_{cw}[Z_{Win}]\{J_{Win}\}=[V_{Win}]\\
&S_{in,cu}[I]\{J_{Uin}\}+S_{in,cu}[I]\{J_{Vin}\}+S_{in,cu}[I]\{J_{Win}\}=0\\
&[C_2]\{A\}-S_{co}[Z_{out}]\{J_{out}\}=[V_o]
\end{aligned}\right\} \tag{4-24}$$

where the applied three-phase voltage sources $\{V_u\}$, $\{V_v\}$, $\{V_w\}$, and $\{V_{NN'}\}$ and $\{V_{out}\}$ are the unknown voltage at the neutral point as expressed below:

$$\{V_u\}=\{\{V_{u1s}\}\ \{V_{u1c}\}\ \{V_{u3s}\}\ \{V_{u3c}\}\ \cdots\}^T \tag{4-25}$$

$$\{V_v\}=\{\{V_{v1s}\}\ \{V_{v1c}\}\ \{V_{v3s}\}\ \{V_{v3c}\}\ \cdots\}^T \tag{4-26}$$

$$\{V_w\}=\{\{V_{w1s}\}\ \{V_{w1c}\}\ \{V_{w3s}\}\ \{V_{w3c}\}\ \cdots\}^T \tag{4-27}$$

$$\{V_{NN'}\}=\{\{V_{NN1s}\}\ \{V_{NN1c}\}\ \{V_{NN3s}\}\ \{V_{NN3c}\}\ \cdots\}^T \tag{4-28}$$

$$\{V_o\}=\{\{V_{o1s}\}\ \{V_{o1c}\}\ \{V_{o3s}\}\ \{V_{o3c}\}\ \cdots\}^T \tag{4-29}$$

and $[C_{ck}]$ is obtained from:

$$[C_{ck}]=\frac{\omega d_0\Delta}{3S_{ck}}[N\ \ N\ \ N] \tag{4-30}$$

$\{J_\mathrm{u}\}$, $\{J_\mathrm{v}\}$, $\{J_\mathrm{w}\}$ and $\{J_\mathrm{out}\}$ are magnetizing current density and output current density, respectively. It can be expressed in a general form as below:

$$\{J^k\}=\{\{J^k_{1s}\}\ \{J^k_{1c}\}\ \{J^k_{3s}\}\ \{J^k_{3c}\}\ \cdots\}^T \tag{4-31}$$

The generalized input voltage matrix equation taking account of circuit can be expressed as:

$$\{V_{ink}\} = \{V_k\} + S_{ck}[Z_k]\{J_k\} \tag{4-32}$$

where matrix $[Z_k]$ is the circuit impedance matrix, which includes the resistance and leakage inductance of windings corresponding to the harmonics, and S_{ck} is the area of windings.

$$[Z_{nk}] = [Z_{Rnk}] + [Z_{Lnk}] = \begin{bmatrix} R + n\omega L & 0 \\ 0 & R + n\omega L \end{bmatrix} \tag{4-33}$$

The input voltage $\{V_{ink}\}$, including all harmonic components which have a known value, are expressed as follows:

$$\{V_{ink}\} = \{V_{0k}, V_{1sk}, V_{1ck}, V_{2sk}, V_{2ck}, \cdots\}^T \tag{4-34}$$

where V_0, V_{3sk} and V_{3ck} are zero components; only fundamental component exists in the input voltage.

The above system matrix, including all input and output circuits, can be expressed as:

$$[H]\{A\} - [G_w]\{J_w\} - [G_v]\{J_v\} - [G_w]\{J_w\} \cdots - [G_{out}]\{J_{out}\} = \{0\} \tag{4-35}$$

where $[Gk]$ is obtained from a single element, that is $[G^e] = \Delta^e/3$.

Combining Equations (4-24) and (4-35), the global system matrix equations for multiple input and output are obtained from following system equation.

$$\left.\begin{aligned} &[H]\{A\} - [G_U]\{J_{Uin}\} - [G_V]\{J_{Vin}\} - [G_W]\{J_{Win}\} - [G_{2out}]\{J_{out}\} = \{0\} \\ &[C_{Uin}]\{A\} + [I]\{V_{NNin}\} + S_{cu}[Z_{Uin}]\{J_{Uin}\} = [V_{Uin}\} \\ &[C_{Vin}]\{A\} + [I]\{V_{NNin}\} + S_{cv}[Z_{Vin}]\{J_{Vin}\} = [V_{Vin}\} \\ &[C_{Win}]\{A\} + [I]\{V_{NNin}\} + S_{cw}[Z_{Win}]\{J_{Win}\} = [V_{Win}\} \\ &[C_{out}]\{A\} - S_{co}[Z_{out}]\{J_{out}\} = [V_{out}\} \\ &S_{in,cu}[I]\{J_{Uin}\} + S_{in,cu}[I]\{J_{Vin}\} + S_{in,cu}[I]\{J_{Win}\} = 0 \end{aligned}\right\} \tag{4-36}$$

The harmonic balance FEM matrix equations for voltage source excitation can, therefore, be expressed as the following system matrix equation:

$$\begin{bmatrix} [H] & -[G_u] & -[G_v] & -[G_w] & -[G_{out}] & 0 \\ [C_{Uin}] & S_{cu}[Z_u] & 0 & 0 & 0 & [I] \\ [C_{Vin}] & 0 & S_{cv}[Z_v] & 0 & 0 & [I] \\ [C_{Win}] & 0 & 0 & S_{cw}[Z_w] & 0 & [I] \\ [C_{Out}] & 0 & 0 & 0 & -S_{co}[Z_{out}] & 0 \\ 0 & S_{cu}[I] & S_{cv}[I] & S_{cw}[I] & 0 & 0 \end{bmatrix} \begin{Bmatrix} \{A\} \\ \{J_u\} \\ \{J_v\} \\ \{J_w\} \\ \{J_{out}\} \\ \{V_{NN}\} \end{Bmatrix} = \begin{Bmatrix} 0 \\ \{V_u\} \\ \{V_v\} \\ \{V_w\} \\ \{0\} \\ \{0\} \end{Bmatrix} \tag{4-37}$$

where $\{A\}$ and $\{J_k\}$ are unknown, and can be calculated by solving the system matrix equation, $[H]$ is the matrix obtained from $([S] + [M])$, $[C_k]$ represents the geometric co-efficient related to transformer windings. $[Z_{ink}]$, $[Z_{out}]$ and $S_{c,in}$, $S_{c,out}$ are external circuit impedances and cross-sectional areas of windings, respectively, $[I]$ is unit matrix, V_{NN} is the voltage for neutral point when it is not grounded, and $[C_{in}]$ and $[C_{out}]$ are geometric coefficients related to transformer windings. Current density J_k can be presented as:

$$\{J_k^e\} = \{\, J_0 \;\; J_{1s} \;\; J_{1c} \;\; J_{2s} \;\; J_{2c} \;\; J_{3s} \;\; J_{3c} \;\; \cdots \}^T \tag{4-38}$$

$[H]$ is the system matrix, and the detailed definitions can be obtained from:

$$\begin{aligned}
[H^e] &= \frac{1}{4\Delta^e}\begin{bmatrix} (b_1b_1+c_1c_1)D & (b_1b_2+c_1c_2)D & (b_1b_3+c_1c_3)D \\ (b_2b_1+c_2c_1)D & (b_2b_2+c_2c_2)D & (b_2b_3+c_2c_3)D \\ (b_3b_1+c_3c_1)D & (b_3b_2+c_3c_2)D & (b_3b_3+c_3c_3)D \end{bmatrix}\begin{Bmatrix} A_1^e \\ A_2^e \\ A_3^e \end{Bmatrix} \\
&+ \frac{\sigma\omega\Delta^e}{12}\begin{bmatrix} 2N & N & N \\ N & 2N & N \\ N & N & 2N \end{bmatrix}\begin{Bmatrix} A_1^e \\ A_2^e \\ A_3^e \end{Bmatrix}
\end{aligned} \tag{4-39}$$

where matrices D and N are:

$$D = \begin{bmatrix} d_{11} & d_{12} & d_{13} & d_{14} & \cdots \\ d_{21} & d_{22} & d_{23} & d_{24} & \cdots \\ d_{31} & d_{32} & d_{33} & d_{34} & \cdots \\ d_{41} & d_{42} & d_{43} & d_{44} & \cdots \\ \cdots & \cdots & \cdots & \cdots & \ddots \end{bmatrix} = \frac{1}{2}\begin{bmatrix} 2\nu_0-\nu_{2c} & \nu_{2s} & \nu_{2c}-\nu_{4c} & -\nu_{2s}+\nu_{4s} & \cdots \\ & 2\nu_0+\nu_{2c} & \nu_{2s}+\nu_{4s} & \nu_{2c}+\nu_{4c} & \cdots \\ & & 2\nu_0-\nu_{6c} & \nu_{6s} & \cdots \\ & Symmetry & & 2\nu_0+\nu_{6c} & \cdots \\ & & & & \ddots \end{bmatrix} \tag{4-40}$$

and

$$N = \begin{bmatrix} 0 & -1 & 0 & 0 & 0 & 0 & \cdots \\ 1 & 0 & 0 & 0 & 0 & 0 & \cdots \\ 0 & 0 & 0 & -3 & 0 & 0 & \cdots \\ 0 & 0 & 3 & 0 & 0 & 0 & \cdots \\ 0 & 0 & 0 & 0 & 0 & -5 & \cdots \\ 0 & 0 & 0 & 0 & 5 & 0 & \cdots \\ & \cdots & & & \cdots & & \ddots \end{bmatrix} \tag{4-41}$$

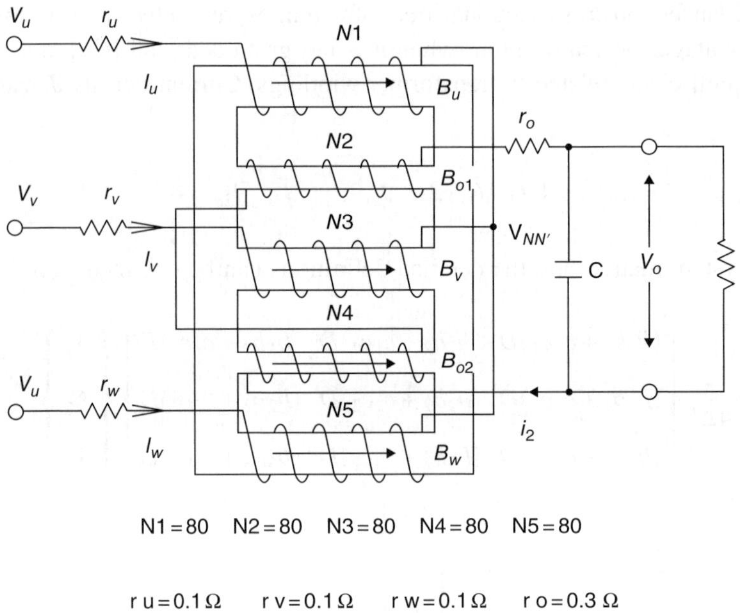

Figure 4.15 A three-phase magnetic tripler problem as a voltage-driven source connected to the magnetic system

Figure 4.15 shows a real application model with the physical structure shown in Figure 4.16. This is a three-phase circuit with magnetic field-coupled problem, which can be solved by HBFEM taking account of electric circuit.

The output voltage waveform can be calculated from numerical result i ($V = iR$), and the simulation result is compared with the experimental result, as illustrated in Figure 4.17.

The magnetic flux distribution for each harmonic component calculated from HBFEM is presented as below. From Figure 4.18, we can see that only the third harmonic component can pass through the lag 2 and lag 4, which will generate the induced voltage with three times fundamental frequency on secondary windings. The input and output currents with harmonic distortion in each winding can be calculated from the HBFEM matrix equation. The calculation results are also compared with experimental results. A good agreement has been achieved in this calculation.

The waveforms of input and output currents, and input voltage (phase U) and neutral voltage $V_{NN''}$ are calculated from HBFEM. The simulation results compared with experiment results are presented in Figure 4.19.

The waveforms of the magnetic flux density distribution for each phase of magnetic leg, and output side of magnetic legs, are calculated from HBFEM. The simulation results, compared with experiment results, are presented in Figure 4.20.

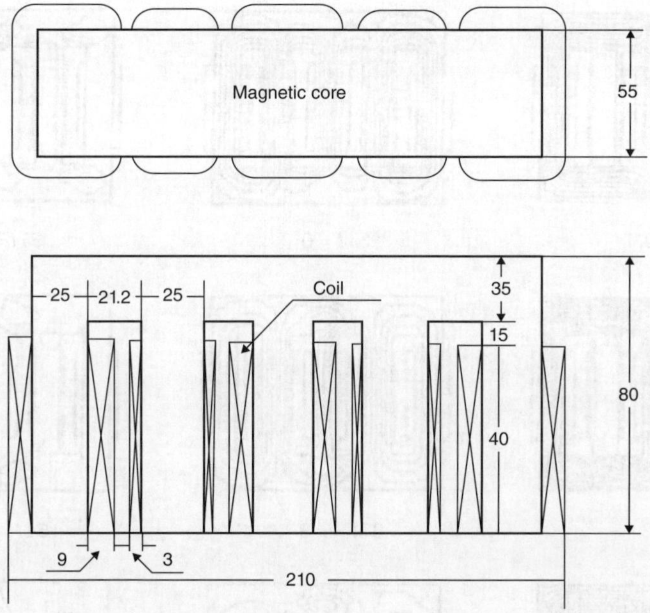

Figure 4.16 Geometric size ½, a configuration of the magnetic frequency tripler ½ for numerical computation model

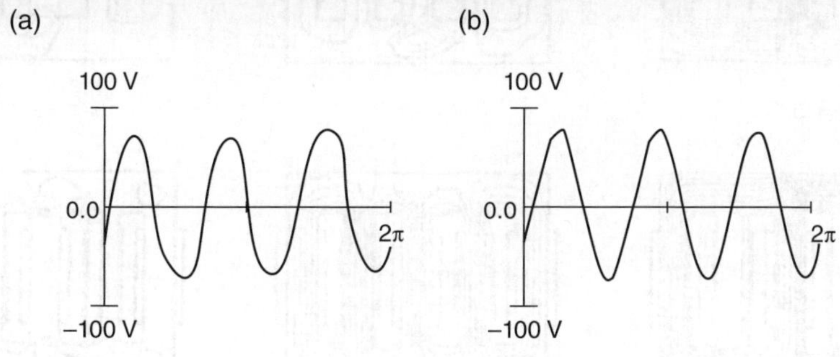

Figure 4.17 Output voltage waveform: (a) experimental result, (b) simulation result

Figure 4.21 illustrates characteristics of output current against load, and phase U input current against input voltage, respectively. The output current of fundamental component is very small, as shown in Figure 4.21(a). When the input voltage reaches to the rated voltage (100 V), the input current will be increased significantly.

4.4 Design of High-Speed and Hybrid Induction Machine using HBFEM

4.4.1 Construction of High-Speed and Hybrid Induction Machine

The high-speed hybrid induction motor consists of a three-phase to two-phase magnetic frequency tripler, as shown in Figure 4.22(a) [9]. The induction motor is

Figure 4.18 Flux distribution of fundamental, third and fifth harmonics. Source: Mohan *et al.* [3]. Reprinted with permission from John Wiley & Sons

constructed by two pairs of magnetic frequency triplers and four magnetic poles, with the air gap in the middle leg of cores. The three-phase magnetizing windings are connected as Scott-connection, and four additional coils, connected with the capacitors for increasing output power, are put in the poles, as shown in Figure 4.22(b). When a 60 Hz commercial source is applied to the hybrid induction motor, the rotation speed (10 800 rpm) will be gained between the poles. The principle of the three-phase to two-phase magnetic frequency tripler is that two single-phase triplers, composed of three-legged

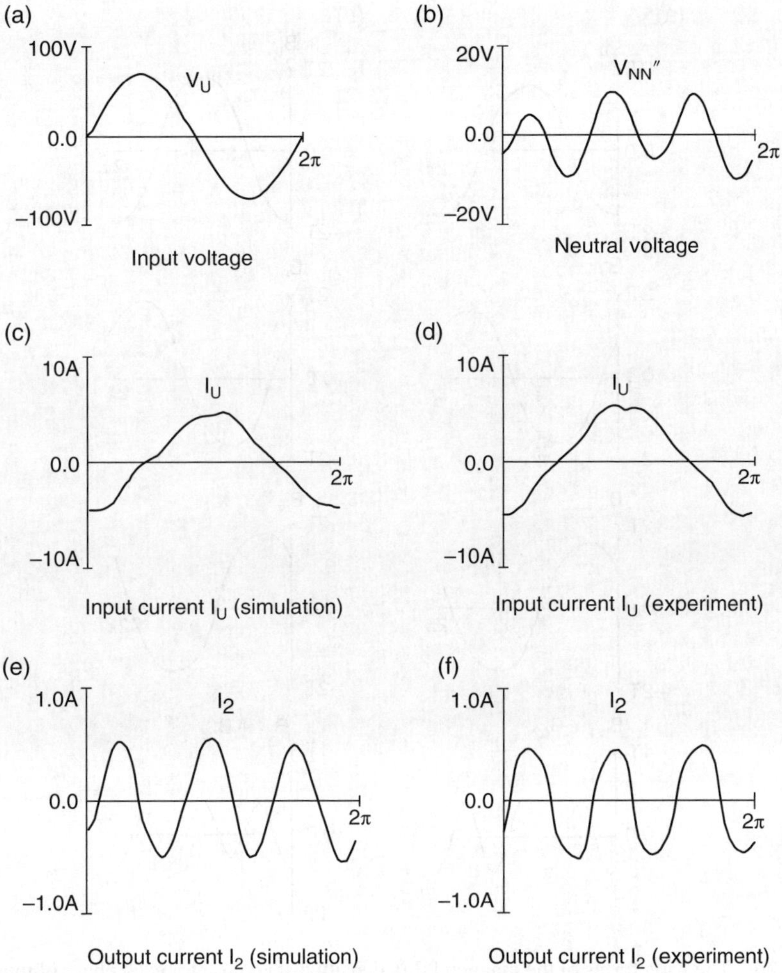

Figure 4.19 Comparison between computation and experiment results of waveforms of input and output currents, and input voltage (phase U) and neutral voltage $V_{NN''}$

cores, are connected in Scott-connection. Therefore, the input voltages, shifted at 90° are applied to two triplers.

Figure 4.23 shows the compact structure of a high-speed and hybrid induction motor with three-phase to two-phase magnetic frequency tripler, three-phase magnetizing windings connected as Scott-connection, and four additional coils connected with the capacitors for increasing the output power.

4.4.2 Numerical Model of High-Speed and Hybrid Induction Machine using HBFEM, Taking Account of Motion Effect

The electric machines are usually excited by three-phase voltage sources and connected with an external circuit, as shown in Figure 4.24. In this case, the magnetic field and

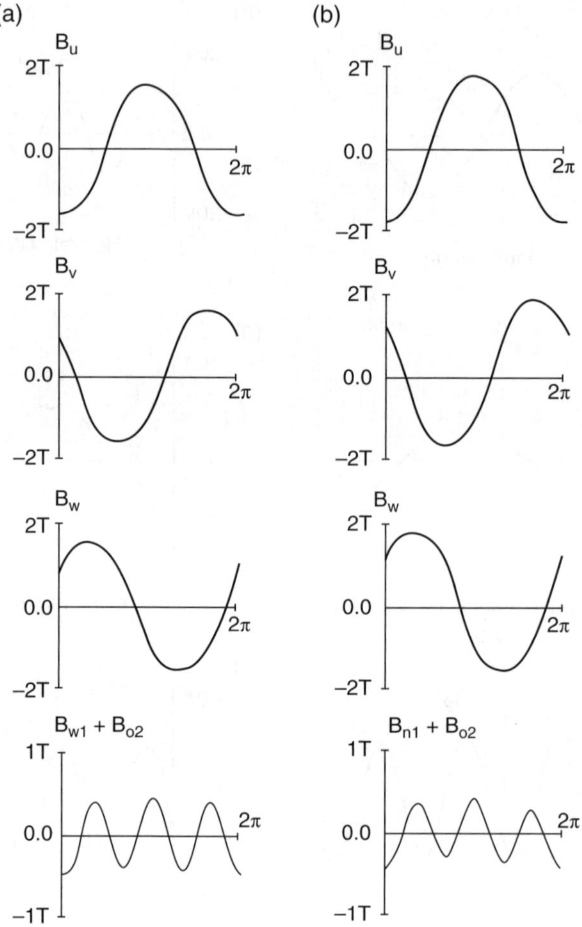

Figure 4.20 The waveforms of the magnetic flux density distribution for each phase of magnetic leg and output side of magnetic legs

external circuits including both primary and secondary should be considered in the numerical analysis.

Since the numerical analysis model should include rotating or moving parts, according to Maxwell's equations, the formulation of a two-dimensional magnetic field, related to the vector magnetic potential A, can be given by:

$$\frac{\partial}{\partial x}\left(\nu\frac{\partial A}{\partial x}\right)+\frac{\partial}{\partial y}\left(\nu\frac{\partial A}{\partial y}\right)=-J_0+\sigma\frac{\partial A}{\partial t}+\sigma V_x\frac{\partial A}{\partial x}+\sigma V_y\frac{\partial A}{\partial y} \tag{4-42}$$

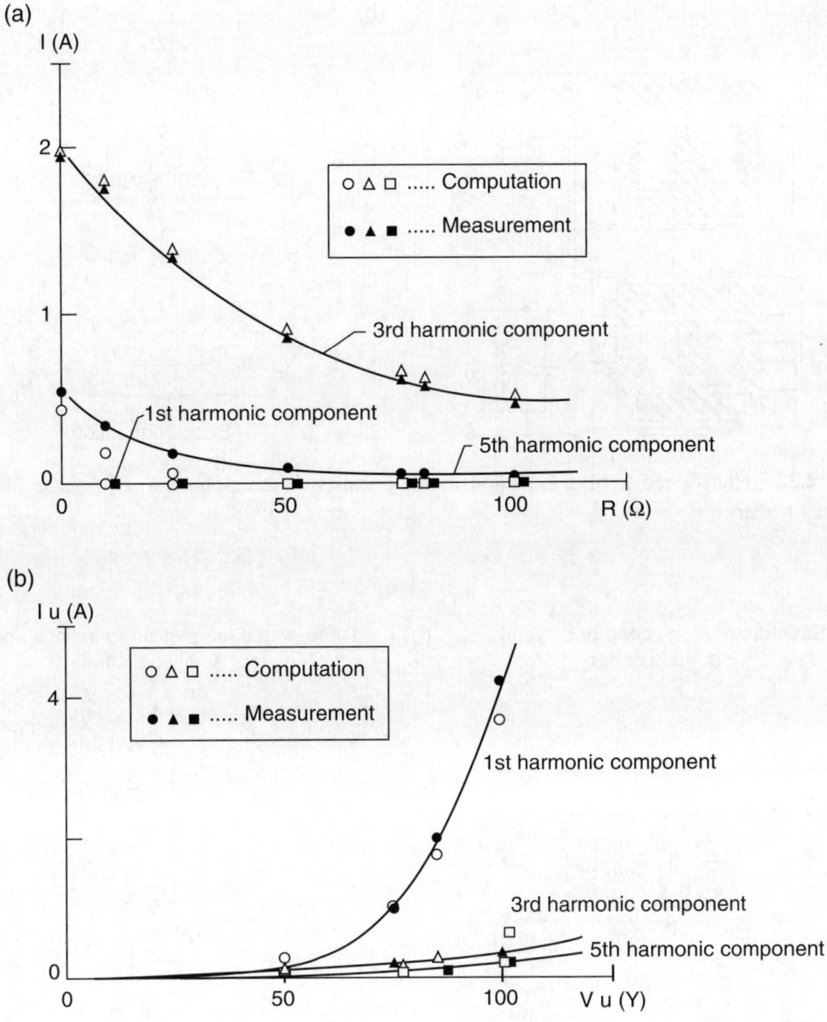

Figure 4.21 Characteristics of output current against load and phase U input current against input voltage. (a) Output current against load; (b) Phase U input current against input voltage

By using Galerkin's procedure, Equation (4-42) can be re-written as Equation (4-43):

$$\iint_{\Omega}\left\{\frac{\partial N_i}{\partial x}\left(\nu\frac{\partial A}{\partial x}\right)+\frac{\partial N_i}{\partial y}\left(\nu\frac{\partial A}{\partial y}\right)\right\}dxdy-\iint_{\Omega}\left\{J_0-\sigma\frac{\partial A}{\partial t}-\sigma V_x\frac{\partial A}{\partial x}-\sigma V_y\frac{\partial A}{\partial y}\right\}N_i dxdy=0 \tag{4-43}$$

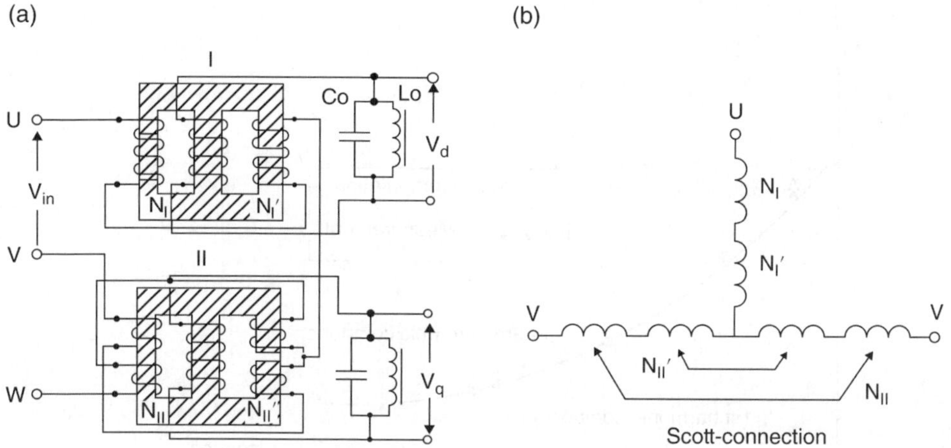

Figure 4.22 High-speed hybrid induction motor consists of three-phase to two-phase magnetic frequency tripler

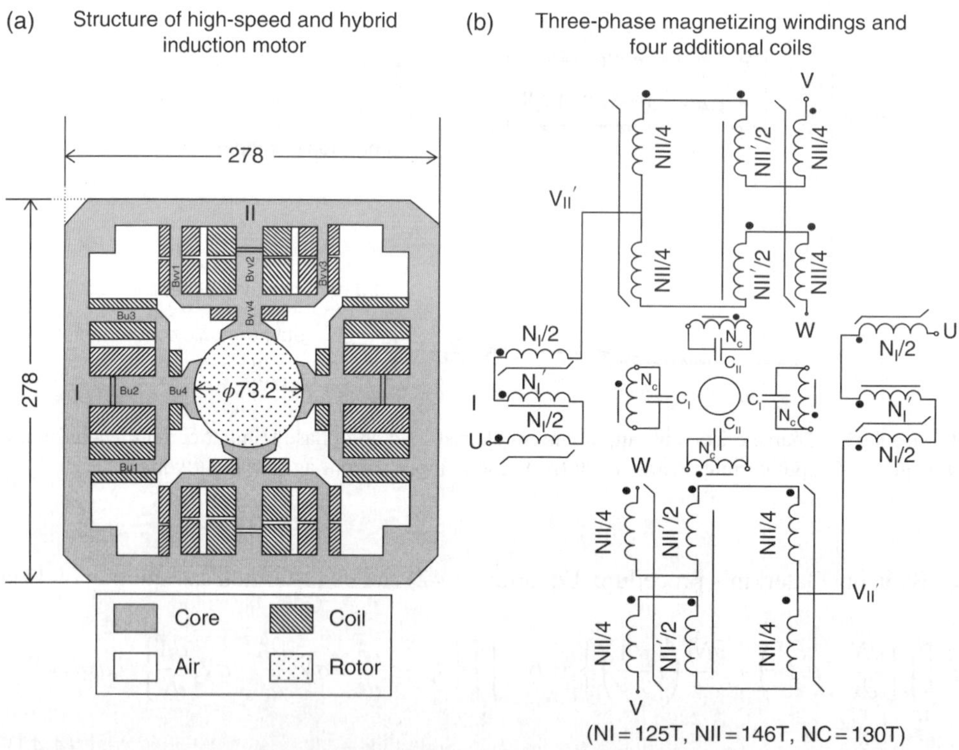

Figure 4.23 The compact structure of high-speed and hybrid induction motor and three-phase magnetizing windings and four additional coils connected with the capacitors

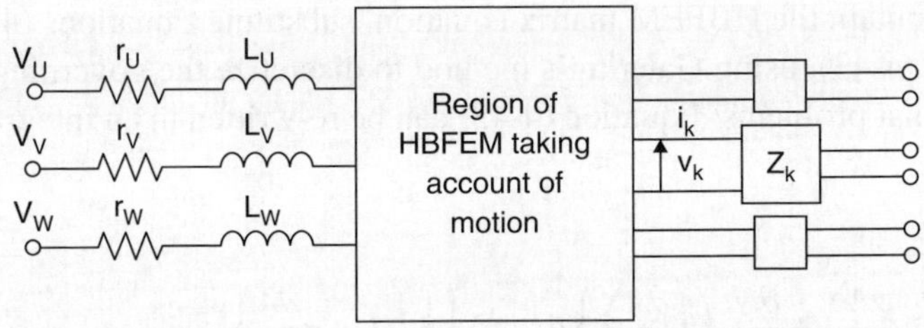

Figure 4.24 Numerical model of electric machine taking account of motion

where N is the shape function of the first-order triangular element as a weighting function. The speeds of x- and y-direction $(Vx, Vy,)$ are assumed to be constant. ν and σ are the magnetic reluctivity and the conductivity respectively. Variables such as the vector potentials A, the magnetizing current density J, and the nonlinear magnetic reluctivity ν $(=1/\mu)$ can be approximately expressed in harmonic solutions. Then:

$$A^i = \sum_{k=2n-1}^{\infty} \{A^i_{ks}\sin(k\omega t) + A^i_{kc}\cos(k\omega t)\} \tag{4-44}$$

$$J_s = \sum_{k=2n-1}^{\infty} \{J_{ks}\sin(k\omega t) + J_{ks}\cos(k\omega t)\} \tag{4-45}$$

$$\nu(B(t)) = H\{B(t)\}/B(t) = \nu_0 + \sum_{k=2n-2}^{\infty} \{\nu_{ks}\sin(k\omega t) + \nu_{kc}\cos(k\omega t)\} \tag{4-46}$$

where:

$$\nu_0 = \frac{1}{T}\int_0^T \nu(t)dt \tag{4-47}$$

$$\nu_{ks} = \frac{2}{T}\int_0^T \nu(t)\cdot\sin(k\omega t)dt \tag{4-48}$$

$$\nu_{kc} = \frac{2}{T}\int_0^T \nu(t)\cdot\cos(k\omega t)dt \tag{4-49}$$

ω is the fundamental angular frequency and k is the harmonic number. $H(B)$ is the magnetizing characteristic of the core which is expressed by an approximate function.

$$H(B) = aB + bB^3 + cB^5 + \cdots + jB^{(2n-1)} + \cdots \tag{4-50}$$

In order to obtain the HBFEM matrix equation, substitute Equations (4-44) to (4-46) into Equation (4-43), using Galerkin's method to discretize the governing equation for two-dimensional problems. Equation (4-43) can be re-written in an integral form that is given as:

$$\iint_e \left\{ \frac{\partial N_i}{\partial x}\left(\nu\frac{\partial A}{\partial x}\right) + \frac{\partial N_i}{\partial y}\left(\nu\frac{\partial A}{\partial y}\right) \right\} dxdy - \iint_e \left\{ J_0 - \sigma\frac{\partial A}{\partial t} \right\}$$
$$+ \iint_e \{\sigma V_x (A_1 b^1 + A_2 b^2 + A_3 b^3)/2\Delta + \sigma V_y (A_1 c^1 + A_2 c^2 + A_3 c^3)/2\Delta\} N_i dxdy = 0 \tag{4-51}$$

where the interpolating function $N_i(x, y)$ is $N_i = (a_i + b_i x + c_i y)/2\Delta$ and

$$a_i = x_i y_k - x_k y_i,\ b_i = y_j - y_k,\ c_i = x_k - x_j \tag{4-52}$$

Δ is the cross-section of the element; $x_{i,j,k}$, $y_{i,j,k}$ are coordinates of the node.
The matrix expression of HBFEM for a single element is obtained as follows:

$$[S^e]\{A^e\} + [N^e]\{A^e\} + [M^e]\{A^e\} - \{K^e\} = 0 \tag{4-53}$$

The magnetic vector potential $\{A_i^e\}$ and current density $\{K_i^e\}$ are expressed as:

$$\{A^e\} = \{ A^1_{1s}\ A^1_{1c}\ A^1_{3s}\ A^1_{3c},\ A^2_{1s}\ A^2_{1c}\ A^2_{3s}\ A^2_{3c},\ A^3_{1s}\ A^3_{1c}\ A^3_{3s}\ A^3_{3c} \} \tag{4-54}$$

$$\{K_i^e\} = \frac{\Delta^e}{3}\{J_{1s}\ J_{1c}\ J_{3s}\ J_{3c}\ \cdots,\ \cdots\}^T \tag{4-55}$$

The magnetic reluctivity coefficient matrix, D, can be derived as:

$$[S^e] = \frac{1}{4\Delta^e}\begin{bmatrix} (b_1b_1 + c_1c_1)D & (b_1b_2 + c_1c_2)D & (b_1b_3 + c_1c_3)D \\ (b_2b_1 + c_2c_1)D & (b_2b_2 + c_2c_2)D & (b_2b_3 + c_2c_3)D \\ (b_3b_1 + c_3c_1)D & (b_3b_2 + c_3c_2)D & (b_3b_3 + c_3c_3)D \end{bmatrix} \tag{4-56}$$

where D and N are the reluctivity and the harmonic matrices, as shown in reference [1]:

$$D=\begin{bmatrix} d_{11} & d_{12} & d_{13} & d_{14} & \cdots \\ d_{21} & d_{22} & d_{23} & d_{24} & \cdots \\ d_{31} & d_{32} & d_{33} & d_{34} & \cdots \\ d_{41} & d_{42} & d_{43} & d_{44} & \cdots \\ \cdots & \cdots & \cdots & \cdots & \ddots \end{bmatrix} = \frac{1}{2}\begin{bmatrix} 2\nu_0-\nu_{2c} & \nu_{2s} & \nu_{2c}-\nu_{4c} & -\nu_{2s}+\nu_{4s} & \cdots \\ & 2\nu_0+\nu_{2c} & \nu_{2s}+\nu_{4s} & \nu_{2c}+\nu_{4c} & \cdots \\ & & 2\nu_0-\nu_{6c} & \nu_{6s} & \cdots \\ & Symmetry & & 2\nu_0+\nu_{6c} & \cdots \\ & & & & \ddots \end{bmatrix} \tag{4-57}$$

and

$$[N^e]=\frac{\sigma\omega\Delta^e}{12}\begin{bmatrix} 2N & N & N \\ N & 2N & N \\ N & N & 2N \end{bmatrix} \tag{4-58}$$

The matrix $[M^e]$ is related to motion, and can be given by:

$$[M^e]=\frac{\sigma V_x}{6}\begin{bmatrix} b_1I & b_2I & b_3I \\ b_1I & b_2I & b_3I \\ b_1I & b_2I & b_3I \end{bmatrix}+\frac{\sigma V_y}{6}\begin{bmatrix} c_1I & c_2I & c_3I \\ c_1I & c_2I & c_3I \\ c_1I & c_2I & c_3I \end{bmatrix} \tag{4-59}$$

where I is the Unix matrix.

According to Faraday's law, the relation between magnetic vector potential and terminal voltage can be given by the following equation:

$$V=-\frac{d}{dt}\left(\oint A d\ell\right) \tag{4-60}$$

where the integration is calculated along the magnetizing coil.

Substituting magnetic vector potential A in Equation (4-44) into Equation (4-60), the excitation voltage related to magnetic field can be obtained as:

$$V_{ck}=\sum_{coil}\left\{\sum_{k=2n-1}^{\infty}\left\{kA_{ks}^{i}\cos(k\omega t)+kA_{kc}^{i}\sin(k\omega t)\right\}\right\}\omega d_0\Delta/3S_{ck} \tag{4-61}$$

where ck denotes the number of circuits. S_{ck}, and d_0 are the area for a coil and the depth in the z-direction, respectively.

Based on the harmonic balance method, the terminal voltage $\{V_{ck}\}$ for a single element in the *ck*-th circuit is expressed as:

$$\{V_{ck}\} = \frac{\omega d_0 \Delta^e}{3S_{ck}} [N \;\; N \;\; N] \begin{Bmatrix} \{A^1\} \\ \{A^2\} \\ \{A^3\} \end{Bmatrix} = \left[C_{ck}^e\right] \begin{Bmatrix} \{A^1\} \\ \{A^2\} \\ \{A^3\} \end{Bmatrix} \tag{4-62}$$

According to Kirchhoff's law, the matrix equations of magnetizing voltage sources and external circuits related to the field are obtained by:

$$\left.\begin{array}{l} [C_{Uin}]\{A\} + [I]\{V_{NNin}\} + S_{cu}[Z_{Uin}]\{J_{Uin}\} = [V_{Uin}\} \\ [C_{Vin}]\{A\} + [I]\{V_{NNin}\} + S_{cv}[Z_{Vin}]\{J_{Vin}\} = [V_{Vin}\} \\ [C_{Win}]\{A\} + [I]\{V_{NNin}\} + S_{cw}[Z_{Win}]\{J_{Win}\} = [V_{Win}\} \\ [C_{out}]\{A\} - S_{co}[Z_{out}]\{J_{out}\} = [V_{out}\} \\ S_{in,cu}[I]\{J_{Uin}\} + S_{in,cu}[I]\{J_{Vin}\} + S_{in,cu}[I]\{J_{Win}\} = 0 \end{array}\right\} \tag{4-63}$$

where the applied three-phase voltage sources $\{V_u\}$, $\{V_v\}$, $\{V_w\}$ and $\{V_{NN'}\}$ represent the unknown voltage at the neutral point, expressed as below:

$$\{V_u\} = \{\{V_{u1s}\} \;\; \{V_{u1c}\} \;\; \{V_{u3s}\} \;\; \{V_{u3c}\} \;\; \cdots\}^T \tag{4-64}$$

$$\{V_v\} = \{\{V_{v1s}\} \;\; \{V_{v1c}\} \;\; \{V_{v3s}\} \;\; \{V_{v3c}\} \;\; \cdots\}^T \tag{4-65}$$

$$\{V_w\} = \{\{V_{w1s}\} \;\; \{V_{w1c}\} \;\; \{V_{w3s}\} \;\; \{V_{w3c}\} \;\; \cdots\}^T \tag{4-66}$$

$$\{V_{NN'}\} = \{\{V_{NN1s}\} \;\; \{V_{NN1c}\} \;\; \{V_{NN3s}\} \;\; \{V_{NN3c}\} \;\; \cdots\}^T \tag{4-67}$$

and $[C_{ck}]$ is obtained from:

$$[C_{ck}] = \frac{\omega d_0 \Delta}{3S_{ck}} [N \;\; N \;\; N] \tag{4-68}$$

$\{J_u\}$, $\{J_v\}$, $\{J_w\}$ and $\{J_{out}\}$ are magnetizing current density and output current density, respectively. These variables have the same expression as Equations (4-64) to (4-67).

The matrix $[Z_{ck}]$ is the circuit impedance, including the resistance of windings and leakage inductance corresponding to the harmonics, and:

$$[Z_{ck}] = \begin{bmatrix} [Z_{c1}] & 0 & \cdots \\ 0 & [Z_{c3}] & \cdots \\ \vdots & \vdots & \ddots \end{bmatrix} \tag{4-69}$$

The system equation of magnetic field for the whole region can be written as below:

$$[H]\{A\}-[G_w]\{J_w\}-[G_v]\{J_v\}-[G_w]\{J_w\}\cdots-[G_k]\{J_k\}=\{0\} \tag{4-70}$$

The coefficient matrix $[H]$ of HBFEM has the form of:

$$[H]=\sum_e([S^e]+[N^e]+[M^e]) \tag{4-71}$$

$[G_k]$ is the constant matrix related to the currents, that is:

$$[G_k]=\sum_e[G_k^e] \tag{4-72}$$

where $[G_k^e]$ is obtained from a single element, that is $[G^e]=\Delta^e/3$. $\{K^e\}$ is defined as:

$$\{K^e\}=[G_k^e]\{J_k\} \tag{4-73}$$

$\{A\}$ is the magnetic vector potential for each harmonic component, expressed as:

$$\{A^e\}=\left\{\{A^1\}^T\ \{A^2\}^T\ \{A^3\}^T\right\}^T \tag{4-74}$$

where $[A^i]$ includes all harmonic components, that is:

$$\{A^i\}=\{A_{1s}^i\ A_{1c}^i\ A_{2c}^i\ A_{3c}^i\ \cdots\}^T \tag{4-75}$$

Combining (4-63) and (4-70), the system matrix equation for whole magnetic field, voltage sources and external circuits is obtained from the following system equations:

$$\left.\begin{aligned}
&[H]\{A\}-[G_U]\{J_{Uin}\}-[G_V]\{J_{Vin}\}-[G_W]\{J_{Win}\}\\
&-[G_{Uout}]\{J_{Uout}\}-[G_{Vout}]\{J_{Vout}\}-[G_{Wout}]\{J_{Wout}\}=\{0\}\\
&[C_{Uin}]\{A\}+[I]\{V_{NNin}\}+S_{cu}[Z_{Uin}]\{J_{Uin}\}=[V_{Uin}\}\\
&[C_{Vin}]\{A\}+[I]\{V_{NNin}\}+S_{cu}[Z_{Vin}]\{J_{Vin}\}=[V_{Vin}\}\\
&[C_{Win}]\{A\}+[I]\{V_{NNin}\}+S_{cu}[Z_{Win}]\{J_{Win}\}=[V_{Win}\}\\
&[C_{out}]\{A\}-S_{cu}[Z_{out}]\{J_{out}\}=[V_{out}\}\\
&S_{in,cu}[I]\{J_{Uin}\}+S_{in,cu}[I]\{J_{Vin}\}+S_{in,cu}[I]\{J_{Win}\}=0
\end{aligned}\right\} \tag{4-76}$$

Thus, the system matrix equation can be obtained as:

$$\begin{bmatrix} [H] & -[G_u] & -[G_v] & -[G_w] & -[G_{out}] & 0 \\ [C_{Uin}] & S_{cu}[Z_u] & 0 & 0 & 0 & [I] \\ [C_{Vin}] & 0 & S_{cv}[Z_v] & 0 & 0 & [I] \\ [C_{Win}] & 0 & 0 & S_{cw}[Z_w] & 0 & [I] \\ [C_{out}] & 0 & 0 & 0 & -S_{co}[Z_{out}] & 0 \\ 0 & S_{cu}[I] & S_{cv}[I] & S_{cw}[I] & 0 & 0 \end{bmatrix} \begin{Bmatrix} \{A\} \\ \{J_u\} \\ \{J_v\} \\ \{J_w\} \\ \{J_{out}\} \\ \{V_{NN}\} \end{Bmatrix} = \begin{Bmatrix} 0 \\ \{V_u\} \\ \{V_v\} \\ \{V_w\} \\ \{0\} \\ \{0\} \end{Bmatrix} \tag{4-77}$$

Since $[H]$ is a large-scale band matrix, the system matrix equation (4-76) becomes a doubly-bordered band diagonal which can be solved by the Gaussian elimination method.

4.4.3 Numerical Analysis of High Speed and Hybrid Induction Machine using HBFEM

Since the high-speed hybrid motor has a very complex configuration, the optimal design is requested in the design of an electric machine. The good design prevents the flux of fundamental harmonic component from passing through the poles and rotor, and only the three-times frequency flux passes through the poles and moves the rotor. In our numerical analysis, the half model is used as a analysis area. The magnetizing windings are applied with 60 Hz commercial three-phase voltage source. The harmonic components will be generated in the core when the core becomes saturated. Figure 4.25(a) shows the distributions of magnetic flux in the core at excitation voltage, 213 V. The comparison made between the different slips is shown in Figure 4.25(b), where we find that the motion has had some influence on the distribution of magnetic flux.

Figure 4.26 shows the waveform of the flux density at different legs of the magnetic tripler indicated in Fig 4.22(a), where the magnetic flux of the third harmonic component has been generated in the leg B_{u4} and B_{uw4}.

The magnetic flux distributions at different rotating angles are shown in Figure 4.27. It is clear to see that the third harmonic component is far larger than the fundamental harmonic component in the rotor. The condition of generating a high rotation speed is:

$$|B_3| > |B_1| \tag{4-78}$$

Therefore, the rotation speed (10 800 rpm) of the high speed and hybrid electric machine can be achieved by the third harmonic magnetic flux.

Figure 4.27 shows the distribution of the fundamental and third harmonic components of magnetic flux in the magnetic core at different rotating angles. Figure 4.28

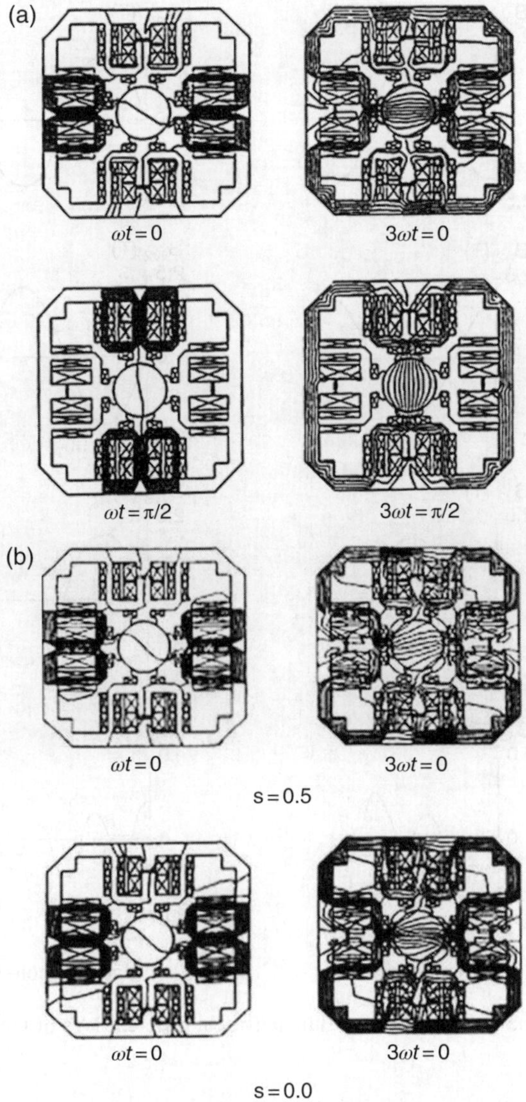

Figure 4.25 The distributions of magnetic flux in the core at excitation voltage 213 V. (a) Magnetic flux distributions of different harmonic component; (b) The impact of slip on the distributions of magnetic flux

shows magnetic flux distributions of the high-speed and hybrid induction motor at normal rotating case (a) and stopping case (b).

When the input voltage (fundamental frequency) is increased, the output voltage of the third harmonic component is also increased significantly at the saturation point, while fundamental component voltage decreased at the same point (213 V), as shown in Figure 4.29.

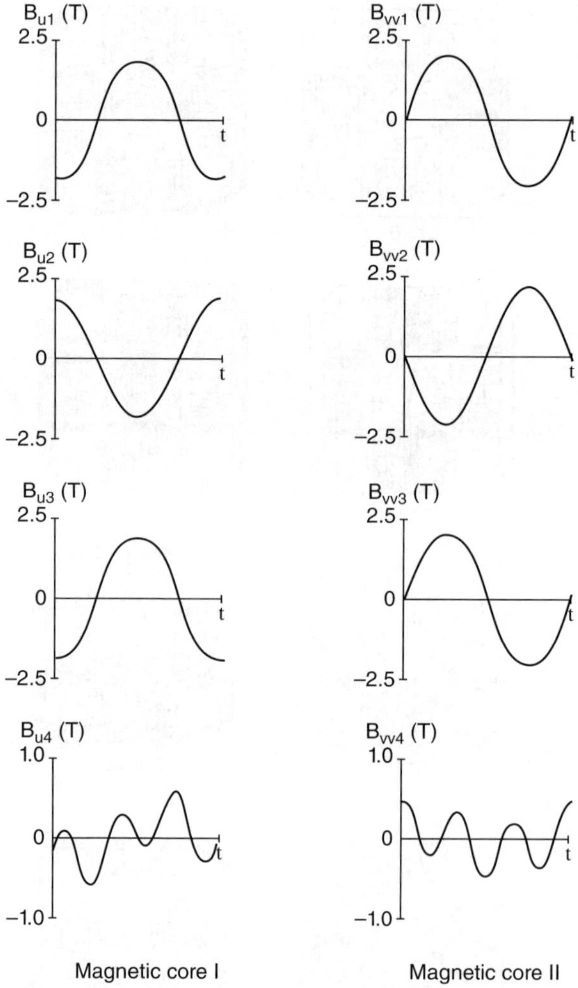

Figure 4.26 The waveform of the flux density at different legs of the magnetic tripler

Figure 4.30 shows various waveforms, including input voltages with frequency f_1 (50 Hz), output voltage with frequency f_2 (150 Hz) and voltage accrued at neutral point. Obviously, the phase input voltages have some harmonic during the operation, while the output voltage is mainly dominated by the third harmonic component. Neutral voltage is very high in this case. The HBFEM, taking account of external circuits and motion, can be used effectively to solve the high-speed and hybrid induction motor problem. A comparison is made between experimental and numerical results for the static model, and has shown a good agreement.

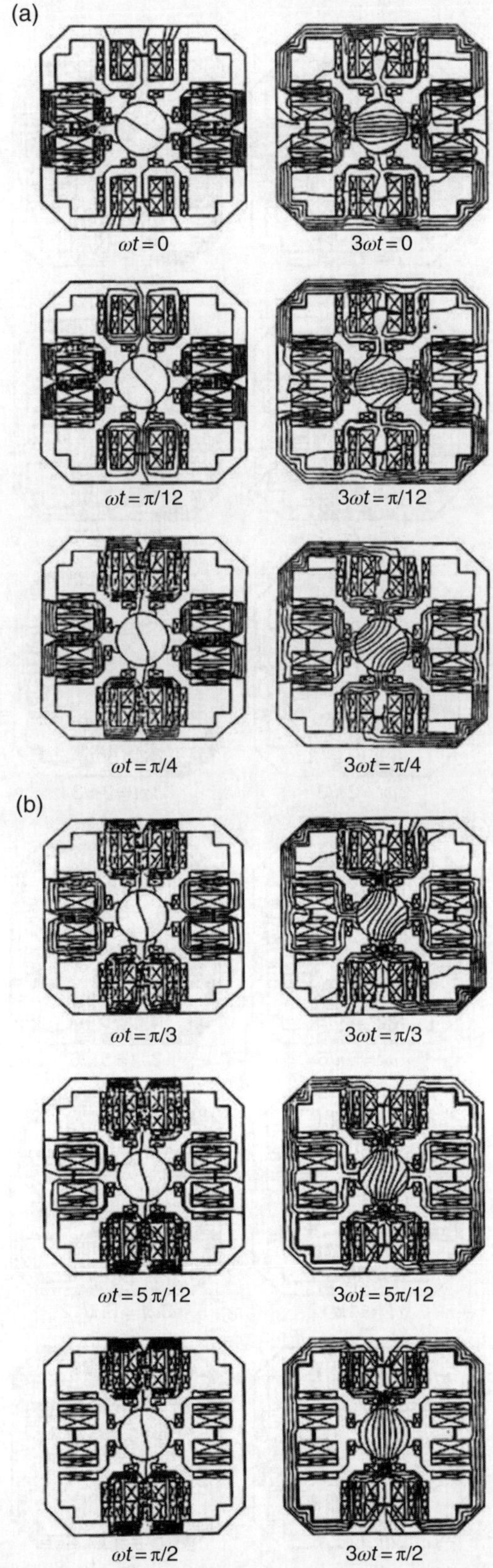

Figure 4.27 The fundamental and third harmonic components of magnetic flux distributions at different rotating angles

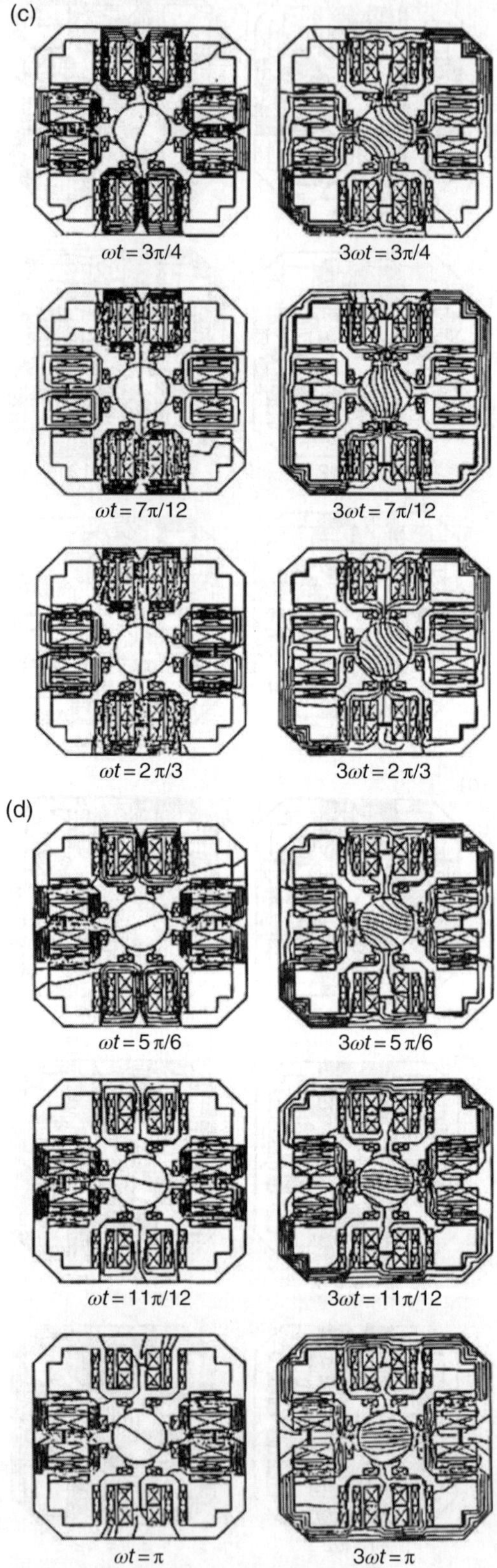

Figure 4.27 (*Continued*)

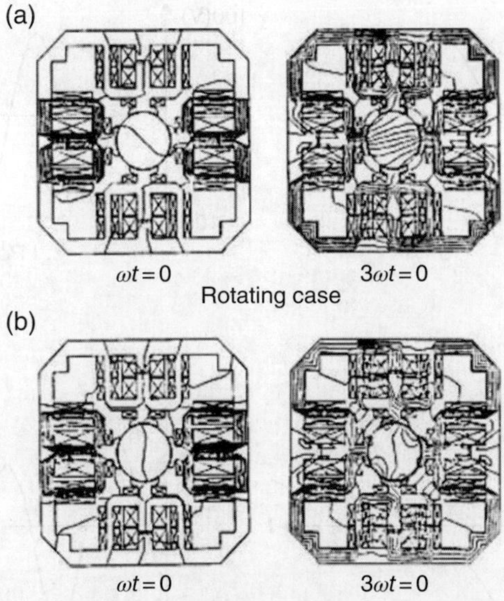

Figure 4.28 Magnetic flux distributions of high-speed and hybrid induction motor: (a) at normal rotating case; (b) at stopping case

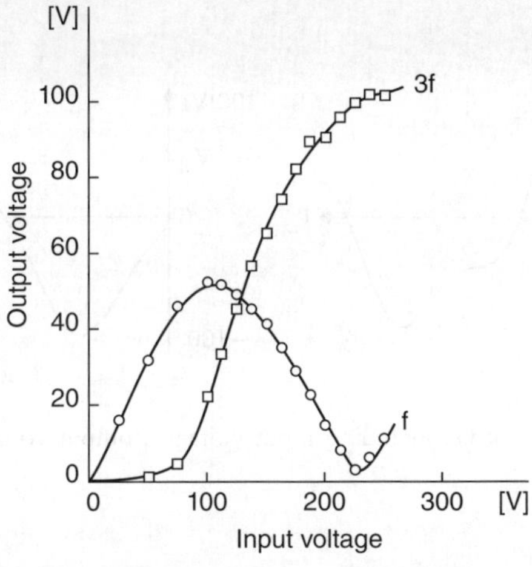

Figure 4.29 Input voltage vs output voltage for both fundamental and third harmonics

4.5 Three-Dimensional Axi-Symmetrical Transformer with DC-Biased Excitation

In high-frequency switching power supplies, leakage inductance, skin and proximity effects, winding self-capacitance and inter-winding capacitance can cause some serious problems in high-frequency transformers. Detailed information about the distribution of

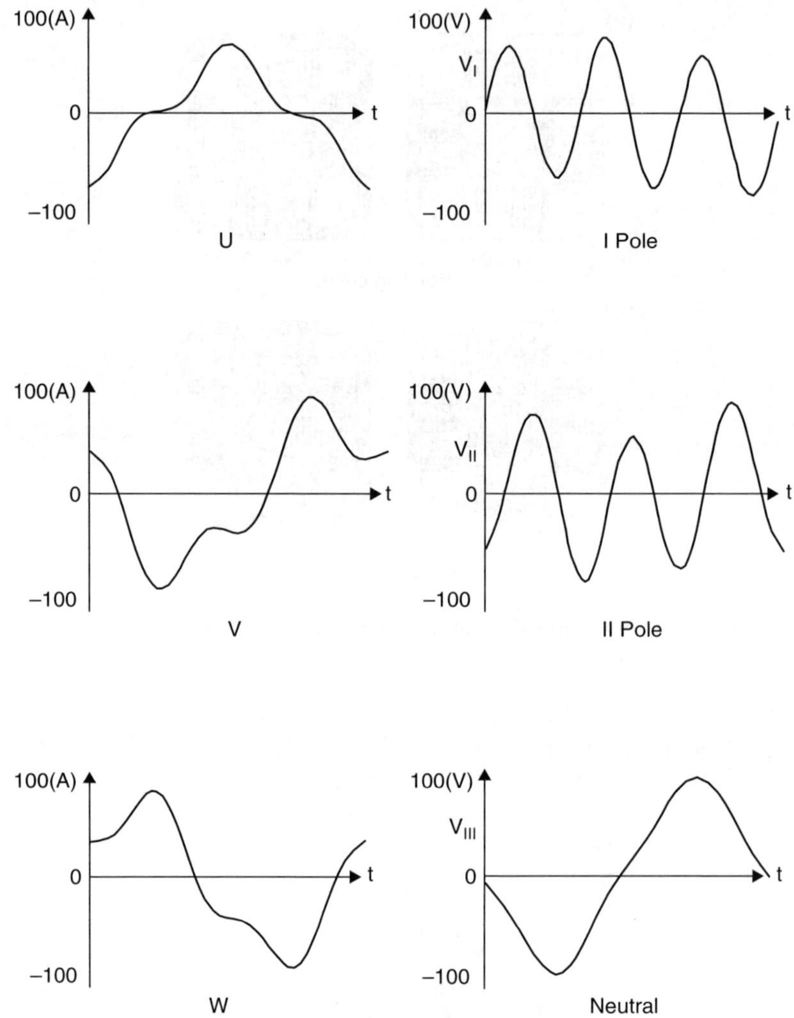

Figure 4.30 Various waveforms, including input voltages, output voltage and voltage accrued at neutral point

eddy currents, flux density and harmonics distribution in the magnetic core and windings has to be known when designing a high-frequency transformer. Furthermore, the nonlinear nature and hysteresis of the core material can cause waveform distortion. These distortions cause further harmonics, which will increase power losses in both the winding and magnetic core, resulting in a loss of efficiency, as well as the possibility of parasitic resonance in the system [5].

A typical port core transformer has an axi-symmetrical structure and a B-H hysteresis curve, as shown in Figure 4.31. The transformer is excited by quasi-sinusoidal

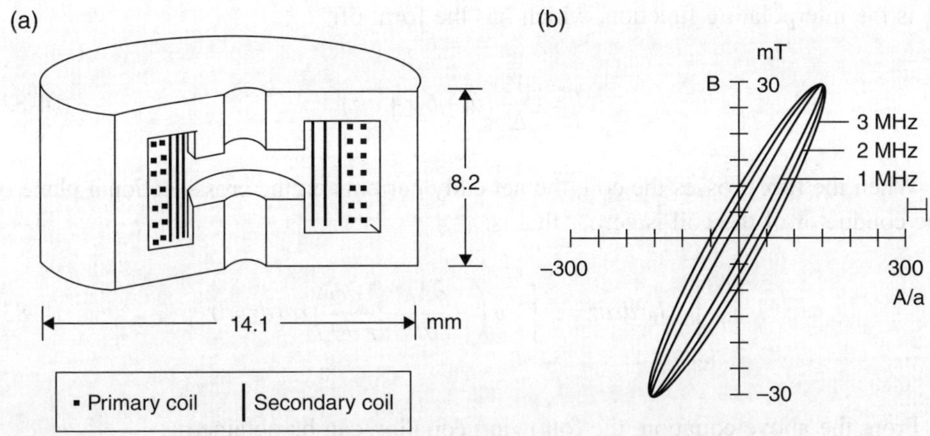

Figure 4.31 (a) Simulation model of a switching mode transformer. (b) B-H curve of the magnetic core

waveform, which includes AC fundamental, harmonics and DC components. When the magnetic core becomes saturated, the waveforms will be distorted, and harmonics will be generated in the magnetic field and circuits. The hysteresis loss in winding is also increased due to the effect of high-frequency harmonics.

4.5.1 *Numerical Simulation of 3-D Axi-Symmetrical Structure*

Since the 3-D port core transformer has an axi-symmetrical structure, the numerical model can be reduced to a 3-D axi-symmetrical structure problem. Thus, Maxwell's equation, related to the vector potential A, can be written as:

$$\frac{\partial}{\partial r}\left(\frac{\nu}{r}\frac{\partial}{\partial r}(rA_\theta)\right)+\frac{\partial}{\partial z}\left(\nu\frac{\partial A_\theta}{\partial z}\right)=-j_s+\sigma\frac{\partial A_\theta}{\partial t}+\sigma\frac{1}{r}\frac{\partial\phi}{\partial\theta} \tag{4-79}$$

According to Galerkin's method and weighted function, the integral equation then becomes:

$$\begin{aligned}&2\pi\int\left\{\left(\frac{N_i}{r}+\frac{\partial N_i}{\partial r}\right)\nu\left(\frac{A_\theta}{r}+\frac{\partial A_\theta}{\partial r}\right)+\frac{\partial N_i}{\partial z}\nu\frac{\partial A_\theta}{\partial z}\right\}rdrdz\\&-2\pi\int\left(J_s-\sigma\frac{\partial A_\theta}{\partial t}-\sigma\frac{1}{r}\frac{\partial\phi}{\partial\theta}\right)N_r rdrdz=0\end{aligned} \tag{4-80}$$

where vector potential A_θ is:

$$A_\theta=\sum_{i-1,2,3}N_iA_\theta^i \tag{4-81}$$

N_i is the interpolating function, which has the form of:

$$N_i = \frac{1}{2\Delta^e}(a_i + b_i r + c_i z) \tag{4-82}$$

When the flux crosses the coil, the net eddy currents on the cross-sectional plane of the conductor of the coil is zero – that is:

$$\int_{S(coil)} J_e r dr dz = \int_{S(coil)} \sigma\left(-\frac{\partial A_\theta}{\partial t} - \frac{1}{r}\frac{\partial \phi}{\partial \theta}\right) r dr dz = 0 \tag{4-83}$$

From the above equation, the following equation can be obtained:

$$R = \frac{\partial \phi}{\partial \theta} = -\frac{\int_{S(coil)} \left(\frac{\partial A_\theta}{\partial t}\right) r dr dz}{\int_{S(coil)} \left(\frac{1}{r}\right) r dr dz} = -\frac{1}{S}\sum_{e}^{coil} \int_{S(e)} \left(\frac{\partial A_\theta}{\partial t}\right) r dr dz \tag{4-84}$$

where S is the cross-section of the coil.

Combining harmonic balance techniques with the FEM matrix equation and considering the DC-biased condition for a single element can be obtained as below:

$$G^e = \frac{1}{4\Delta^e}\begin{bmatrix} (b_1b_1 + c_1c_1)D & (b_1b_2 + c_1c_2)D & (b_1b_3 + c_1c_3)D \\ (b_2b_1 + c_2c_1)D & (b_2b_2 + c_2c_2)D & (b_2b_3 + c_2c_3)D \\ (b_3b_1 + c_3c_1)D & (b_3b_2 + c_3c_2)D & (b_3b_3 + c_3c_3)D \end{bmatrix}\begin{Bmatrix} A_1^e \\ A_2^e \\ A_3^e \end{Bmatrix}$$
$$+\frac{1}{6}\begin{bmatrix} (b_1 + b_1)D & (b_1 + b_2)D & (b_1 + b_3)D \\ (b_2 + b_1)D & (b_2 + b_2)D & (b_2 + b_3)D \\ (b_3 + b_1)D & (b_3 + b_2)D & (b_3 + b_3)D \end{bmatrix}\begin{Bmatrix} A_1^e \\ A_2^e \\ A_3^e \end{Bmatrix} + \frac{\Delta^e}{9r_c}\begin{bmatrix} D & D & D \\ D & D & D \\ D & D & D \end{bmatrix} \tag{4-85}$$
$$+\frac{r_c\sigma\omega\Delta^e}{12}\begin{bmatrix} 2N & N & N \\ N & 2N & N \\ N & N & 2N \end{bmatrix}\begin{Bmatrix} A_1^e \\ A_2^e \\ A_3^e \end{Bmatrix} - \begin{Bmatrix} K_1^e \\ K_2^e \\ K_3^e \end{Bmatrix} + \begin{Bmatrix} R_1^e \\ R_2^e \\ R_3^e \end{Bmatrix}$$

D and N have different details when the DC component is considered.

$$D = \begin{bmatrix} d_{11} & d_{12} & d_{13} & d_{14} & \cdots \\ d_{21} & d_{22} & d_{23} & d_{24} & \cdots \\ d_{31} & d_{32} & d_{33} & d_{34} & \cdots \\ d_{41} & d_{42} & d_{43} & d_{44} & \cdots \\ \cdots & \cdots & \cdots & \cdots & \ddots \end{bmatrix} = \frac{1}{2}\begin{bmatrix} 2\nu_0 & \nu_{1s} & \nu_{1c} & \nu_{2s} & \cdots \\ 2\nu_{1s} & 2\nu_0 - \nu_{2c} & \nu_{2s} & \nu_{1c} - \nu_{3c} & \cdots \\ 2\nu_{1c} & & 2\nu_0 + \nu_{2c} & \nu_{1s} + \nu_{3s} & \cdots \\ 2\nu_{2s} & symmetry & & 2\nu_0 - \nu_{4c} & \cdots \\ \vdots & \vdots & \vdots & \vdots & \ddots \end{bmatrix} \tag{4-86}$$

and the harmonic matrix, including both odd and even harmonics, is expressed as below:

$$N = \begin{bmatrix} 0 & 0 & 0 & 0 & 0 & 0 & 0 & \cdots \\ 0 & 0 & -1 & 0 & 0 & 0 & 0 & \cdots \\ 0 & 1 & 0 & 0 & 0 & 0 & 0 & \cdots \\ 0 & 0 & 0 & 0 & -2 & 0 & 0 & \cdots \\ 0 & 0 & 0 & 2 & 0 & 0 & 0 & \cdots \\ 0 & 0 & 0 & 0 & 0 & 0 & -3 & \cdots \\ 0 & 0 & 0 & 0 & 0 & 3 & 0 & \cdots \\ \vdots & \vdots & \vdots & \vdots & \vdots & \vdots & \vdots & \ddots \end{bmatrix} \tag{4-87}$$

The system equations for the eddy current problem, taking account of scalar potential φ and the DC component, are obtained from (4-85) – that is:

$$[S]\{A\} + [N]\{A\} + \{R\} - \{K\} = 0 \tag{4-88}$$

where $[S]$ and $[N]$ are system coefficient matrix and harmonic matrix, respectively. For the single element, the other terms can be expressed as:

$$\left\{A_\theta^e\right\} = \left\{\left\{A_\theta^1\right\}^T \left\{A_\theta^2\right\}^T \left\{A_\theta^3\right\}^T\right\}^T \tag{4-89}$$

$$\left\{A^i\right\} = \left\{A_0 \;\; A_{1s} \;\; A_{1c} \;\; A_{2s} \;\; A_{2c} \;\; \cdots\right\}^T \tag{4-90}$$

and

$$\left\{K_i^e\right\} = \left\{\left\{K^1\right\}^T \left\{K^2\right\}^T \left\{K^3\right\}^T\right\}^T \tag{4-91}$$

$$\left\{K^i\right\} = \frac{\Delta}{4}\left(r_c + \frac{r_i}{3}\right)\left\{J_0 \;\; J_{1s} \;\; J_{1c} \;\; J_{2s} \;\; J_{2c} \;\; \cdots\right\}^T \tag{4-92}$$

and

$$\{R_i^e\} = \left\{\{R^1\}^T \ \{R^2\}^T \ \{R^3\}^T\right\}^T \tag{4-93}$$

$$\{R^i\}^T = \frac{\sigma\Delta^e}{3}\{R_0 \ R_{1s} \ R_{1c} \ R_{2s} \ R_{2c} \ \cdots, \ R_{ns}, \ R_{nc}\}^T \tag{4-94}$$

where the eddy current matrix $\{R\}$ for a single element can be calculated from:

$$R_{ns} = \sum_e \frac{n\omega\Delta^e}{4S}\sum_{1,2,3}\left(r_c + \frac{r_i}{3}\right)A^i_{nc} \tag{4-95}$$

and

$$R_{nc} = \sum_e \frac{n\omega\Delta^e}{4S}\sum_{1,2,3}\left(r_c + \frac{r_i}{3}\right)A^i_{ns} \tag{4-96}$$

Therefore, the system matrix equation of (4-88) can be rewritten as:

$$[S]\{A\} + ([N] + [R])\{A\} - \{K\} = 0 \tag{4-97}$$

where the eddy current loss matrix $[R]$ for a single element can be calculated as:

$$[R] = \frac{(\Delta^e)^2 n\omega\sigma}{12S} \tag{4-98}$$

where S is the cross-section of the coil and Δ is the area of the element.

4.5.2 Numerical Analysis of the Three-Dimensional Axi-Symmetrical Model

Figure 4.32 shows the numerical results for the distributions of magnetic flux calculated by HBFEM. From these numerical results, one can find that, around the air gap, the DC component of magnetic flux can easily cross the windings without inducing any current in the windings, while the AC component of magnetic flux is pushed out of the windings by induced current, called eddy current, in the windings. Figure 4.32 shows the flux distributions for DC, first and second harmonics.

To verify the HB-FEM based solution, a single-phase DC-biased switching transformer with a voltage source driven model is used. The calculated result of flux density is compared with the measured result, as shown in Figure 4.33. In this simulation model,

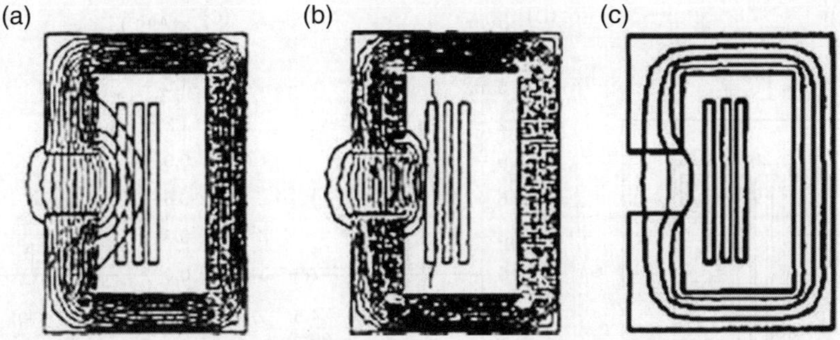

Figure 4.32 Magnetic flux distribution [4]. (a) DC flux; (b) First harmonic flux; (c) Second harmonic flux

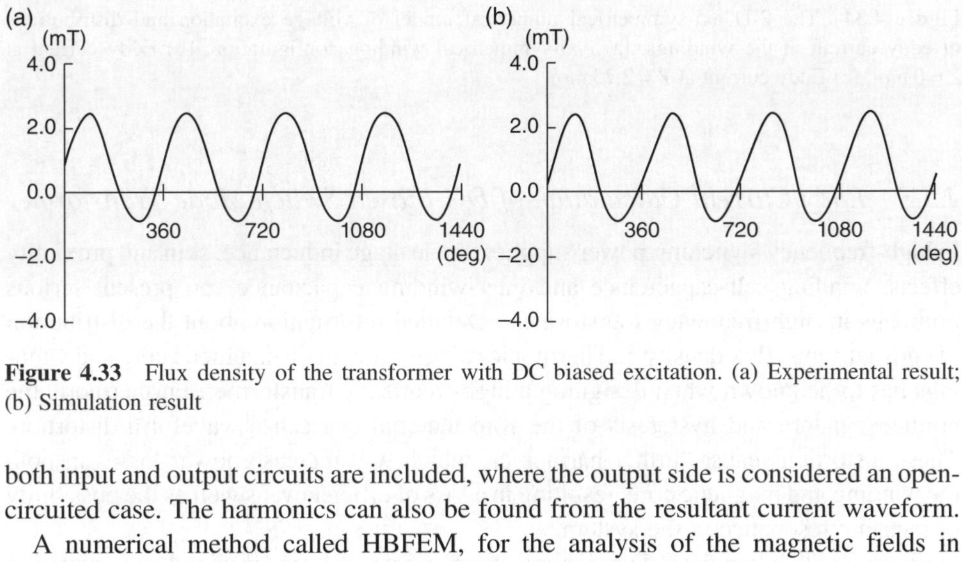

Figure 4.33 Flux density of the transformer with DC biased excitation. (a) Experimental result; (b) Simulation result

both input and output circuits are included, where the output side is considered an open-circuited case. The harmonics can also be found from the resultant current waveform.

A numerical method called HBFEM, for the analysis of the magnetic fields in switching power supplies, has been proposed. This method can be used effectively to solve the harmonic problems in the magnetic field. To design and analyze switching power supplies, numerical techniques are required to determine the leakage inductance, eddy currents and power loss. All of these losses involve electromagnetic field analysis. Therefore, the numerical technique plays an important role in the analysis of switching power supplies. This chapter presents a harmonic balance technique, combined with the finite element method, for solving quasi-static nonlinear magnetic field problems. The magnetic field coupled with the circuits is discussed, and a numerical model described by a system matrix equation is obtained. Several examples show the application of HBFEM to solve harmonic and eddy current problems. The comparison between numerical and experimental results is in good agreement.

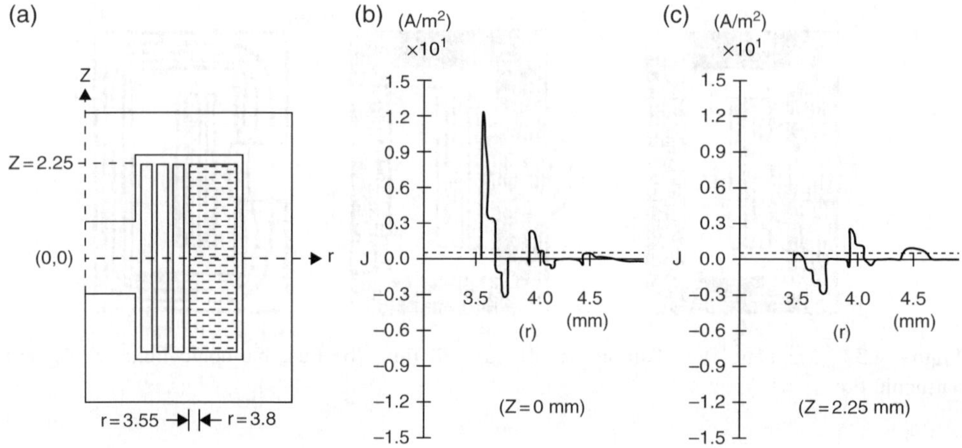

Figure 4.34 The 2-D axi-symmetrical numerical model of voltage excitation and distributions of eddy current in the windings. (a) Axi-symmetrical winding configuration; (b) Eddy current at $Z = 0$ mm, (c) Eddy current at $Z = 2.25$ mm

4.5.3 Eddy Current Calculation of DC-Biased Switch Mode Transformer

In high-frequency switching power supplies, the leakage inductance, skin and proximity effects, winding self-capacitance and inter-winding capacitance can present serious problems in high-frequency transformers. Detailed information about the distribution of eddy currents, flux density and harmonics distribution in the magnetic core and windings has to be known when designing a high-frequency transformer. Furthermore, the nonlinear nature and hysteresis of the core material can cause waveform distortion. These distortions cause further harmonics, which will increase power losses in both the winding and magnetic core, resulting in a loss of efficiency, as well as the possibility of parasitic resonance in the system.

Figure 4.34 shows the 3-D axi-symmetrical numerical model of voltage excitation with a nonlinear magnetic field and the hysteresis curve of the magnetic core [5,7]. The eddy currents in transformer windings are significantly increased by leakage flux of AC components, especially around the air gap of a magnetic core.

The eddy currents induced in the windings are shown in Figure 4.34. The largest current density occurs at the edge of the air gap in the magnetic core. The power losses produced by eddy currents in both the coil and the magnetic core can be easily calculated by using the numerical results of HBFEM. The formula for the total eddy current power loss in the areas of the coil and the magnetic core can be expressed as:

$$P_{total} = P_{coil} + P_{core} \tag{4-99}$$

where P_{coil} is the total eddy current power loss in the coil, given by:

$$P_{coil} = \sum_{e}^{coil} \left(P_{1coil}^{e} + P_{2coil}^{e} + \cdots P_{ncoil}^{e}\right) \tag{4-100}$$

and P_{core} is a summation of the eddy current power loss in the magnetic core – that is:

$$P_{core} = \sum_{e}^{core} \left(P_{1core}^{e} + P_{2core}^{e} + \cdots P_{ncore}^{e}\right) \tag{4-101}$$

where n indicates the *nth* harmonic. Eddy current power loss for each harmonic component can be calculated using the following generalized equation:

$$P_{n}^{e} = \sigma \Delta^{e} \left\{ \sqrt{\left(J_{ens}^{e}\right)^{2} + \left(J_{enc}^{e}\right)^{2}} \right\}^{2} \tag{4-102}$$

where J_{en} can be obtained by:

$$J_{ens} = \frac{n\omega\sigma\Delta^{e}}{4S} \sum_{i=1,2,3} A_{nc}^{i} \tag{4-103}$$

and

$$J_{enc} = \frac{n\omega\sigma\Delta^{e}}{4S} \sum_{i=1,2,3} A_{ns}^{i} \tag{4-104}$$

where A is the vector potential on each node and S is the cross-sectional for each material.

References

[1] Lee, F.C. (1988). High-Frequency Quasi-Resonant Converters Technologies. *Proceedings of the IEEE* **76**(4), 362–376.

[2] Kassakian, J.G. and Schlecht, M.F. (1988). High-Frequency High-Density Converters for Distributed Power Supply Systems. *Proceedings of the IEEE* **76**(4), 362–376.

[3] Mohan, N., Undeland, T.M. and Robbins, W. (1989). *Power Electronics-Converters, Applications and Design*. John Wiley & Sons.

[4] Tihanyi, L. (1995). *Electromagnetic Compatibility in Power Electronics*. IEEE Press.

[5] Lu, J., Yamada, S. and Harrison, H.B. (1996). Application of HB-FEM in the Design of Switching Power Supplies. *IEEE Transactions on Power Electronics* **11**(2), 347–355.

[6] Yamada, S. and Bessho, K. (1988). Harmonic field calculation by the combination of finite element analysis and harmonic balance method. *IEEE Transactions on Magnetics* **24**(6), 2588–2590.
[7] Lu, J., Yamada, S. and Bessho, K. (1990). Development and Application of Harmonic Balance Finite Element Method in Electromagnetic Field. *International Journal of Applied Electromagnetics in Materials* **1**(2–4), 305–316.
[8] Lu, J., Yamada, S. and Bessho, K. (1990). Time-periodic Magnetic Field Analysis with Saturation and Hysteresis Characteristics by Harmonic Balance Finite Element Method. *IEEE Transactions on Magnetics* **26**(2), 995–998.
[9] Lu, J., Yamada, S. and Bessho, K. (1991). Harmonic Balance Finite Element Method Taking Account of External Circuits and Motion. *IEEE Transactions on Magnetics* **27**(5), 4204–4207.
[10] Biringer, B.P. and Slemon, G.R. (1963). Harmonic analysis of the magnetic frequency tripler. *IEEE Transactions on Communication and Electronics* **82**, 327–332.
[11] Bendzsak, G.J. and Biringer, B.P. (1974). The influence of magnetic characteristics upon tripler performance. *IEEE Transactions on Magnetics* **10**(3), 961–964.
[12] Ishikawa, T. and Hou, Y. (2002). Analysis of a Magnetic Frequency Tripler Using the Preisach Model. *IEEE Transactions on Magnetics* **38**(2), 841–844.

5

Advanced Numerical Approach using HBFEM

5.1 HBFEM for DC-Biased Problems in HVDC Power Transformers

5.1.1 DC Bias Phenomena in HVDC

Transformers are key components in the normal operation of power systems. Typically, they are designed for sinusoidal voltage excitation. Occasionally, direct current (DC) flows through the power transformer coils from the directly earthed neutral of the star-connected winding [1]. Generally, there are two major causes of such DC. The first cause is the geomagnetically induced currents (GICs), caused by the interaction of the solar wind and the earth's ionosphere [2–4]. The second cause is the leakage currents of the ground electrodes of a DC transmission system, which operates in the ground-return mode [5], as shown in Figure 5.1. There are actually two main operation modes of the HVDC, ground-return mode and metallic-return mode. The ground-return mode includes monopolar ground-return mode, balanced bipolar ground-return mode and unbalanced bipolar ground-return mode, which all generate direct current in the ground.

The DC intruding into the power transformer can bias the working point of the magnetic field in the limb, making parts of the core saturate in half a cycle. This results in a series of undesirable effects in a power transformer, such as increased noise and loss, and excessive VAR consumption [6–8]. In order to analyze a DC-biased power transformer, the following aspects require special attention: the topology of a core, the nonlinearity of ferromagnetic materials, the coupling and connection of coils, the effects of eddy currents, the terminal load levels, and so on.

Harmonic Balance Finite Element Method: Applications in Nonlinear Electromagnetics and Power Systems, First Edition. Junwei Lu, Xiaojun Zhao and Sotoshi Yamada.

Companion website: www.wiley.com\go\lu\HBFEM

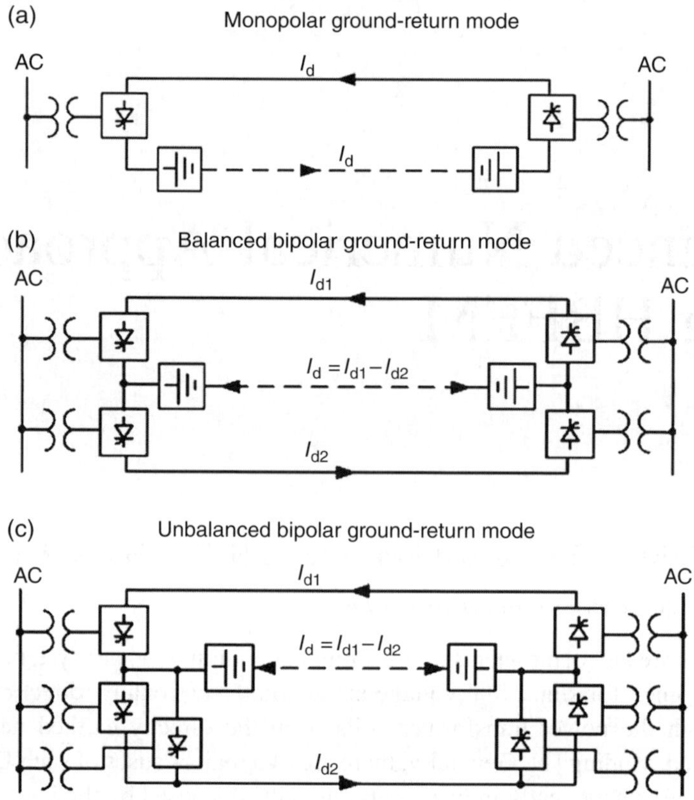

Figure 5.1 Operation modes of the high-voltage direct current transmission system

5.1.2 HBFEM for DC-Biased Magnetic Field

Research on the mechanism of the DC-biased problem contributes to important developments in transformer design. Various methods have been used to study the electromagnetic field in transformers under the DC bias condition [2, 4, 9]. The electric circuit model and the magnetic circuit model [10, 11] were proposed to calculate the excitation current in windings. The time-stepping finite element method [12] was also used by some researchers to compute the magnetizing current and magnetic field. However, it is difficult to obtain accurate results from the electric or magnetic circuit model, especially when magnetic field analysis is required to explore the mechanism of the DC-biased problem. The time-stepping finite element method is an alternative method to calculate the transient magnetic field, but accurate solutions of high order harmonic components in the exciting current and magnetic induction require many more iterations in the time domain, which can reduce the effectiveness of this method.

When a direct current flows additionally through the windings of a transformer, a steady-state nonlinear magnetic field is established in the transformer. Odd and even harmonics exist in the magnetizing current simultaneously, and the total flux in the magnetic core consists of both DC flux and AC flux. Therefore, the DC bias phenomenon is actually a harmonic problem [8], and the effect of the DC bias on each harmonic component of excitation current and magnetic induction should be analyzed.

Sotoshi Yamada and Junwei Lu developed the harmonic balance finite element method (HBFEM) [13, 14], which is an effective method to calculate the nonlinear magnetic field without the DC bias in the harmonic frequency domain. However, when DC bias arises from the HVDC transmission, the nonlinear magnetic field exhibits some unique features, such as the coexistence of odd and even harmonics in the magnetizing current, a DC bias effect on the magnetic induction and the mutual influence between DC flux and AC flux. None of these phenomena have been investigated systematically.

5.1.2.1 A. Basic Theory of HBFEM

The nonlinear magnetic field can be described by the following equation:

$$\nabla \times \nu \nabla \times A - J = 0 \tag{5-1}$$

where A is the magnetic vector potential, ν is the reluctivity and J is the current density, which includes exciting current density and eddy current density.

In the two-dimensional case, the nonlinear magnetic field equation can be written as:

$$\frac{\partial}{\partial x}\left(\nu \frac{\partial A}{\partial x}\right) + \frac{\partial}{\partial y}\left(\nu \frac{\partial A}{\partial y}\right) + \sigma \frac{\partial A}{\partial t} - J_s = 0. \tag{5-2}$$

In light of Galerkin's method, the weighted residual can be obtained from Equation (5-2):

$$\begin{aligned} G = & \iint_{\Omega_e} \left(\frac{\partial N_i^e}{\partial x} \nu \frac{\partial A}{\partial x} + \frac{\partial N_i^e}{\partial y} \nu \frac{\partial A}{\partial y} \right) dxdy + \\ & \iint_{\Omega_e} \left(\sigma \frac{\partial A}{\partial t} N_i^e \right) dxdy - \iint_{\Omega_e} \left(J_s N_i^e \right) dxdy, \end{aligned} \tag{5-3}$$

where N_i is the interpolation function for a linear triangular element, J_s is the exciting current density, and σ is the conductivity.

The time-periodic solutions are focused in the DC bias phenomenon, since it is a harmonic problem with alternating and direct excitations. The magnetic flux density and magnetic vector potentials are both periodic functions in the time domain. According to the Fourier transformation theory, all variables, such as vector potential A, flux density B and exciting current density J_s, can be approximated by a triangular series.

Therefore, a HBFEM matrix equation for a single element can be obtained based on the harmonic balanced method:

$$\begin{aligned}\boldsymbol{G}^e &= \boldsymbol{S}^e\boldsymbol{A}^e + \boldsymbol{M}^e\boldsymbol{A}^e - \boldsymbol{K}^e \\ &= \frac{1}{4\Delta^e}\begin{bmatrix}(b_1b_1+c_1c_1)\boldsymbol{D} & (b_1b_2+c_1c_2)\boldsymbol{D} & (b_1b_3+c_1c_3)\boldsymbol{D} \\ (b_2b_1+c_2c_1)\boldsymbol{D} & (b_2b_2+c_2c_2)\boldsymbol{D} & (b_2b_3+c_2c_3)\boldsymbol{D} \\ (b_3b_1+c_3c_1)\boldsymbol{D} & (b_3b_2+c_3c_2)\boldsymbol{D} & (b_3b_3+c_3c_3)\boldsymbol{D}\end{bmatrix}\begin{Bmatrix}\boldsymbol{A}_1^e \\ \boldsymbol{A}_2^e \\ \boldsymbol{A}_3^e\end{Bmatrix} \\ &+ \frac{\sigma\omega\Delta^e}{12}\begin{bmatrix}2N & N & N \\ N & 2N & N \\ N & N & 2N\end{bmatrix}\begin{Bmatrix}\boldsymbol{A}_1^e \\ \boldsymbol{A}_2^e \\ \boldsymbol{A}_3^e\end{Bmatrix} - \begin{Bmatrix}\boldsymbol{K}_1^e \\ \boldsymbol{K}_2^e \\ \boldsymbol{K}_3^e\end{Bmatrix} = \boldsymbol{0},\end{aligned} \tag{5-4}$$

in which $b_i = y_j - y_k$, $c_i = x_k - x_j$. The harmonic forms of the magnetic vector potential $\boldsymbol{A}_i^e$ and the exciting term $\boldsymbol{K}_i^e$ are expressed in Equations (5-5) and (5-6), respectively:

$$\boldsymbol{A}_i^e = \{A_{i0}^e \ \ A_{i1s}^e \ \ A_{i1c}^e \ \ A_{i2s}^e \ \ A_{i2c}^e \ \ \ldots\} \tag{5-5}$$

$$\boldsymbol{K}_i^e = \{K_{i0}^e \ \ K_{i1s}^e \ \ K_{i1c}^e \ \ K_{i2s}^e \ \ K_{i2c}^e \ \ \ldots\}. \tag{5-6}$$

5.1.2.2 B. Coupling Between Electric Circuits and the Magnetic Field

When electromagnetic devices, such as transformers under DC bias, are excited by voltage, the exciting currents are unknown. Therefore, Equation (5-2) is no longer applicable to solve the coupled problem. In that case, the coupling between the electric circuit and the magnetic field should be taken into account [15–17].

According to Kirchhoff's Law, the applied voltage on the external port of the electric circuit can be defined as follows:

$$V_{ink} = V_k + R_kI_k + L_k(dI_k/dt) + (1/C_k)\int I_k dt, \tag{5-7}$$

where V_{ink} is the input voltage of circuit k, and V_k is the corresponding induced electromotive force. C_k and L_k are the capacitance and inductance in circuit k, respectively.

Combined with the finite element method, the equations of electric circuits coupled to the magnetic field can be expressed in the form of matrix,

$$V_{ink} = C_k\boldsymbol{A} + S_{ck}\boldsymbol{Z}_k\boldsymbol{J}_k \tag{5-8}$$

where $\boldsymbol{V}_{ink}$ is the harmonic vector of the input voltage, $\boldsymbol{C}_k$ is the coupling matrix, and $\boldsymbol{Z}_k$ is the corresponding impedance matrix.

Finally, the new system matrix equation considering the applied voltage can be obtained by combining Equation (5-4) with Equation (5-8):

$$\begin{bmatrix} \boldsymbol{H} & -\boldsymbol{G}_1 & -\boldsymbol{G}_2 & \cdots & -\boldsymbol{G}_k & \cdots \\ \boldsymbol{C}_1 & S_{c1}\boldsymbol{Z}_1 & \boldsymbol{0} & \boldsymbol{0} & \boldsymbol{0} & \cdots \\ \boldsymbol{C}_2 & \boldsymbol{0} & S_{c2}\boldsymbol{Z}_2 & 0 & \boldsymbol{0} & \cdots \\ \vdots & \boldsymbol{0} & \boldsymbol{0} & \ddots & \boldsymbol{0} & \cdots \\ \boldsymbol{C}_k & \boldsymbol{0} & \boldsymbol{0} & \boldsymbol{0} & S_{ck}\boldsymbol{Z}_k & \cdots \\ \vdots & \vdots & \vdots & \vdots & \vdots & \ddots \end{bmatrix} \begin{Bmatrix} \boldsymbol{A} \\ \boldsymbol{J}_1 \\ \boldsymbol{J}_2 \\ \vdots \\ \boldsymbol{J}_k \\ \vdots \end{Bmatrix} = \begin{Bmatrix} \boldsymbol{0} \\ \boldsymbol{V}_{in1} \\ \boldsymbol{V}_{in2} \\ \vdots \\ \boldsymbol{V}_{ink} \\ \vdots \end{Bmatrix}. \tag{5-9}$$

where the DC voltage component is also included in the following triangular series:

$$V_{ink} = V_{k0} + \sum_{n=1}^{\infty} \{V_{kns}\sin(n\omega t) + V_{knc}\cos(n\omega t)\}. \tag{5-10}$$

The vector potentials and current densities can be solved simultaneously by the above equation.

5.1.2.3 C. Epstein Frame-Like Core Model

An Epstein frame-like core model made by the Tianwei Group, Baoding, China, has been tested under different DC bias conditions. Harmonic analysis of the magnetizing current and magnetic field is carried out, based on the consistency between the computation and the measurement.

The Epstein frame-like core model for the DC-biased test is shown in Figure 5.2. The iron core is made up of silicon steel lamination (model number 30Q140). Figure 5.3 shows the fitted magnetizing curve of the silicon steel. There are two windings on the ferromagnetic core: the exciting coil (fed by an alternating voltage) and the measuring coil.

The peak value of the excitation current without a DC bias is selected as a reference. This reference current causes the flux density in the silicon steel to reach the rated value (1.7 T) in the transformer's no-load operation. The DC bias, in the form of direct current, is then applied in proportion to the reference current to the exciting coil.

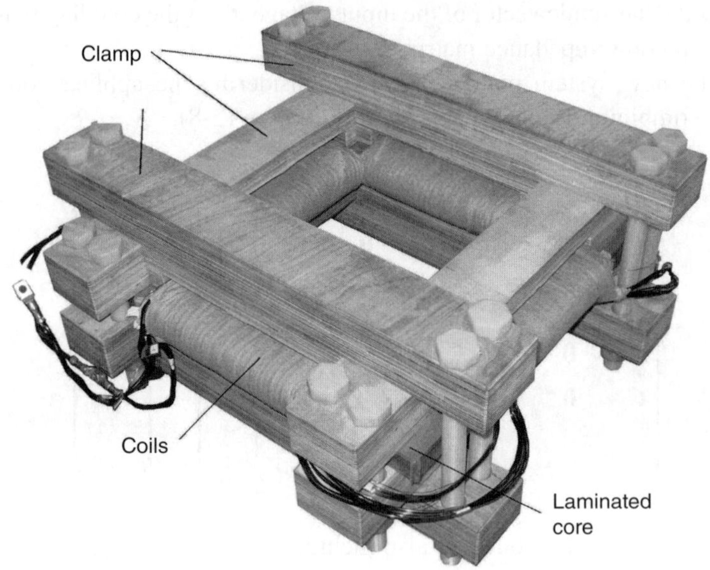

Figure 5.2 Epstein frame-like core model

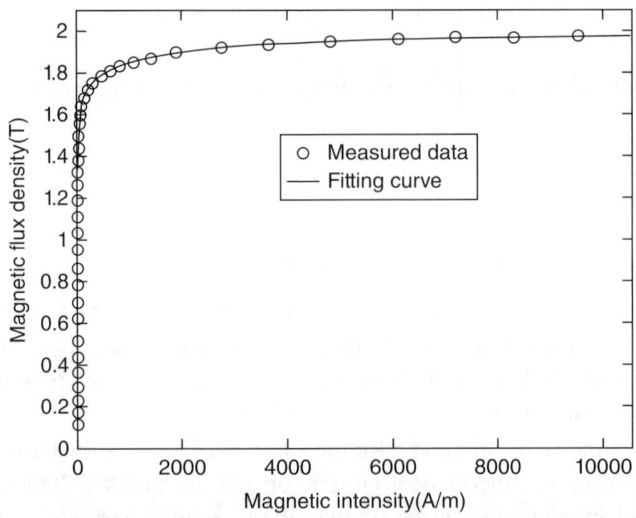

Figure 5.3 Magnetizing curve of the silicon-steel sheet

The DC bias current should be integrated with the input voltage of the corresponding circuit for the implementation of computation in Equation (5-9). The input voltage V_{ink} in Equation (5-10) is actually a harmonic vector that includes DC and AC components. The DC component of V_{ink} is connected with the DC bias current through the resistance of the transformer's winding.

Table 5.1 Different DC bias conditions specified by quantity in the magnetic field

	DC bias			AC excitation	
Cases (i/j)	P_i ($\%I_0$)	$I_{dc,i}$(A)	$H_{dc,i}$(A/m)	$U_{m,j}$(V)	$B_{m,j}$(T)
1	25	0.4256	105.68	26	0.09
2	50	0.847	213.12	133	0.49
3	75	1.273	320.30	240	0.88
4	100	1.697	425.23	370	1.37
5	150	2.530	636.58	420	1.57
6				495	1.82

5.1.2.4 D. Calculated and Measured Results

The value of reference current I_0 measured on the square ferromagnetic core model is 1.68 amperes. The DC bias is applied in incremental proportions of the reference current, which are represented by P_i ($i = 1, 2, 3, 4$) in Table 5.1. The AC excitation is also applied in four different cases, indicated by the subscript j ($j = 1, 2, 3, 4, 5, 6$). I_{dc} represents the DC bias current that corresponds to different proportions of the reference current I_0, while H_{dc} is the subsequently generated magnetic intensity. The peak value of alternating flux density B_m in the magnetic core varies with the step-increase of alternating voltage U_m (peak value) [18], which is also shown in the same table.

The calculated results are compared with the experimental data in different excitation cases, as shown in Figures 5.4 to 5.7. It is observed that there is consistency in the computational and measured results obtained from the magnetizing current waveforms.

A quantitative comparison between the calculated and measured results is necessary to estimate the calculated errors by HBFEM. There are two main causes of inaccuracy in computation. The first is the hysteresis effect of the magnetic core in the model, which is neglected here for the simplicity of computation. The second cause is a truncation error that plays a key role in the calculation of exciting currents and magnetic fields under DC bias. The higher the harmonic order considered, the more accurate are the results. The truncated harmonic order depends on the DC and AC excitation. A deficiency of considered harmonic numbers in computation will result in ripples in waveforms of exciting currents.

Because the waveform of the exciting current is non-sinusoidal under DC bias conditions, the root-mean-square value is selected to carry out the quantitative comparison between the calculated and measured results [18]. $I_{c,rms}$ and $I_{m,rms}$ represent the root-mean-square values of the exciting current obtained from calculation and measurement, respectively. The error data reflects the inaccuracy resulting mainly from neglect of the hysteresis effects and truncation in computation. The error analysis, as shown in Table 5.2, is performed in different cases related to DC and AC excitations, which can be combined with the waveform comparison in Figures 5.4 to 5.7 to evaluate the validity of the computation by HBFEM.

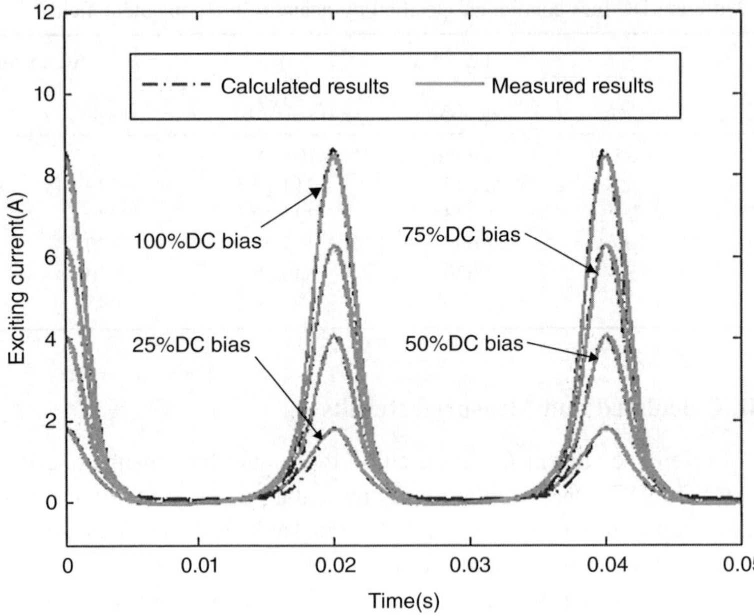

Figure 5.4 Exciting current under different DC bias ($U_m = U_{m,2} = 133$ V; $B_m = B_{m,2} = 0.49$ T)

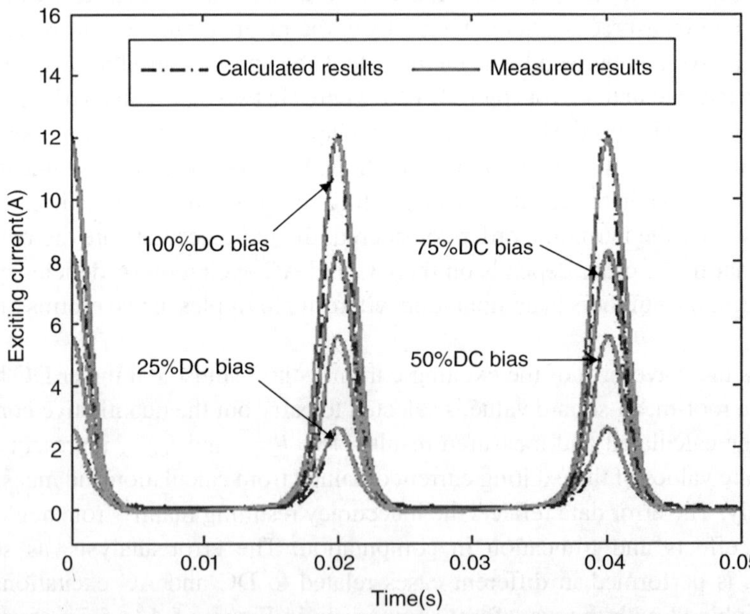

Figure 5.5 Exciting current under different DC bias ($U_m = U_{m,3} = 240$ V; $B_m = B_{m,3} = 0.88$ T)

Table 5.2 Errors between calculated and measured results in exciting current

H_{dc}(A/m)	B_m(T)	$I_{c,rms}$(A)	$I_{m,rms}$(A)	Error (%)
105.68	0.49	0.8677	0.9088	4.5258
105.68	0.88	1.0861	1.1449	5.1419
105.68	1.37	1.2943	1.3736	5.7773
105.68	1.82	2.9859	3.0984	3.630
213.12	0.49	1.9141	1.9312	0.8869
213.12	0.88	2.3463	2.3695	0.9794
213.12	1.37	2.7315	2.7613	1.0789
213.12	1.82	4.9705	5.3355	6.8412
320.30	0.49	2.9749	2.9878	4.3294
320.30	0.88	3.4717	3.5846	3.1513
320.30	1.37	4.2376	4.5009	5.8500
320.30	1.82	10.042	10.513	4.4700
425.23	0.49	4.0284	4.0519	0.5790
425.23	0.88	4.9335	5.2631	6.2600

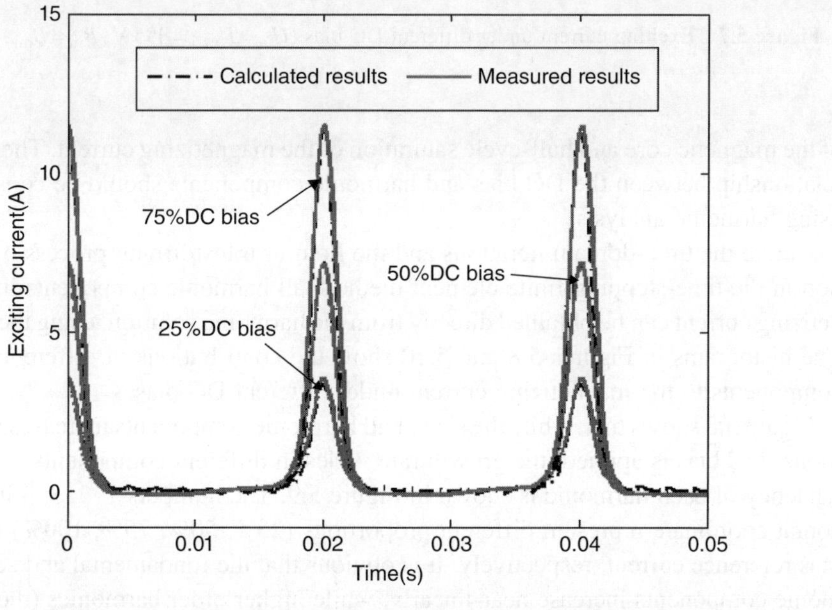

Figure 5.6 Exciting current under different DC bias ($U_m = U_{m,4} = 370$ V; $B_m = B_{m,4} = 1.37$ T)

5.1.2.5 E. Harmonic Analysis of the Magnetizing Current

There are only odd harmonics in the magnetizing current when the transformer is fed by AC excitation. However, additional harmonics appear when direct current invades the transformer windings. The generation of large harmonics results in significant saturation

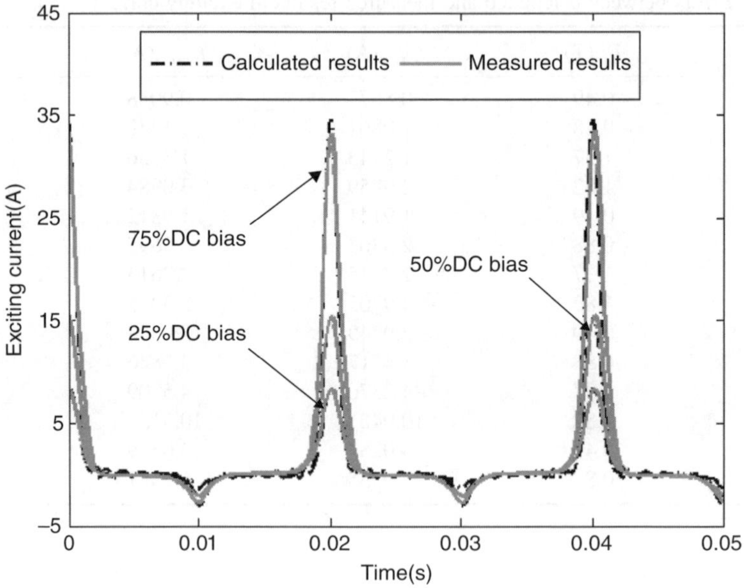

Figure 5.7 Exciting current under different DC bias ($U_m = U_{m,6} = 495$ V; $B_m = B_{m,6} = 1.82$ T)

of the magnetic core and half-cycle saturation of the magnetizing current. Therefore, the relationship between the DC bias and harmonic components should be considered by using harmonic analysis.

Unlike the time-domain iterations and the Fourier transforming process of the solution in the time-stepping finite element method, all harmonic components in the magnetizing current can be obtained directly from the harmonic solution, using the HBFEM. The histograms in Figures 5.8 and 5.10 show the contributions of different harmonic components to the magnetizing current under different DC biases.

Figure 5.8 shows that, while the size of all harmonic components increases when additional DC bias is applied, the growth rate varies in different components. The growth tendency of each harmonic is shown in Figure 5.9. The numbers 1, 2, 3, 4 in the horizontal coordinate represent different proportions (25%, 50%, 75%, 100%) of the DC bias reference current, respectively. It is obvious that the fundamental and second harmonic components increase near-linearly, while higher order harmonics (the third and fourth) grow faster, rather than linearly.

The contribution of each harmonic component is different when the peak value of alternating voltage is increased up to 495 volts, which is given in Figure 5.10. Odd harmonics are greater than even order components under 25% and 50% DC bias respectively. It is implied that the growth of odd harmonic components is related to the increased AC excitation.

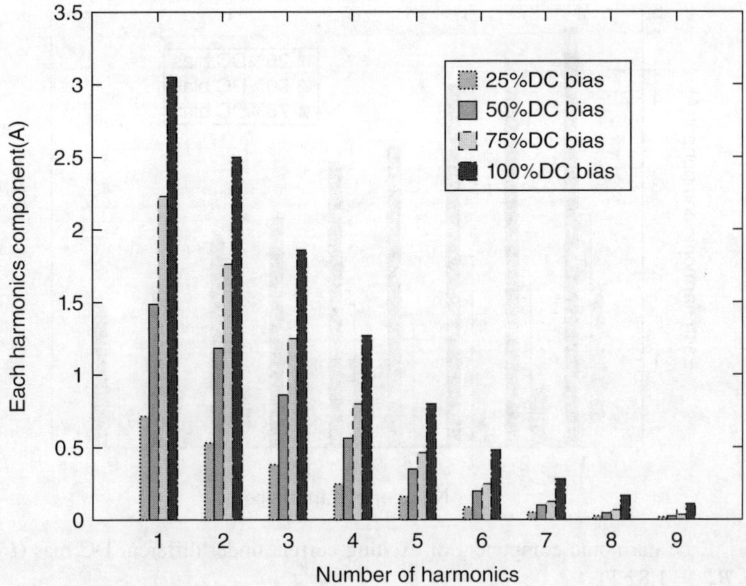

Figure 5.8 Each harmonic component of exciting current under different DC bias ($U_m = U_{m,3} = 240$ V; $B_m = B_{m,3} = 0.88$ T)

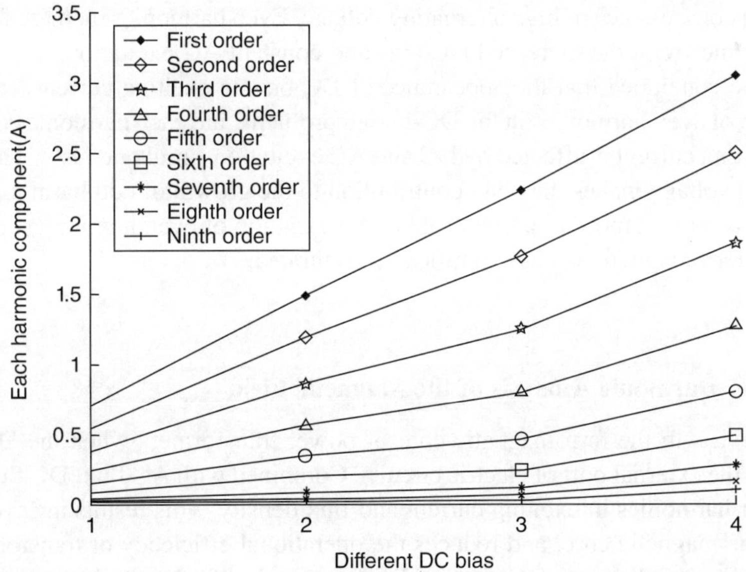

Figure 5.9 DC bias effect on different harmonics ($U_m = U_{m,3} = 240$ V; $B_m = B_{m,3} = 0.88$ T)

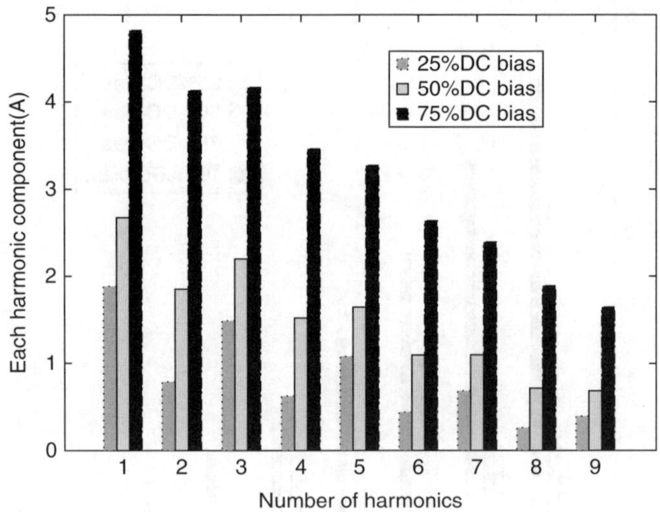

Figure 5.10 Each harmonic component of exciting current under different DC bias ($U_m = U_{m,6} =$ 495 V; $B_m = B_{m,6} = 1.82$ T)

Curves in Figure 5.11 display a relationship between odd harmonics and AC excitation. With increased alternating voltage, odd harmonics grow faster (and are greater in size) than even harmonics. On the other hand, the negative influence of DC bias on each harmonic is analyzed in Figure 5.12, when the ferromagnetic core is significantly saturated as a consequence of high alternating voltage. Even harmonics increase faster than odd harmonics with the increased DC bias and constant AC excitation.

It can be concluded that the appearance of DC bias in exciting current leads to the generation of even harmonics in the DC-biased problem, and each harmonic component in the exciting current is affected by DC and AC excitation simultaneously. The applied alternating voltage makes the main contribution to the growth of odd harmonics, while the DC bias plays a more important role in the variation of even harmonics, especially when the ferromagnetic core is significantly saturated.

5.1.2.6 F. Harmonic Analysis of the Magnetic Field

DC flux exists in the ferromagnetic core of power transformers when the DC bias is applied to the external port of electric circuits. Combined with AC flux, DC flux creates high-order harmonics in exciting current and flux density. This results in severe saturation of the magnetic core, and reduces the operational efficiency of transformers.

The relationship between DC bias and DC flux requires further study, because the DC flux is not affected linearly by the DC bias. DC and AC harmonic components of the magnetic flux density can be computed directly by the HBFEM. Analysis of the DC

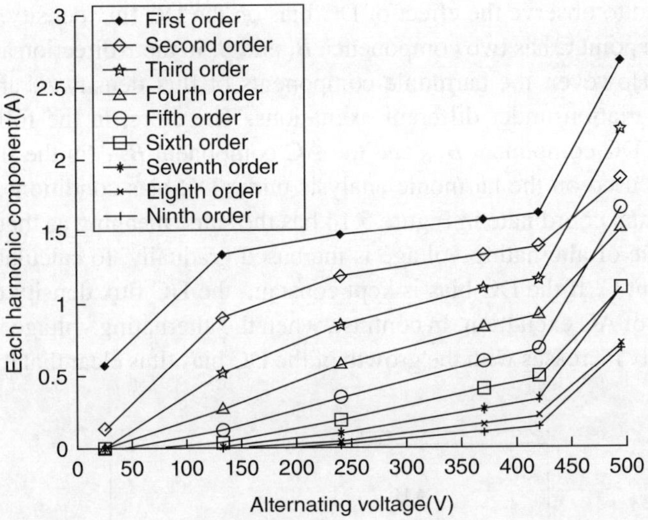

Figure 5.11 AC voltage (peak value) effect on each harmonic component under 50% DC bias ($I_{dc} = I_{dc,2} = 0.847$ A; $H_{dc} = H_{dc,2} = 213.12$ A/m)

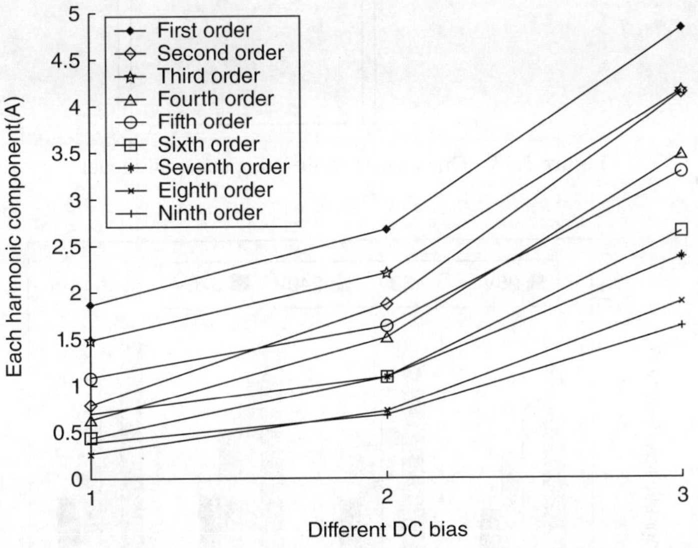

Figure 5.12 DC bias effect on different harmonics ($U_m = U_{m,6} = 495$ V; $B_m = B_{m,6} = 1.82$ T)

component of the magnetic flux density is carried out through the calculated harmonic solutions.

A quarter of the Epstein frame-like core model in Figure 5.13 is computed, considering its structural symmetry. One point on the cross-section of the core, such as point C, can be selected to observe the effect of DC bias on the DC flux density. The magnetic flux density in point C has two components: B_x and B_y in the x-direction and y-direction, respectively. However, the harmonic components of flux density B_y are too small to analyze the variation under different excitations. Therefore, in the following part of the paper, the DC component $B_{x,0}$ and the AC component $B_{x,1}$ (in the flux density B_x) are mainly focused on the harmonic analysis under DC bias conditions.

The horizontal coordinate in Figure 5.14 has the same meaning as that in Figure 5.9. The peak value of alternating voltage is increased gradually, to calculate the DC flux density in point C. If the DC bias is kept constant, the DC flux density decreases with the increment of AC excitation. In contrast, when the alternating voltage is constant, the DC flux density increases with the growth of the DC bias. It is clear that the DC bias and

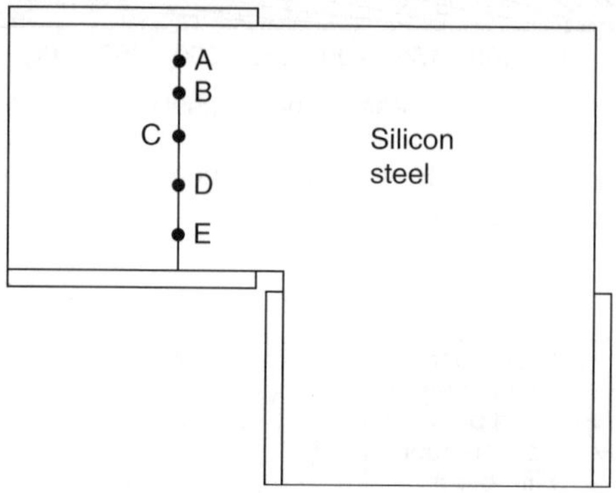

Figure 5.13 One quarter of the computational model

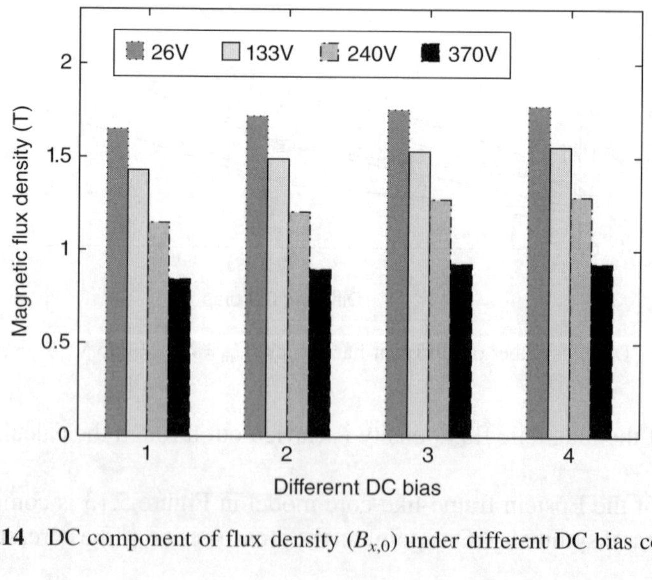

Figure 5.14 DC component of flux density ($B_{x,0}$) under different DC bias conditions

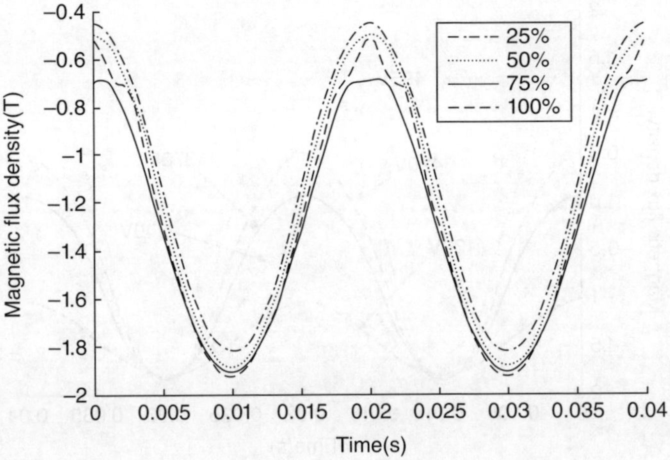

Figure 5.15 Magnetic flux density (B_x) under different DC bias when the AC excitation is constant ($U_m = U_{m,3} = 240$ V)

AC excitation affect the DC flux density at the same time. In fact, the DC component of flux density is due to the balanced effect of DC and AC excitation.

The waveforms of flux density in point C are analyzed under different DC bias conditions. In Figure 5.15, there are four waveforms of magnetic flux density, corresponding to 25%, 50%, 75% and 100% DC bias, respectively (under the condition that the alternating voltage is 240 V). The waveform is raised by the increased DC bias, and the negative peak value of the magnetic flux density approaches –2.0 Tesla. This leads to the rapid saturation of the ferromagnetic core.

As shown in Figure 5.16, 50% DC bias is selected to observe the effect of alternating voltage on the waveform of the magnetic flux density under DC bias conditions. When the applied AC voltage is low, the magnetic core is saturated only at the negative peak value of the magnetic flux density, which corresponds to the half-cycle saturation of the magnetizing current in Figures 5.4 to 5.6. An increase in alternating voltage causes the positive peak value of the flux density to increase so rapidly that the magnetic core is also saturated in the other half-cycle. This is consistent with the appearance of a negative peak of magnetizing current, as shown in Figure 5.7.

The detailed harmonic analysis of DC and AC components in the magnetic flux density is given in Figures 5.17 to 5.22. Of the AC components of flux density, the fundamental component is dominant and much larger than other high order components (see Table 5.3). Therefore, in Figures 5.17 to 5.22, the AC flux density refers to the first harmonic component in the magnetic flux density, neglecting high-order components in the analysis.

Figures 5.17-5.19 show that the DC flux density decreases with higher AC voltage while the AC flux density increases at the same time. This variation is even more apparent when the AC voltage is very high because the core has been significantly saturated.

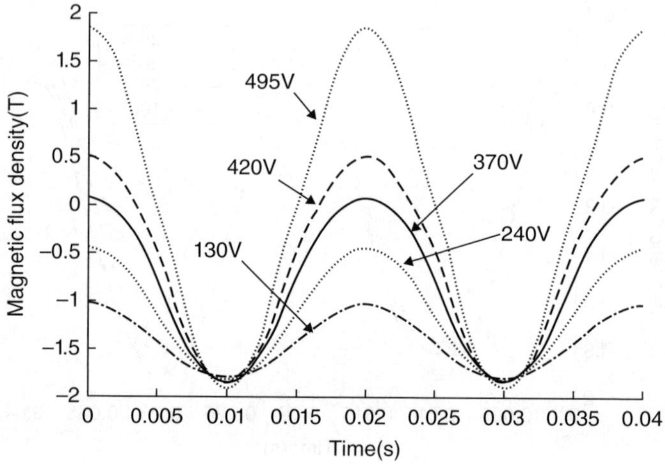

Figure 5.16 AC voltage effect on waveforms of magnetic flux density (B_x) under 50% DC bias ($I_{dc} = I_{dc,2} = 0.847$ A)

Table 5.3 Each harmonic component of magnetic induction in one element in the silicon steel region under DC bias conditions

Harmonic order	Harmonic component (T)
0	0.8410
1	0.9774
2	0.0612
3	0.0295
4	0.0143
5	0.0113
6	0.0019
7	0.0002
8	0.0022
9	0.0014

Compared with Figures 5.17 to 5.19, Figures 5.20 to 5.22 reflect the influence of different DC bias on DC and AC flux density. The DC flux density increases slowly and tends to remain constant (the value of the DC bias is 2.5 amperes). In that case, the peak value of the magnetic flux density approaches 2.0 Tesla, which means significant saturation of the ferromagnetic core. The AC flux density varies little with the dramatic increase in the DC bias.

It can be predicted that a nonlinear relationship exists between the DC bias and the DC flux, and between the AC voltage and the AC flux. The DC flux is a necessary result of the DC bias, but it is actually affected by the simultaneous excitation of the DC and AC. The AC flux depends mainly on the alternating excitation.

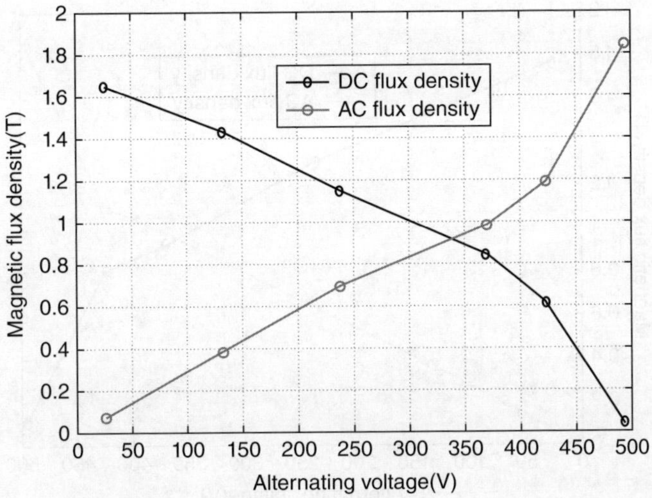

Figure 5.17 AC voltage effect on DC and AC components of the magnetic flux density ($B_{x,0}$ and $B_{x,1}$) under 25% DC bias

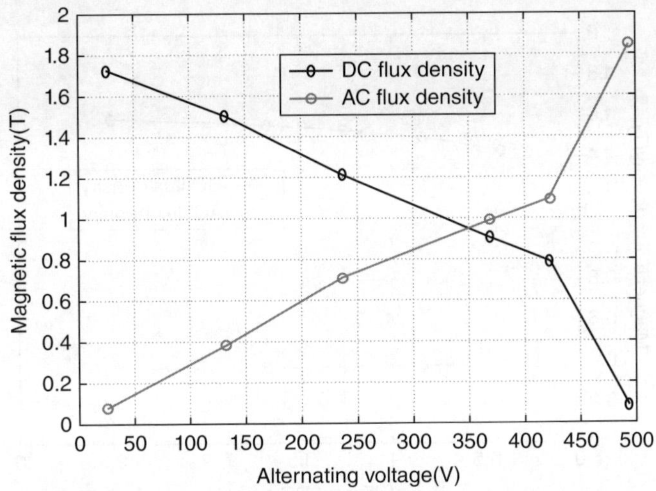

Figure 5.18 AC voltage effect on DC and AC components of the magnetic flux density ($B_{x,0}$ and $B_{x,1}$) under 50% DC bias

5.1.2.7 G. Harmonic Analysis of Flux Distribution

The magnetic flux density in each element is represented in the form of a harmonic component, and can be calculated directly from Equation (5-9). Therefore, the harmonic flux distribution in the steel region can be given directly through the solution of the magnetic field.

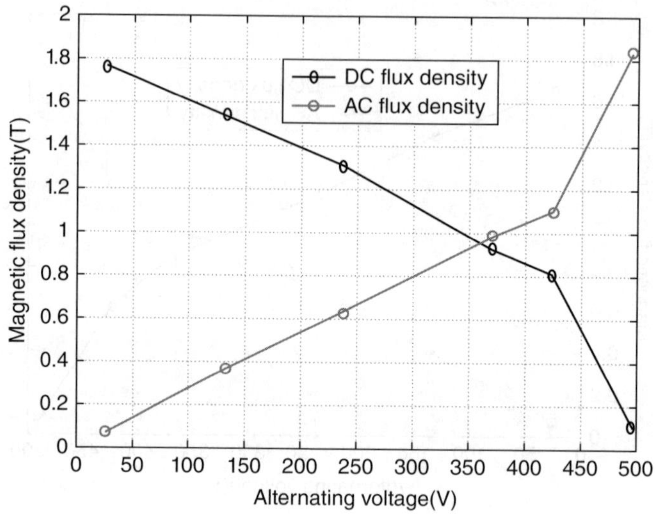

Figure 5.19 AC voltage effect on DC and AC components of the magnetic flux density ($B_{x,0}$ and $B_{x,1}$) under 75% DC bias

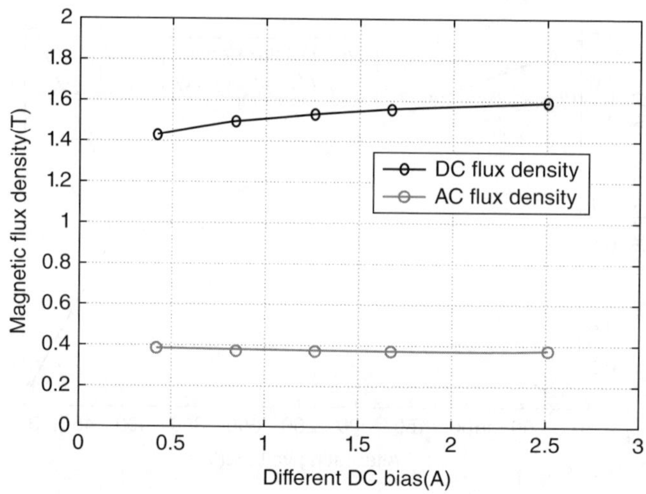

Figure 5.20 DC bias effect on DC and AC components of the magnetic flux density ($B_{x,0}$ and $B_{x,1}$) with alternating voltage of 133 V

Figure 5.23 shows the calculated distribution of total flux, which is the superposition of all harmonic components in the ferromagnetic core under DC bias. The sub-graphics (a) to (j) in Figure 5.24 represent the computational flux distribution of each harmonic component in the moment $\omega t = \pi/3$ ($\omega = 100\pi$). The harmonic flux distribution varies with time and excitation.

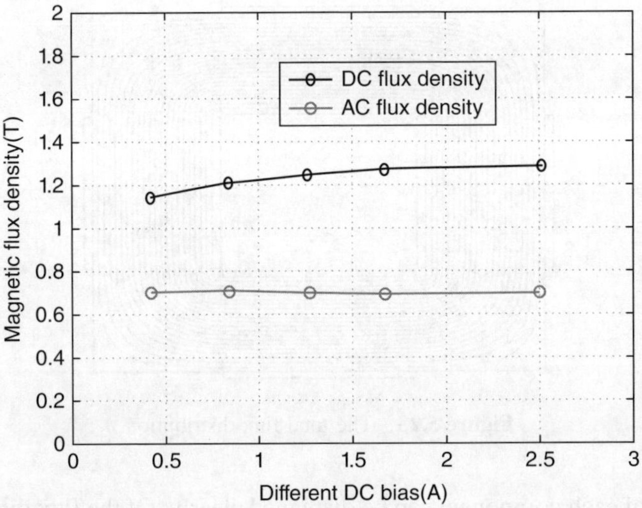

Figure 5.21 DC bias effect on DC and AC components of the magnetic flux density ($B_{x,0}$ and $B_{x,1}$) with alternating voltage of 240 V

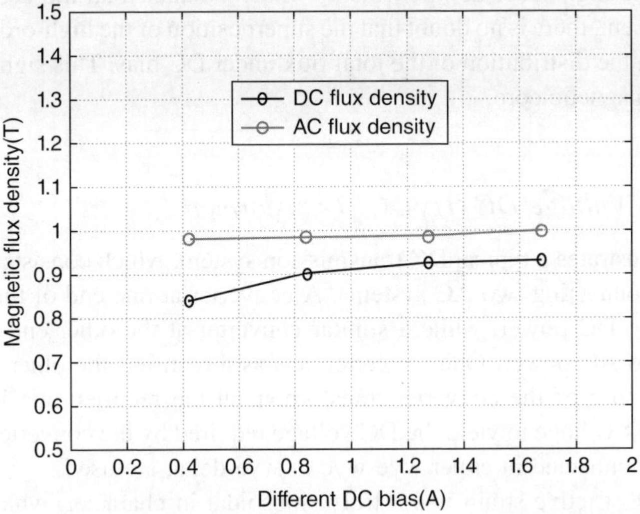

Figure 5.22 DC bias effect on DC and AC components of the magnetic flux density ($B_{x,0}$ and $B_{x,1}$) with alternating voltage of 370 V

In the ferromagnetic core, high-order components are much smaller than the DC component and the first harmonic component (all harmonic components of flux density in one element region belonging to point C are given in Table 5.3). The flux distribution of high-order components is magnified to be many times larger, so that the distribution

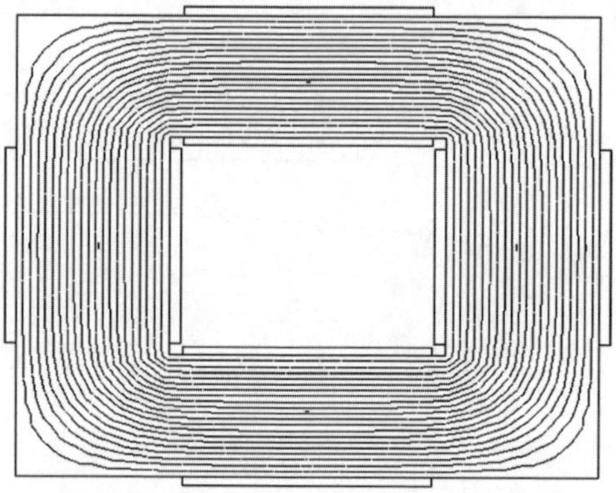

Figure 5.23 The total flux distribution

characteristic of each component can be displayed clearly. Of the flux distribution of all harmonics, the second to ninth harmonics exhibit different characteristics of local vertex. The local vertex characteristic will affect the waveform of flux density in the ferromagnetic core under DC bias conditions.

Although the high-order components are much smaller than the DC and first harmonic component, there is no doubt that the superposition of the high-order components interferes with the distribution of the total flux under DC bias. This significantly saturates the ferromagnetic core.

5.1.3 High-Voltage DC (HVDC) Transformer

Figure 5.25 illustrates a typical DC transmission system, which consists of a DC transmission line connecting two AC systems. A converter at one end of the line converts AC power into DC power, while a similar converter at the other end reconverts the DC power into AC power. One converter acts as a rectifier, the other as an inverter. The basic purpose of the converter transformer on the rectifier side is to transform the AC network voltage to yield the DC voltage required by the converter. Three-phase transformers, connected in either wye-wye or wye-delta, are used.

The magnetostrictive strain is not truly sinusoidal in character, which leads to the introduction of the harmonics. With a DC-biased transformer, magnetizing saturation will also cause some harmonics. The harmonics in transformer noise may have a substantial effect on an observer, even though their level is 10 dB or more lower than the 100 Hz fundamental. In fact, the most striking point is the strength of the component at 100 Hz, or twice the normal operating frequency of the transformer.

Deviation from a "square-law" magnetostrictive characteristic would result in even harmonics (at 200, 400, 600 Hz, etc.), while the different values of the magnetostrictive strain for increasing and decreasing flux densities – a pseudo-hysteresis effect – lead to

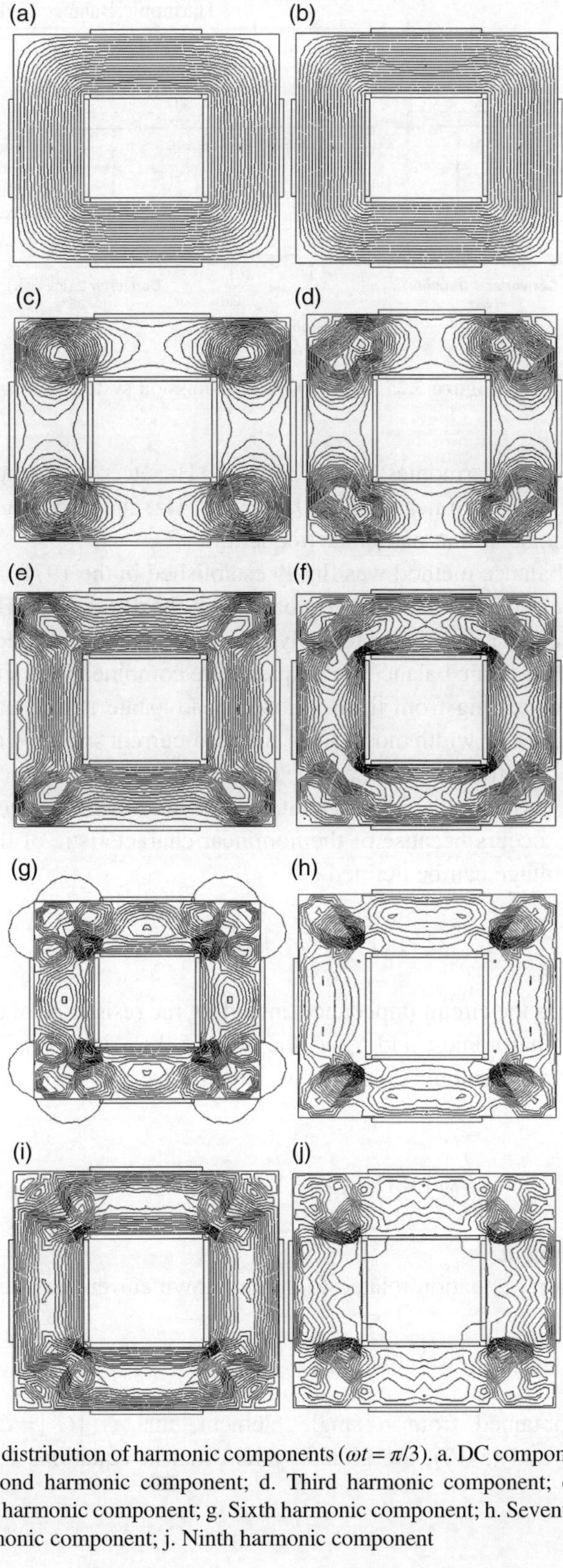

Figure 5.24 Flux distribution of harmonic components ($\omega t = \pi/3$). a. DC component; b. Fundamental component; c. Second harmonic component; d. Third harmonic component; e. Fourth harmonic component; f. Fifth harmonic component; g. Sixth harmonic component; h. Seventh harmonic component; i. Eighth harmonic component; j. Ninth harmonic component

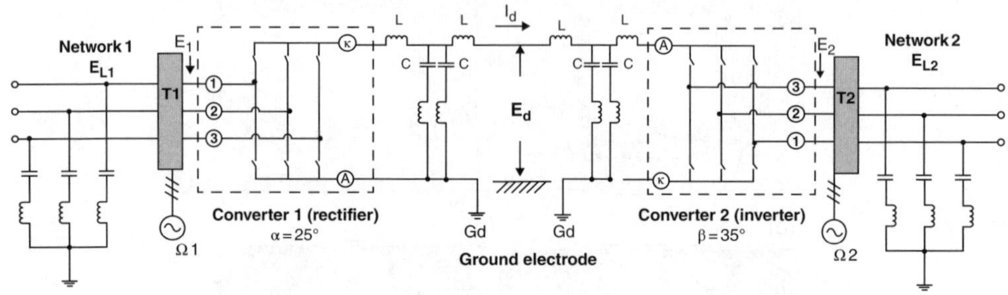

Figure 5.25 An HVDC transmission system

the introduction of odd harmonics (at 300, 500, 700 Hz, etc.). If any part of the structure has a natural frequency at or near 100, 200, 300, 400 Hz, and so on, the result will be an amplification of noise at that particular frequency.

The harmonic balance method was firmly established in the 1970s, and widely used in solving nonlinear microwave circuit problems in the 1980s [19].The harmonic balance technique was first introduced to analyze low frequency EM field problems in the later 1980s [13]. Harmonic balance techniques were combined with FEM to accurately solve the problems arising from time-periodic steady-state nonlinear magnetic fields [20]. In most cases, pulse width modulation and zero-current switched resonant converters, as well as HVDC transformers, can be considered as a voltage-source to the magnetic system. The current in the input circuits will be unknown, and the saturation of the current waveform occurs because of the nonlinear characteristic of the magnetic core [20]. The input voltage can be defined as:

$$\{V_{ink}\} = \{V_k\} + S_{ck}[Z_k]\{J_k\} \tag{5-11}$$

where matrix $[Z_k]$ is the circuit impedance, including the resistance of the windings corresponding to the harmonics, and S_{ck} is the area of the windings. The input voltage $\{V_{ink}\}$, including all harmonic components which have a known value, is expressed as follows:

$$\{V_{ink}\} = \{V_{0k}, V_{1sk}, V_{1ck}, V_{2sk}, V_{2ck}, \cdots\}^T \tag{5-12}$$

V_0 is a DC component.

The system matrix equation related to the unknown current can be rewritten as:

$$([S] + [M])\{A\} - [G_k]\{J_k\} = 0 \tag{5-13}$$

where $[G_k]$ is obtained from a single element, that is $[G^e] = \Delta^e/3$. Combining Equations (5-11) and (5-13), the global system matrix equations for multiple input and output are obtained. The harmonic balance FEM matrix equations for voltage source excitation can therefore be expressed as:

$$\begin{bmatrix} [H] & -[G_1] & -[G_2] & \cdots & -[G_k] & \cdots \\ [C_1] & S_{c1}[Z_1] & 0 & 0 & 0 & \cdots \\ [C_2] & 0 & S_{c2}[Z_2] & 0 & 0 & \cdots \\ \vdots & 0 & 0 & \ddots & 0 & \cdots \\ [C_k] & 0 & 0 & 0 & S_{ck}[Z_k] & \cdots \\ \vdots & \vdots & \vdots & \vdots & \vdots & \ddots \end{bmatrix} \begin{Bmatrix} \{A\} \\ \{J_1\} \\ \{J_2\} \\ \vdots \\ \{J_k\} \\ \vdots \end{Bmatrix} = \begin{Bmatrix} 0 \\ \{V_{in1}\} \\ \{V_{in2}\} \\ \vdots \\ \{V_{ink}\} \\ \vdots \end{Bmatrix} \tag{5-14}$$

where $\{A\}$ and $\{J_k\}$ are unknown, and can be calculated by solving the system matrix equation. $[H]$ is the matrix obtained from $([S]+[M])$, and $C_k = (\omega d_O \Delta/3S_{ck})[N\,N\,N]$ is the geometric coefficient related to transformer windings.

The model has a voltage-driven source connected to the magnetic system, which is always coupled to the external circuits. Considering a three-phase transformer connected in wye-wye, a computer simulation model, with a neutral *NN* and external circuits for both primary and secondary windings, is obtained using the HBFEM technique. According to the Galerkin procedure, system matrix equations of HBFEM for the HVDC transformer can be obtained through Faraday's and Kirchhoff's laws for the external circuit. That is:

$$\left.\begin{aligned} &[H]\{A\}-[G_U]\{J_{Uin}\}-[G_V]\{J_{Vin}\}-[G_W]\{J_{Win}\} \\ &-[G_{Uout}]\{J_{Uout}\}-[G_{Vout}]\{J_{Vout}\}-[G_{Wout}]\{J_{Wout}\}=\{0\} \\ &[C_{Uin}]\{A\}+[I]\{V_{NNin}\}+S_{cu}[Z_{Uin}]\{J_{Uin}\}=[V_{Uin}\} \\ &[C_{Vin}]\{A\}+[I]\{V_{NNin}\}+S_{cu}[Z_{Vin}]\{J_{Vin}\}=[V_{Vin}\} \\ &[C_{Win}]\{A\}+[I]\{V_{NNin}\}+S_{cu}[Z_{Win}]\{J_{Win}\}=[V_{Win}\} \\ &[C_{Uout}]\{A\}+[I]\{V_{NNout}\}-S_{cu}[Z_{Uout}]\{J_{Uout}\}=[V_{Uout}\} \\ &[C_{Vout}]\{A\}+[I]\{V_{NNout}\}-S_{cu}[Z_{Vout}]\{J_{Vout}\}=[V_{Vout}\} \\ &[C_{Wout}]\{A\}+[I]\{V_{NNout}\}-S_{cu}[Z_{Wout}]\{J_{Wout}\}=[V_{Wout}\} \\ &S_{in,cu}[I]\{J_{Uin}\}+S_{in,cu}[I]\{J_{Vin}\}+S_{in,cu}[I]\{J_{Win}\}=0 \\ &S_{out,cu}[I]\{J_{Uout}\}+S_{out,cu}[I]\{J_{Vout}\}+S_{out,cu}[I]\{J_{Wout}\}=0 \end{aligned}\right\} \tag{5-15}$$

where $[G_k]$ is obtained from a single element, that is $[G^e] = \Delta^e/3$. $[Z_{in}]$, $[Z_{out}]$ $S_{in,cu}$, and $S_{out,cu}$ are external circuit impedances and cross-sectional areas of windings respectively, $[I]$ is the unit matrix, V_{NN} is the voltage for primary or secondary neutral points when it is not grounded, and $[C_{in}]$ and $[C_{out}]$ are geometric coefficients related to transformer windings. Current density J can be presented as:

$$\{J_k^e\} = \{\, J_0 \;\; J_{1s} \;\; J_{1c} \;\; J_{2s} \;\; J_{2c} \;\; J_{3s} \;\; J_{3c} \;\; \cdots \}^T \tag{5-16}$$

$[H]$ is the system matrix and the detailed definitions can be obtained from:

$$[H^e] = \frac{1}{4\Delta^e}\begin{bmatrix} (b_1b_1+c_1c_1)D & (b_1b_2+c_1c_2)D & (b_1b_3+c_1c_3)D \\ (b_2b_1+c_2c_1)D & (b_2b_2+c_2c_2)D & (b_2b_3+c_2c_3)D \\ (b_3b_1+c_3c_1)D & (b_3b_2+c_3c_2)D & (b_3b_3+c_3c_3)D \end{bmatrix}\begin{Bmatrix} A_1^e \\ A_2^e \\ A_3^e \end{Bmatrix} + \frac{\sigma\omega\Delta^e}{12}\begin{bmatrix} 2N & N & N \\ N & 2N & N \\ N & N & 2N \end{bmatrix}\begin{Bmatrix} A_1^e \\ A_2^e \\ A_3^e \end{Bmatrix} \tag{5-17}$$

where matrices D and N are:

$$D = \frac{1}{2}\begin{bmatrix} 2\nu_0 & \nu_{1s} & \nu_{1c} & \nu_{2s} & \nu_{2c} & \nu_{3s} & \nu_{3c} & \cdots \\ 2\nu_{1s} & 2\nu_0-\nu_{2c} & \nu_{2s} & \nu_{1c}-\nu_{3c} & -\nu_{1s}+\nu_{3s} & \nu_{2c}-\nu_{4c} & -\nu_{2s}+\nu_{4s} & \cdots \\ 2\nu_{1c} & & 2\nu_0+\nu_{2c} & \nu_{1s}+\nu_{3s} & \nu_{1c}+\nu_{3c} & \nu_{2s}+\nu_{4s} & \nu_{2c}+\nu_{4c} & \cdots \\ 2\nu_{2s} & & & 2\nu_0-\nu_{4c} & \nu_{4s} & \nu_{1c}-\nu_{5c} & -\nu_{1s}+\nu_{5s} & \cdots \\ 2\nu_{2c} & & & & 2\nu_0+\nu_{4c} & \nu_{2s}+\nu_{4s} & \nu_{1c}+\nu_{5c} & \cdots \\ 2\nu_{3s} & & \text{Symmetry} & & & 2\nu_0-\nu_{6c} & \nu_{6s} & \cdots \\ 2\nu_{3c} & & & & & & 2\nu_0+\nu_{6c} & \cdots \\ \vdots & \vdots & \vdots & \vdots & \vdots & \vdots & \vdots & \ddots \end{bmatrix} \tag{5-18}$$

and

$$N^e = \begin{bmatrix} 0 & 0 & 0 & 0 & 0 & 0 & 0 & \cdots \\ 0 & 0 & -1 & 0 & 0 & 0 & 0 & \cdots \\ 0 & 1 & 0 & 0 & 0 & 0 & 0 & \cdots \\ 0 & 0 & 0 & 0 & -2 & 0 & 0 & \cdots \\ 0 & 0 & 0 & 2 & 0 & 0 & 0 & \cdots \\ 0 & 0 & 0 & 0 & 0 & 0 & -3 & \cdots \\ 0 & 0 & 0 & 0 & 0 & 3 & 0 & \cdots \\ \vdots & \vdots & \vdots & \vdots & \vdots & \vdots & \vdots & \ddots \end{bmatrix} \tag{5-19}$$

The full matrix equations of the HB-FEM model in a compact form can be obtained as:

$$\begin{bmatrix} [H] & -[G_{in}] & -[G_{out}] & 0 & 0 \\ [C_{in}] & S_{in}[Z_{in}] & 0 & [I] & 0 \\ [C_{out}] & 0 & S_{out}[Z_{out}] & 0 & [I] \\ 0 & S_{in}[I] & 0 & 0 & 0 \\ 0 & & S_{out}[I] & 0 & 0 \end{bmatrix} \begin{Bmatrix} \{A_n\} \\ \{J_{kin}\} \\ \{J_{kout}\} \\ V_{NNin} \\ V_{NNout} \end{Bmatrix} = \begin{Bmatrix} 0 \\ \{V_{kin}\} \\ \{V_{kout}\} \\ 0 \\ 0 \end{Bmatrix} \tag{5-20}$$

5.2 Decomposed Algorithm of HBFEM

5.2.1 *Introduction*

Electromagnetic devices such as electrical machines and power transformers are generally operated under steady-state conditions. Owing to the nonlinearity of magnetic material in electromagnetic devices, there are often high-order harmonics in the exciting current and magnetic field. The HBFEM has been proposed to solve the steady-state magnetic field with eddy current problems [21, 22] and to design the switching power supplies [20]. A modified method via block decomposition of the system equation has been presented to reduce the memory requirement [23], although neither the DC component nor harmonics are considered in the computation.

Based on the harmonic balance technique, a generalized parametric formulation of the nonlinearity and the Jiles-Atherton hysteresis model has been employed to analyze the saturating and hysteresis characteristics of the inductor [24]. The HBFEM has been further developed by introducing differential reluctivity tensor [25] and transmission-line modeling techniques [26], respectively. The magnetic field of a three-phase transformer was also calculated. The fixed-point technique has been introduced to calculate the nonlinear eddy current problems stepping through one period only, and allows all harmonics to be solved in parallel [27, 28].

In recent years, more attention has been paid to the DC-biased problem of power transformers [29, 30]. The magnetic storm and HVDC transmission system may generate large quasi-direct and direct current, respectively, on the earth [31, 32]. Harmonic analysis of exciting current and flux density has been done by using the conventional HBFEM to investigate the DC bias phenomena [33]. However, the conventional approach generates a large-size system matrix to solve all harmonic solutions simultaneously. As a result of the large memory requirement, the widespread application of conventional HBFEM has been limited in large-scale computation.

This section introduces the decomposed HBFEM to calculate the DC-biasing magnetic field of LCM in the harmonic domain, considering the harmonic coupling between the external circuits and magnetic field. The magnetic reluctivity matrix can be decomposed in the harmonic domain, so each harmonic solution is solved sequentially. A modified resolution scheme for the nonlinear equation set is developed at the same time. A more detailed analysis of the flux density and flux distribution under different DC bias conditions is carried out through the harmonic solutions of the magnetic field.

5.2.2 *Decomposed Harmonic Balanced System Equation*

The following vector potential equation can be used to describe the two dimensional magnetic field, where A and J are magnetic vector potential and current density respectively, ν is the magnetic reluctivity, and σ is the conductivity.

$$\nabla \times (\nu \nabla \times A) + \sigma(\partial A/\partial t) = J. \tag{5-21}$$

Due to the periodic characteristics of the electromagnetic field under DC bias conditions, the steady-state variables, such as exciting current density J, magnetic vector potential A, and flux density B, can be approximated by a summation of trigonometric function, as follows:

$$P(t) = P_0 + \sum_{i=1}^{\infty} (P_{2i-1} \sin i\omega t + P_{2i} \cos i\omega t), \tag{5-22}$$

where $P(t)$ can be replaced by J, A, B_x and B_y. The symbols i and ω represent the harmonic number and the fundamental angular frequency, respectively.

The magnetic reluctivity ν, which is a function of magnetic flux density B, can also be expressed by:

$$\nu(t) = H(t)/B(t) = \nu_0 + \sum_{i=1}^{\infty} (\nu_{2i-1} \sin i\omega t + \nu_{2i} \cos i\omega t). \tag{5-23}$$

Galerkin's method and finite element method can be applied to discretize the governing equation for two-dimensional problems. This is written as:

$$\iint_{\Omega_e} \left(\frac{\partial N_m}{\partial x} \nu \frac{\partial A}{\partial x} + \frac{\partial N_m}{\partial y} \nu \frac{\partial A}{\partial y} \right) dxdy + \iint_{\Omega_e} \left(\sigma \frac{\partial A}{\partial x} \right) dxdy = \iint_{\Omega_e} (J \cdot N_m) dxdy \tag{5-24}$$

where N_m represents the interpolation functions on node m in the finite element region Ω_e.

The decomposed finite element equation can be obtained by substituting the periodic variables expressed in Equation (5-22) into Equation (5-24) and equating the coefficients of sin($i\omega t$) and cos($i\omega t$) on both sides, according to the harmonic balance method:

$$\left(\boldsymbol{S}^e * \mathfrak{R}^e_{i,i} + \boldsymbol{T}^e * \boldsymbol{h}_i\right)\boldsymbol{A}^e_i = -\sum_{j=1,2,3\ldots}^{j\neq i}\left(\boldsymbol{S}^e * \mathfrak{R}^e_{i,j}\right)\boldsymbol{A}^e_j + \boldsymbol{K}^e_i \quad (i=1,2\ldots N) \tag{5-25}$$

$$\boldsymbol{A}^e_i = \begin{cases} \{A^1_0, A^1_1, A^1_2, A^2_0, A^2_1, A^2_2, \cdots, A^q_0, A^q_1, A^q_2\}^T & (i=1) \\ \{A^1_i, A^1_{i+1}, A^2_i, A^2_{i+1}, \cdots, A^q_i, A^q_{i+1}\}^T & (i=2,\ldots N) \end{cases}, \tag{5-26}$$

$$\begin{aligned} \boldsymbol{K}^e_i &= \iint_{\Omega^e} (\boldsymbol{J}^e \cdot N_m)\,dxdy \quad (m=1,2,\cdots q) \\ &= \begin{cases} \{K^1_0, K^1_1, K^1_2, K^2_0, K^2_1, K^2_2, \cdots, K^q_0, K^q_1, K^q_2\}^T & (i=1) \\ \{K^1_i, K^1_{i+1}, K^2_i, K^2_{i+1}, \cdots, K^q_i, K^q_{i+1}\}^T & (i=2,\ldots N) \end{cases}, \end{aligned} \tag{5-27}$$

where N is the truncated harmonic number and q is the total node number in an element. $\boldsymbol{J}^e$ has the same vector form with $\boldsymbol{A}^e_i$. $\boldsymbol{K}_i$ is obtained from the spatial distribution of the i-th harmonic component of current density.

The operator $*$ in Equation (5-25) can be defined by:

$$\boldsymbol{S}^e * \mathfrak{R}^e_{i,i} + \boldsymbol{T}^e * \boldsymbol{h}_i = \begin{bmatrix} S_{11}\mathfrak{R}^e_{i,i} & S_{12}\mathfrak{R}^e_{i,i} & \cdots \\ S_{21}\mathfrak{R}^e_{i,i} & S_{22}\mathfrak{R}^e_{i,i} & \cdots \\ \vdots & \vdots & \ddots \end{bmatrix} + \begin{bmatrix} T_{11}\boldsymbol{h}_i & T_{12}\boldsymbol{h}_i & \cdots \\ T_{21}\boldsymbol{h}_i & T_{22}\boldsymbol{h}_i & \cdots \\ \vdots & \vdots & \ddots \end{bmatrix}, \tag{5-28}$$

where $S_{m,n}$ and $T_{m,n}$ are the elements in $\boldsymbol{S}^e$ and $\boldsymbol{T}^e$ respectively. The two terms can be defined by the following expressions:

$$S_{m,n} = \int_{\Omega_e} \nabla N_m \cdot \nabla N_n d\Omega, \tag{5-29}$$

$$T_{m,n} = \int_{\Omega_e} \sigma N_m \cdot N_n d\Omega, \tag{5-30}$$

where the subscripts m and n represent the node number in an element ($m, n = 1, 2, \ldots q$).

The matrices $\mathfrak{R}^e_{i,i}$ and $\mathfrak{R}^e_{i,j}$ are determined by the Fourier coefficients in equation (5-23). They act as a reluctivity, and are called the reluctivity matrix, which is derived from the multiplication of magnetic vector potential A in Equation (5-22) and reluctivity

ν in Equation (5-23) [23]. The DC component of the magnetic vector potential is coupled with the fundamental harmonic in computation. Therefore, the expression of matrix $\mathfrak{R}^e_{i,i}$ $(i=1)$ differs from $\mathfrak{R}^e_{i,i}$ $(i>1)$ shown in Equation (5-31) and matrix $\mathfrak{R}^e_{i,j}$, which is expressed by Equations (5-32), (5-33) and (5-34):

$$\mathfrak{R}^e_{i,i}=\begin{cases}\dfrac{1}{2}\begin{bmatrix}2\nu_0 & \nu_1 & \nu_2\\ 2\nu_1 & 2\nu_0-\nu_4 & \nu_3\\ 2\nu_2 & \nu_3 & 2\nu_0+\nu_4\end{bmatrix} & (i=1)\\ \dfrac{1}{2}\begin{bmatrix}2\nu_0-\nu_{4i} & \nu_{4i-1}\\ \nu_{4i-1} & 2\nu_0+\nu_{4i}\end{bmatrix} & (i>1)\end{cases}, \tag{5-31}$$

$$\mathfrak{R}^e_{1,i}=\frac{1}{2}\begin{bmatrix}\nu_{2i-1} & \nu_{2i}\\ \nu_{2i-2}-\nu_{2i+2} & \nu_{2i+1}-\nu_{2i-3}\\ \nu_{2i+1}+\nu_{2i-3} & \nu_{2i-2}+\nu_{2i+2}\end{bmatrix}(i>1), \tag{5-32}$$

$$\left(\mathfrak{R}^e_{i,1}\right)^T=\frac{1}{2}\begin{bmatrix}2\nu_{2i-1} & 2\nu_{2i}\\ \nu_{2i-2}-\nu_{2i+2} & \nu_{2i+1}-\nu_{2i-3}\\ \nu_{2i+1}+\nu_{2i-3} & \nu_{2i-2}+\nu_{2i+2}\end{bmatrix}(i>1), \tag{5-33}$$

$$\mathfrak{R}^e_{i,j}=\left(\mathfrak{R}^e_{j,i}\right)^T=\frac{1}{2}\begin{bmatrix}v_{2(m'-n')}-v_{2(m'+n')} & -v_{2(m'-n')-1}+v_{2(m'+n')-1}\\ v_{2(m'-n')-1}+v_{2(m'+n')-1} & v_{2(m'-n')}+v_{2(m'+n')}\end{bmatrix}\quad (j>i>1) \tag{5-34}$$

where the subscripts m' and n' are relevant to the harmonic number i and j $(i, j>1$ and $i\neq j)$. m' is defined by $Max(i, j)$ and n' by $Min(i, j)$.

The matrix $\boldsymbol{h}_i$ is a constant concerned with harmonic orders, and is called the harmonic matrix. It is presented as follows:

$$\boldsymbol{h}_i=\begin{cases}\omega\begin{bmatrix}0 & 0 & 0\\ 0 & 0 & -1\\ 0 & 1 & 0\end{bmatrix} & (i=1)\\ \omega\begin{bmatrix}0 & -i\\ i & 0\end{bmatrix} & (i>1)\end{cases}. \tag{5-35}$$

By assembling all finite elements in the computational region, Equation (5-25) can be written as:

$$\boldsymbol{Q}_i\boldsymbol{A}_i = \boldsymbol{K}_i + \boldsymbol{F}_i \quad (i=1,2,\ldots,N), \tag{5-36}$$

Where $\boldsymbol{F}_i = -\sum_{j=1,2,3\ldots}^{j\neq i} (\boldsymbol{S} * \boldsymbol{\Re}_{i,j})\boldsymbol{A}_j$ and $\boldsymbol{Q}_i = \boldsymbol{S} * \boldsymbol{\Re}_{i,i} + \boldsymbol{T} * \boldsymbol{h}_i$.

5.2.3 *Magnetic Field Coupled with Electric Circuits*

When electromagnetic devices are excited by voltage sources, the electric potential difference in the coil region can be obtained from the following equation [34]:

$$U_k = R_k I_k + L_k \frac{dI_k}{dt} + \frac{1}{C_k}\int I_k dt + \frac{N_k l_k}{S_k}\int_{\Omega_c} (\partial A/\partial t) d\Omega, \tag{5-37}$$

where U_k is the input voltage of circuit k, and I_k is the corresponding exciting current. R_k and S_k represent the resistance and cross-sectional area of the k-th winding, respectively. L_k is the inductance and C_k is the capacitance of the external circuit k. N_k is the turn number of the k-th winding, and l_k is the thin wire's length in the z-direction. In the DC-biasing case, the input voltage can be written as follows:

$$U_k = U_{k,0} + \sum_{i=1}^{\infty} (U_{k,2i-1}\sin i\omega t + U_{k,2i}\cos i\omega t). \tag{5-38}$$

Hence the harmonics in voltages, currents and magnetic vector potentials can be expressed separately by the following equation:

$$\boldsymbol{U}_{k,i} = \boldsymbol{Z}_{k,i}\boldsymbol{J}_{k,i}S_k + \boldsymbol{C}_{k,i}\boldsymbol{A}_i \tag{5-39}$$

$$\boldsymbol{C}_{k,i} = \sum_{n=1}^{N_c} \frac{\omega N_k l_k}{3S_k}[\boldsymbol{h}_i \;\; \boldsymbol{h}_i \;\; \boldsymbol{h}_i] \tag{5-40}$$

$$\boldsymbol{Z}_{k,i} = \begin{bmatrix} R_k & -i\omega L_k + (i\omega C_k)^{-1} \\ i\omega L_k - (i\omega C_k)^{-1} & R_k \end{bmatrix}. \tag{5-41}$$

where the subscripts i indicate the harmonic number. $\boldsymbol{U}_{k,i}$ and $\boldsymbol{J}_{k,i}$ are the i-th harmonic vectors of the input voltage and the current density in circuit k, respectively. $\boldsymbol{C}_{k,i}$ is the coupled matrix linking the electric circuit k with the magnetic field. Nonlinear components in external circuits can be expressed by harmonic impedance matrices $\boldsymbol{Z}_{k,i}$. N_c is the finite element number in the coil region.

The harmonic solutions of the magnetizing currents and magnetic vector potentials are obtained by solving Equations (5-36) and (5-39) together, which is shown by the following equation:

$$\begin{bmatrix} \boldsymbol{Q}_i & \boldsymbol{G}_{k,i} \\ \boldsymbol{C}_{k,i} & \boldsymbol{Z}_{k,i} \end{bmatrix} \begin{Bmatrix} \boldsymbol{A}_i \\ \boldsymbol{J}_{k,i} \end{Bmatrix} = \begin{Bmatrix} \boldsymbol{F}_i \\ \boldsymbol{U}_{k,i} \end{Bmatrix} \quad (i=1,2,\ldots,N) \tag{5-42}$$

where the matrix $\boldsymbol{G}_{k,i}$ is related to the spatial distribution of the i-th harmonic component of the current density. Finally, N separated equations, shown in Equation (5-42), constitute the system equation when N harmonics are considered in the computation.

5.2.4 Computational Procedure Based on the Block Gauss-Seidel Algorithm

5.2.4.1 Previous Calculation Procedure

In [23], a calculation procedure to solve the equation (5-36) has been proposed:

1. First, the fundamental harmonic $\boldsymbol{A}_1$ is calculated based on the assumption of setting $\boldsymbol{A}_2$ to $\boldsymbol{A}_N$ to zero. If the convergence of $\boldsymbol{A}_1$ is satisfied, turn to step **2**. If not, update $\boldsymbol{A}_1$ and repeat step **1**.
2. The other harmonics, $\boldsymbol{A}_2$ to $\boldsymbol{A}_N$, are calculated respectively with the known $\boldsymbol{A}_1$. If the convergence of $\boldsymbol{A}_2$ to $\boldsymbol{A}_N$ is satisfied, the calculation procedure stops. If not, turn to step **1**.

However, the computational procedure above is not dependable nor efficient. The convergent criterion in each harmonic computation and the lack of full use of updated solutions in nonlinear iterations leads to convergence uncertainty and the subsequent inaccuracy of harmonic solutions, especially when the nonlinearities of the magnetic material are strong. Yamada [23] also pointed out that there is convergence uncertainty in the iterative approach. In fact, a compulsory stop is usually done, by setting a maximum number of iterative steps.

5.2.4.2 Improved Computational Procedure

The nonlinear terms $\boldsymbol{Q}_i$ in Equation (5-36) are obtained from $\boldsymbol{\Re}_{i,i}$ in Equation (5-31). The diagonal elements in matrices $\boldsymbol{\Re}_{i,i}$ and $\boldsymbol{S}$ in Equation (5-28) are dominant. Therefore, a diagonally dominant characteristic exists in the system equation (5-42), and the block Gauss-Seidel algorithm [35] can be used to solve the DC-biasing magnetic field. The new computational procedure is as follows:

1. ($p = 0$): Initialize $\boldsymbol{A}_0$ … and solve the decomposed harmonic balance equation set in [5-42], following the procedure at step **2**.
2. ($p > 0$):
   a. ($i = 1$): Update $\boldsymbol{Q}_1^{\mathrm{p}}$ by $\boldsymbol{A}_1^{\mathrm{p}}, \boldsymbol{A}_2^{\mathrm{p}}, \ldots \boldsymbol{A}_{\mathrm{N}}^{\mathrm{p}}$ and $\boldsymbol{F}_1^{\mathrm{p}}$ by $\boldsymbol{A}_2^{\mathrm{p}}, \boldsymbol{A}_3^{\mathrm{p}}, \ldots \boldsymbol{A}_{\mathrm{N}}^{\mathrm{p}}$. Assemble the matrices $\boldsymbol{C}_1$ and $\boldsymbol{G}_1$. Solve the equations (5-42) to obtain the renewed fundamental harmonic vector $\boldsymbol{A}_1^{\mathrm{p}+1}$.
   b. ($1 < i < N$): Update $\boldsymbol{Q}_{\mathrm{i}}^{\mathrm{p}}$ by $\boldsymbol{A}_1^{\mathrm{p}}, \ldots \boldsymbol{A}_{\mathrm{i}}^{\mathrm{p}}, \ldots \boldsymbol{A}_{\mathrm{N}}^{\mathrm{p}}$ and $\boldsymbol{F}_{\mathrm{i}}^{\mathrm{p}}$ by $\boldsymbol{A}_1^{\mathrm{p}+1}, \boldsymbol{A}_2^{\mathrm{p}+1}, \ldots \boldsymbol{A}_{\mathrm{i-1}}^{\mathrm{p}+1}, \boldsymbol{A}_{\mathrm{i+1}}^{\mathrm{p}}, \ldots \boldsymbol{A}_{\mathrm{N}}^{\mathrm{p}}$. Assemble the matrices $\boldsymbol{C}_{\mathrm{i}}$ and $\boldsymbol{G}_{\mathrm{i}}$. Solve the i-th equation in (5-42) to obtain the renewed i-th harmonic vector $\boldsymbol{A}_{\mathrm{i}}^{\mathrm{p}+1}$.
   c. ($i = N$): Update $\boldsymbol{Q}_{\mathrm{N}}^{\mathrm{p}}$ by $\boldsymbol{A}_1^{\mathrm{p}}, \ldots \boldsymbol{A}_{\mathrm{N-1}}^{\mathrm{p}}, \boldsymbol{A}_{\mathrm{N}}^{\mathrm{p}}$ and $\boldsymbol{F}_{\mathrm{N}}^{\mathrm{p}}$ by $\boldsymbol{A}_1^{\mathrm{p}+1}, \boldsymbol{A}_2^{\mathrm{p}+1}, \ldots \boldsymbol{A}_{\mathrm{N-1}}^{\mathrm{p}+1}$. Assemble the matrices $\boldsymbol{C}_{\mathrm{N}}$ and $\boldsymbol{G}_{\mathrm{N}}$. Solve the N-th equation in (5-42) to obtain the renewed N-th harmonic vector $\boldsymbol{A}_{\mathrm{N}}^{\mathrm{p}+1}$.
3. Check for convergence criterion. If satisfied, the calculation stops, otherwise renew all harmonic solutions with a properly selected relaxation factor, $\boldsymbol{A}^{\mathrm{new}} = (1-\alpha)\boldsymbol{A}^{\mathrm{now}} + \alpha\boldsymbol{A}^{\mathrm{old}}$, and return to step **2** for the next iteration ($p = p + 1$).

The symbol i refers to the harmonic number in the computation. N is the truncated harmonic number and p represents the number of iterative steps. ε is an imposed tolerance. The symbol α is a properly selected relaxation factor aimed at rapid convergence of the harmonic solution in nonlinear iterations. The stopping criterion is defined by:

$$\frac{\left\|\boldsymbol{X}^{p+1}-\boldsymbol{X}^{p}\right\|}{\left\|\boldsymbol{X}^{p}\right\|} < \varepsilon \tag{5-43}$$

where $\boldsymbol{X}$ represents all harmonic solutions that correspond to equation set (5-42), including magnetic vector potential $\boldsymbol{A}$ and current density $\boldsymbol{J}$.

5.2.4.3 Comparison Between the Previous Method and the New Method

Compared with the previous calculation procedure in 5.2.4.1, the improved computational approach in 5.2.4.2 makes the harmonic solutions converge smoothly and unconditionally. The Gauss-Seidel algorithm can be used to solve the nonlinear magnetic field equations more efficiently.

When the DC-biasing magnetic field is calculated, suppose that there are M_{T} nodes in the finite element region. Thus, comparison of the memory required in the computation can be indicated by the largest matrix size, as follows:

$$\frac{MS_N}{MS_T} = \frac{[3 \times M_T]^2}{[(2N+1) \times M_T]^2} = \frac{3 \times 3}{(2N+1)^2}, \tag{5-44}$$

where MS_N represents the matrix size of $\boldsymbol{Q}_1$ in Equation (5-42), while MS_T represents the system matrix size in the traditional HBFEM [33, 36]. Therefore, the memory requirement has been significantly reduced by the decomposed harmonic balance equations, which are suitable for large-scale computations.

5.2.5 DC-Biasing Test on the LCM and Computational Results

The square LCM shown in Figure 5.26(a)–(b) was made by the Tianwei Group, Baoding, China. Two coils entwine the magnetic core, both with 312 turns. One is the exciting coil, and the other is the search coil.

5.2.5.1 B-H Curve of the LCM

The ferromagnetic core is laminated by electric sheets, between which there are minor gaps. Therefore, the LCM may have different magnetizing characteristics, compared with the steel strips. Some measurements are carried out to obtain the accurate magnetizing curve of the LCM.

The exciting coil is fed by different alternating voltages. The induced voltage U_{ac} and the exciting current I are measured by the power analyzer (WT-3000, Japan). The experimental data are manipulated as follows:

$$\begin{aligned} &\varphi = (1/N_{coil}) \int U_{ac} dt \\ &B = \varphi / S \\ &\int H \cdot dl = I \cdot N_{coil} \end{aligned} \tag{5-45}$$

where φ is the magnetic flux in the magnetic core, N_{coil} is the number of turns of the exciting coil, S is the cross-sectional area of the core, and l is the mean length of the magnetic circuit. The B-H curve (shown in Figure 5.26(c)) can be determined by solving the equations above.

5.2.5.2 DC-Biasing Test

The model was tested under different DC bias conditions. The DC bias I_{dc} from a DC current source, and alternating voltage from the voltage regulator, are applied to the

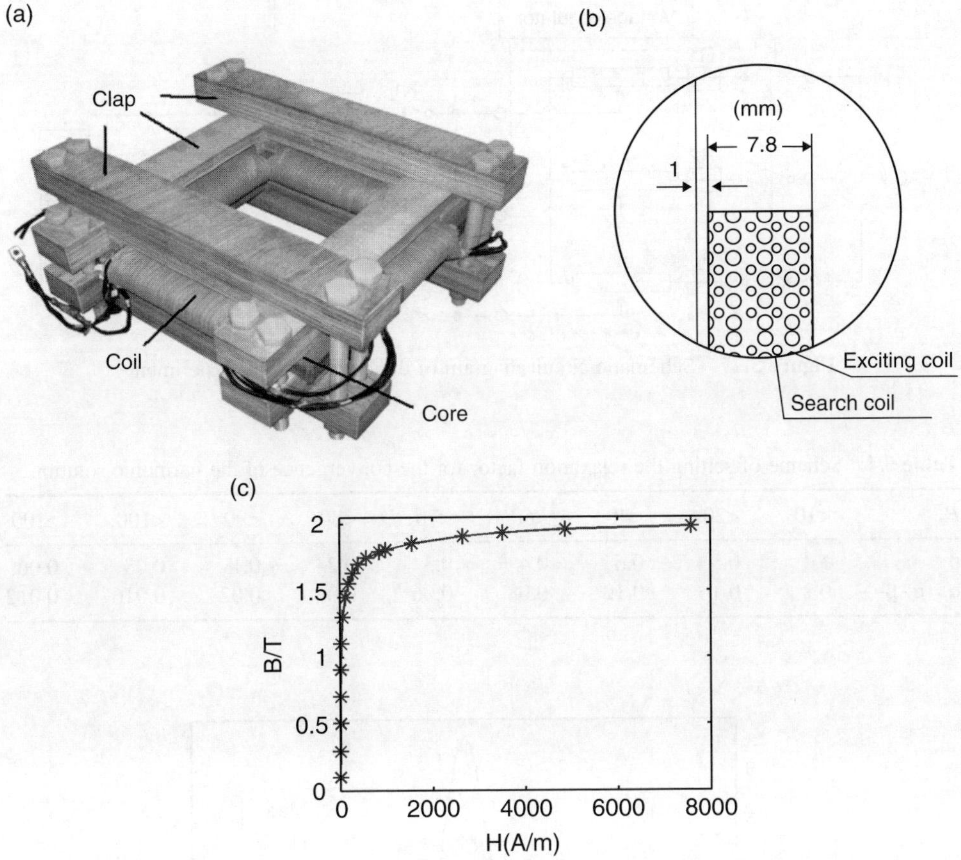

Figure 5.26 LCM and the B-H curve. (a) Laminated core model ($\sigma_{steel} = 2.22 \times 10^6$ S/m); (b) Exciting coil and search coil ($\sigma_{coil} = 5.71 \times 10^7$ S/m); (c) B-H curve of the LCM

exciting coil. Measurements of no-load induced voltage (U_{ac} of 50 Hz) and no-load exciting current have been carried out by the search coil and exciting coil respectively (using the power analyzer WT-3000). In the DC-biasing test, the DC bias I_{dc} is smaller than 3 amperes, because a DC bias that is too large may damage the LCM. Figure 5.27 shows the schematic circuit diagram of the DC-biasing test. The electric circuit of the exciting coil is simulated by the harmonic computation. The DC component of voltage U_0 in Equation (5-38) can be linked with I_{dc} through the resistance R of the exciting coil.

5.2.5.3 Computational Results

The exciting currents in the exciting coil and DC-biasing magnetic field are computed simultaneously by the block Gauss-Seidel algorithm proposed in 5.2.4. The relaxation

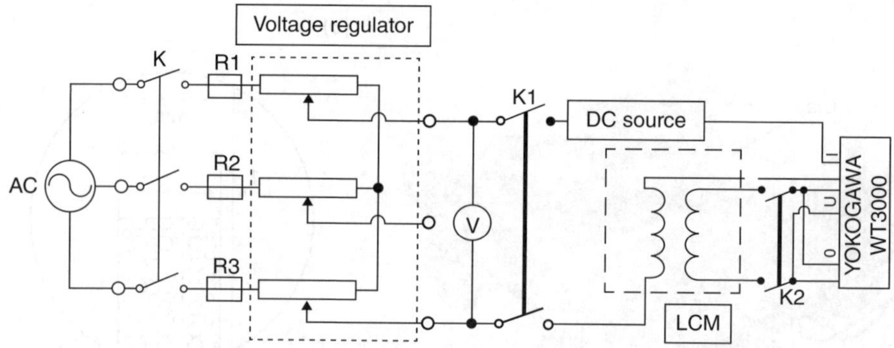

Figure 5.27 Schematic circuit diagram of the core model in experiment

Table 5.4 Scheme of setting the relaxation factor for the convergence of the harmonic solution

P_c	<10	<20	<40	<60	<70	<80	<90	<100	≥100
β	1	0.8	0.6	0.4	0.3	0.2	0.1	0.08	0.06
$\alpha = \alpha \cdot \beta$	0.2	0.16	0.12	0.08	0.06	0.04	0.02	0.016	0.012

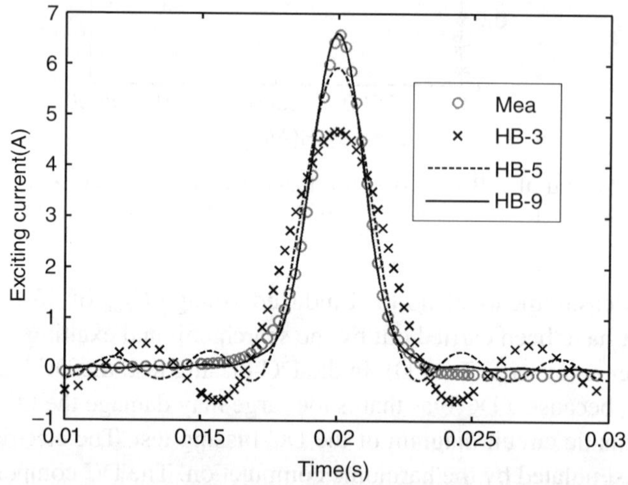

Figure 5.28 Exciting currents under different DC bias conditions ($I_{dc} = 0.847$ A; $U_{ac} = 320$ V)

factor α is initially set to be 0.2 or 0.3 [37], and then varies with the number of iterative steps P_c. A detailed iterative scheme is shown in Table 5.4. In the harmonic computation, 1922 triangular elements and 1014 nodes are used in one-quarter of the square model, which is shown in Figure 5.26.

Table 5.5 Calculation results under different DC bias excitations

I_{dc}(A)	U_{ac}(V)	N	P_c	T_c(s)	$I_{rms,c}$(A)	$I_{rms,m}$(A)	ε	Error(%)
0.4256	370	9	170	5678.5	1.2895	1.3486	0.015	4.38
0.4256	370	7	158	3303.6	1.2813	1.3486	0.01	4.99
0.847	240	5	171	1888.6	2.1202	2.3463	0.01	9.64
0.847	240	9	174	5788.2	2.3516	2.3463	0.015	−0.23

Table 5.6 Comparison of the computational cost between the two methods

				Conventional HB			Decomposed HB		
I_{dc}(A)	U_{ac}(V)	ε	N	$P_{c,1}$	$T_{c,1}$(s)	M_1(Mb)	$P_{c,2}$	$T_{c,2}$(s)	M_2(Mb)
0.4256	370	0.015	8	53	3062.1	10.98	164	4409.1	0.47
0.847	320	0.015	9	72	4287.3	13.61	189	5943.5	0.47

In Figure 5.28 below, the calculated exciting currents are compared with the measured current when the truncated harmonic number is selected differently in the computation. The computational results under different DC bias excitations are listed in Table 5.5. P_c and N represent the number of iterations for convergence and the truncated harmonic number, respectively. $I_{rms,c}$ is the root-mean-square value of calculated exciting current, while $I_{rms,m}$ is the counterpart obtained from the measurement.

From Table 5.5, it can be concluded that at least seven harmonics should be truncated in the computation to control the errors of the exciting current within 5%. The number of iterations for convergence depends on the DC bias excitation, truncated harmonic number N and stopping criterion. The increased (I_{dc} or U_{ac}) indicates a stronger nonlinearity of the magnetic core in computation, and more iterative steps will be required to obtain convergent harmonic solutions. A larger truncated harmonic number also results in more iterative steps. If the stopping criterion is set reasonably, minor errors and fewer iterative steps can be achieved simultaneously.

The decomposed HBFEM is compared with the conventional approach (in terms of memory requirement and computational time) in Table 5.6. The symbol M represents the memory that the system matrix requires in the computation. T_c is the computational time. It is apparent that the conventional HBFEM requires a large memory in the calculation, which may limit its application in large-scale computation, such as 3-D problems. The decomposed method has significantly reduced the memory cost. An increase in computational time is due to the sequential solving scheme of each harmonic solution. In fact, each harmonic solution in equation (5-42) can be solved in parallel, by using the block Jacobi algorithm. In that case, the three-dimensional DC-biasing problems in the power transformer can be solved more efficiently.

5.2.6 *Analysis of the Flux Density and Flux Distribution Under DC Bias Conditions*

One point on the cross-section of the LCM shown in Figure 5.29, such as point B, can be selected to observe the effect of DC bias on the total flux. The flux density B_x in point B is calculated by the decomposed HBFEM and the time domain finite element method (TDFEM), respectively (B_y is too small to be analyzed). The waveforms of flux density and the corresponding exciting current shown in Figures 5.30(a) and 5.30(b) demonstrate that the DC-biasing magnetic fields calculated by the two methods coincide well. The waveform of reluctivity under DC bias conditions is presented in Figure 5.30(c). Due to the existence of DC bias in the exciting coil, the total flux is raised by the DC flux and saturates the magnetic core in one direction. This corresponds to the exciting current distorted in the half-cycle, as shown in Figure 5.28. Table 5.7 presents the calculated harmonic components of flux density (B_x) and reluctivity (ν).

The DC flux is not only affected nonlinearly by the DC bias, but also depends on the alternating voltage. Meanwhile, the AC flux varies with different DC bias conditions. Harmonic components of B_x in point B are observed under different DC bias conditions. As Table 5.7 shows, the DC and fundamental components are dominant in flux density B_x. Therefore, four different DC bias levels, and five different AC excitation cases (shown in Table 5.8 and Table 5.9), are integrated to analyze the variation of the DC component B_{x0} and the fundamental component B_{x1} under DC bias excitation.

Table 5.8 shows that when the alternating voltage U_{ac} is kept constant, the DC flux density increases slowly and tends to be constant with the increment of DC bias I_{dc}. When the DC bias is kept constant, the AC excitation will suppress the DC flux density nonlinearly. It can also be concluded from Table 5.9 that the DC bias has almost no influence on the AC flux density, whereas AC excitation gives rise to variation of AC flux density.

Harmonic flux distributions under DC bias conditions are presented in Figure 5.31. The harmonic flux distributions vary with time (related to phase angle) and excitations (related to DC bias and alternating voltage). The high-order harmonic distribution is apparently different from the fundamental and DC component under DC bias conditions. That may affect the total flux, which is the superposition of all harmonic components.

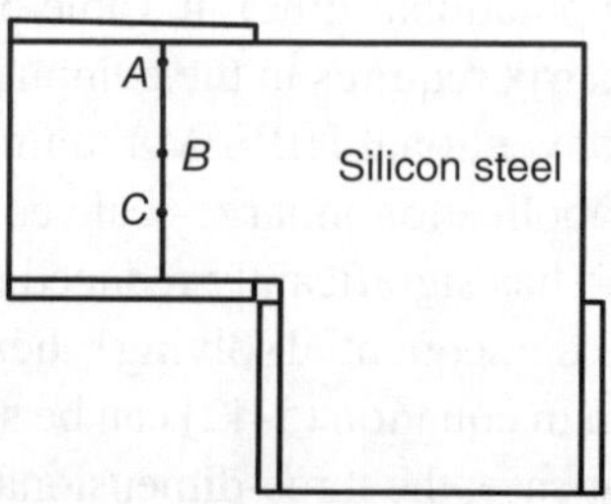

Figure 5.29 Computational region: one quarter of the square laminated core model

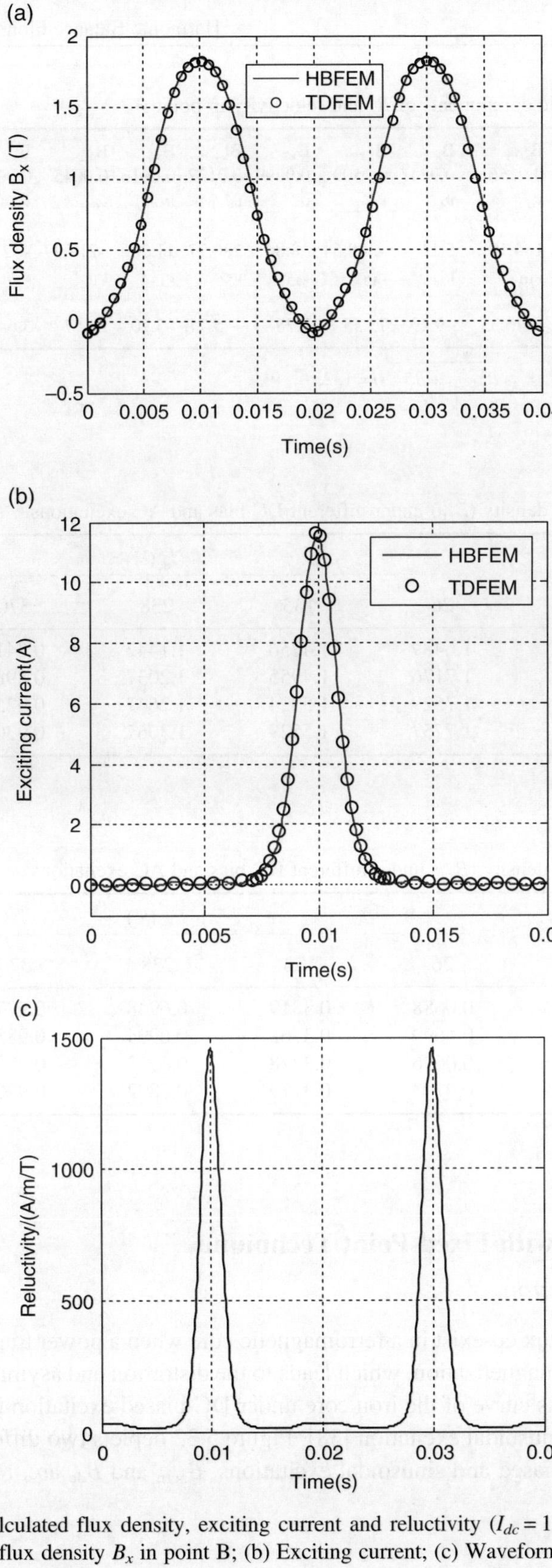

Figure 5.30 Calculated flux density, exciting current and reluctivity ($I_{dc} = 1.27$ A, $U_{ac} = 370$ V). (a) Waveform of flux density B_x in point B; (b) Exciting current; (c) Waveform of reluctivity

Table 5.7 Harmonic components of B_x and reluctivity in point B

Harmonic components	B_{x0}	B_{x1}	B_{x2}	B_{x3}	B_{x4}	B_{x5}	B_{x6}	B_{x7}	B_{x8}	B_{x9}
Flux density (T)	0.9256	0.9847	0.0509	0.0299	0.0122	0.0016	0.0045	0.0015	0.0019	0.0013
Harmonic components	ν_0	ν_1	ν_2	ν_3	ν_4	ν_5	ν_6	ν_7	ν_8	ν_9
Reluctivity (A/(m · T))	184.50	5.11	−308.54	−8.80	280.77	11.36	−218.99	−12.01	165.67	10.93
Harmonic components	ν_{10}	ν_{11}	ν_{12}	ν_{13}	ν_{14}	ν_{15}	ν_{16}	ν_{17}	ν_{18}	ν_{19}
Reluctivity (A/(m · T))	−115.35	−9.49	73.55	7.28	−45.71	−5.10	24.57	3.64	−12.79	−2.18

Note: $B_{x0} = |B_{x,0}|$ and $B_{xi} = \sqrt{B_{x,2i-1}^2 + B_{x,2i}^2}$ $(i = 1, 2, \ldots, 9)$.

Table 5.8 DC flux density (B_{x0}) under different DC bias and AC excitations

		U_{ac}(V)				
B_{dc} (T)		26	133	238	370	420
I_{dc}(A)	0.426	1.6489	1.4256	1.1442	0.8410	0.7201
	0.847	1.7176	1.4955	1.2037	0.8981	0.7846
	1.274	1.7529	1.5324	1.2647	0.9256	0.8126
	1.691	1.7757	1.5529	1.3067	0.9308	0.8298

Table 5.9 AC flux density (B_{x1}) under different DC bias and AC excitations

		U_{ac}(V)				
B_{dc} (T)		26	133	238	370	420
I_{dc}(A)	0.426	0.0688	0.3819	0.6948	0.9774	1.0967
	0.847	0.0699	0.3768	0.6994	0.9835	1.0832
	1.274	0.0696	0.3778	0.6952	0.9846	1.0951
	1.691	0.0703	0.3779	0.6892	0.9902	1.0768

5.3 HBFEM with Fixed-Point Technique

5.3.1 Introduction

DC flux and AC flux co-exist in a ferromagnetic core when a power transformer works under DC-biased magnetization, which leads to the distortion and asymmetry of hysteresis loops. The loss curve of the iron core under DC-biased excitation is also different from that under sinusoidal excitation [38]. Figure 5.32 depicts two different hysteresis loops under DC-biased and sinusoidal excitations. B_{acm} and B_{dc} are, respectively, the

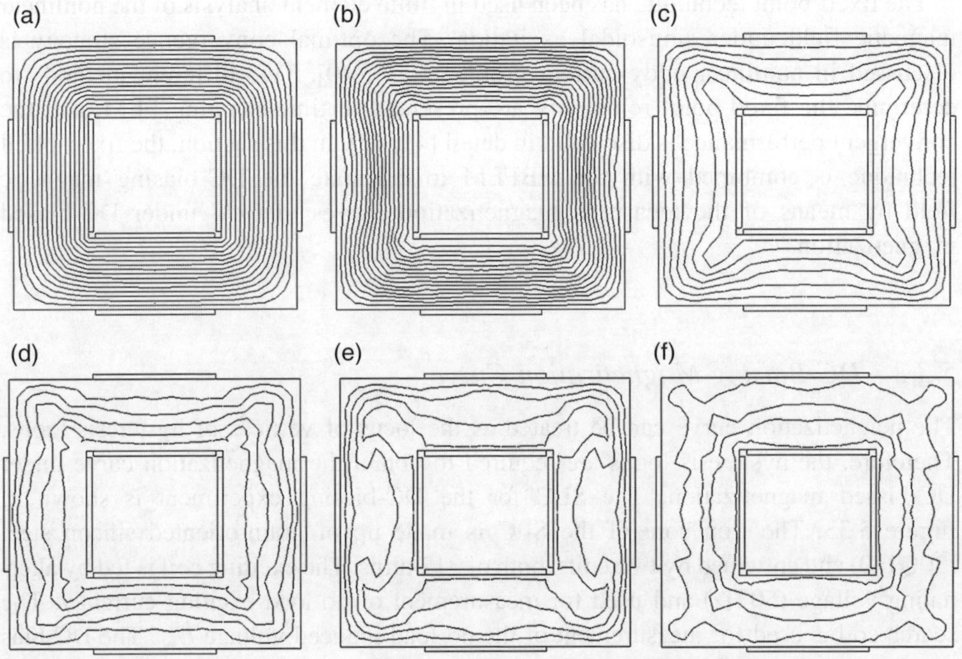

Figure 5.31 Harmonic flux distributions ($I_{dc} = 1.27\text{A}$, $U_{ac} = 370\text{V}$). (a) DC flux; (b) Fundamental ($\omega t = \pi/2$); (c) Second order ($2\omega t = \pi/2$); (d) Third order ($3\omega t = \pi/2$); (e) Fourth order ($4\omega t = \pi/2$); (f) Fifth order ($5\omega t = \pi/2$)

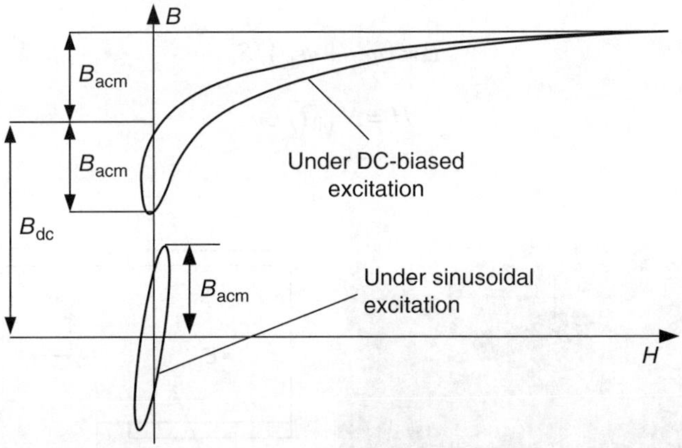

Figure 5.32 Hysteresis loops under DC-biased and sinusoidal excitations

magnitudes of the AC component and the DC component of magnetic flux density. The hysteresis loop is distorted and the ferromagnetic core is saturated due to the DC flux density B_{dc}. As a result, the magnetization curve under DC-biased excitation is different from the basic one.

The fixed-point technique has been used in finite element analysis of the nonlinear magnetic field under sinusoidal excitation. The optimal convergence strategy is discussed in nonlinear eddy current problems [39, 40]. Two different methods to determine the fixed-point reluctivity are presented in time-stepping FEM, and the convergent performance is discussed in detail [41, 42]. In this section, the fixed-point technique is combined with the HBFEM to calculate the DC-biasing magnetic field by means of the measured magnetization curves of SLC under DC-biased magnetization.

5.3.2 DC-Biasing Magnetization Curve

The magnetization curve can be treated as the locus of vertices of hysteresis loops. Therefore, the hysteresis loops are required to obtain the magnetization curve under DC-biased magnetization. The SLC for the DC-biasing experiment is shown in Figure 5.33. The iron core of the SLC is made up of grain-oriented silicon steel (30Q140) and entwined by two coils, both of 312 turns. The exciting coil is fed by alternating voltage (50 Hz) and used for measurement of no-load exciting current I. The search coil is used for measurement of the no-load induced voltage U_{ac}. The DC bias current I_{dc} is applied to the exciting coil by a direct current source. To obtain the DC-biasing hysteresis loops, the measured data are manipulated by:

$$\varphi_{ac} = (1/N_{coil})\int U_{ac}dt \tag{5-46}$$

$$B = (\varphi_{ac} + \varphi_{dc})/S \tag{5-47}$$

$$H = N_{coil}I/L \tag{5-48}$$

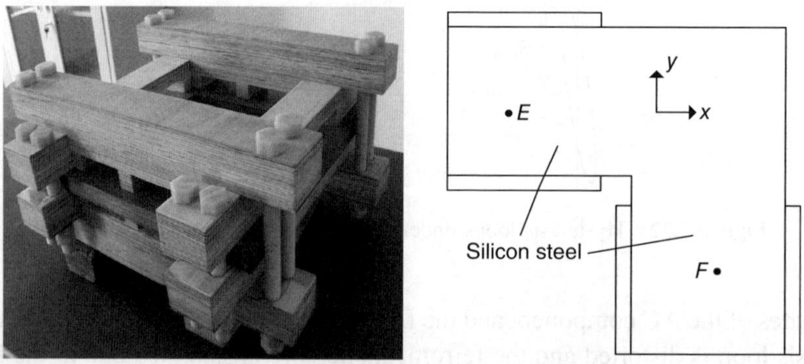

Figure 5.33 The SLC and computational region

where φ_{ac} and φ_{dc} are the AC and DC component of the total flux in the magnetic core, respectively. S is the cross-sectional area of the core, and L is the mean length of the magnetic circuit.

Since the DC flux φ_{dc} cannot be directly measured experimentally, an iterative method has been proposed in [43] to calculate φ_{dc}. First, φ_{ac} is calculated by Equation (5-46), and exciting current I can be captured by the search coil under a given DC-biasing working condition. Second, an initial value of φ_{dc} is assumed. Based on the basic magnetization curve and the total flux φ (sum of φ_{ac} and φ_{dc}), the calculated exciting current I_c is compared with I. Finally, φ_{dc} is adjusted iteratively until I_c matches with I.

B and H can be determined by solving Equations (5-46) to (5-48) with the calculated φ_{dc}, and a series of hysteresis loops under DC-biased magnetization (shown in Figure 5.34) can be obtained. Consequently, the DC-biased B_m-H_b curve [44] can also be obtained. Table 5.10 presents the measured data, which shows the asymmetry of the DC-biasing magnetization curve. The DC component H_{dc} of the magnetic field intensity can be calculated by Equation (5-48) with the given direct current I_{dc} in the exciting coil.

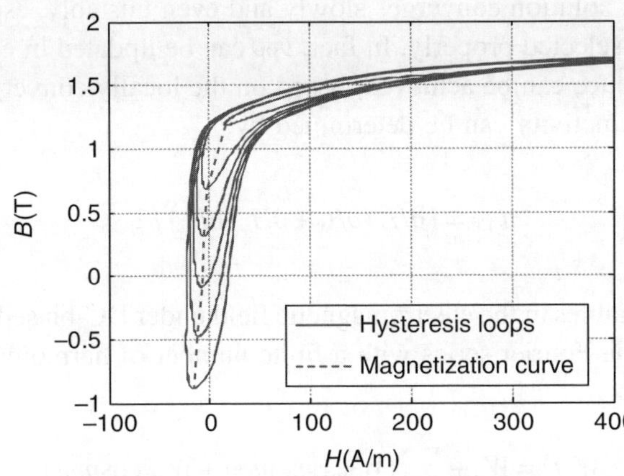

Figure 5.34 Measured hysteresis loops (solid line) and magnetization curve (dashed line) under DC-biased magnetization (H_{dc} = 107 A/m)

Table 5.10 Data of magnetization curve under DC-biased magnetization (H_{dc} = 107 A/m)

B_m(T)	−1.543	−1.244	−0.673	−0.276	−0.079	0.114
H_b(A/m)	−104.77	−32.89	−13.98	−10.14	−8.328	−7.238
B_m(T)	0.314	0.876	1.330	1.647	1.783	1.908
H_b(A/m)	−6.323	−0.747	47.177	325.51	1319.6	2665.8

5.3.3 Fixed-Point Harmonic-Balanced Theory

A relationship between magnetic field intensity $\boldsymbol{H}$ and magnetic flux density $\boldsymbol{B}$ is represented by introducing the fixed-point reluctivity ν_{FP} [45]:

$$\boldsymbol{H}(\boldsymbol{B}) = \nu_{\mathrm{FP}}\boldsymbol{B} - \boldsymbol{M}(\boldsymbol{B}) \tag{5-49}$$

where ν_{FP} is a constant value. $\boldsymbol{M}$ is a magnetization-like quantity which varies nonlinearly with $\boldsymbol{B}$. Therefore, the magnetic field intensity $\boldsymbol{H}$ is divided into two parts – the linear part related to ν_{FP}, and the nonlinear part related to $\boldsymbol{M}$.

The vector potential equation is used to describe the two-dimensional nonlinear magnetic field by substituting (5-49) into Maxwell's equations:

$$\nabla \times \nu_{\mathrm{FP}}(\nabla \times A) + \sigma \frac{\partial A}{\partial t} = J - \nabla \times \boldsymbol{M} \tag{5-50}$$

where A is the magnetic vector potential and σ is the electric conductivity.

The fixed-point reluctivity determines the convergence of solutions. If ν_{FP} is a constant, the harmonic solution converges slowly and even unstably, especially when the value of ν_{FP} is not selected properly. In fact, ν_{FP} can be updated in each iterative step, and a fast convergence can be achieved, based on the locally convergent method. The local fixed-point reluctivity can be determined by:

$$\nu_{\mathrm{FP}} = \left(\partial H_x/\partial B_x + \partial H_y/\partial B_y\right)/2 \tag{5-51}$$

The periodic variables in the electromagnetic field under DC-biased excitation can be approximated by the Fourier series with a finite number of harmonics [11]:

$$W(t) = W_0 + \sum_{n=1}^{\infty}\left(W_{2n-1}\sin n\omega t + W_{2n}\cos n\omega t\right) \tag{5-52}$$

where $W(t)$ can be replaced by current density $\boldsymbol{J}$, vector potential $\boldsymbol{A}$, magnetic flux density $\boldsymbol{B}$, magnetic field intensity $\boldsymbol{H}$, and the magnetization-like quantity $\boldsymbol{M}$. Equation (5-49) can be rewritten in isotropic material by means of a harmonic vector in the harmonic domain:

$$\begin{cases} \boldsymbol{H}_x = \nu_{\mathrm{FP}}\boldsymbol{B}_x - \boldsymbol{M}_x \\ \boldsymbol{H}_y = \nu_{\mathrm{FP}}\boldsymbol{B}_y - \boldsymbol{M}_y \end{cases} \tag{5-53}$$

in which:

$$\begin{cases} \boldsymbol{H}_x = [H_{x,0} \ \ H_{x,1} \ \ H_{x,2} \ \ H_{x,3} \ \ H_{x,4} \ \ \cdots]^T \\ \boldsymbol{H}_y = [H_{y,0} \ \ H_{y,1} \ \ H_{y,2} \ \ H_{y,3} \ \ H_{y,4} \ \ \cdots]^T \end{cases} \tag{5-54}$$

Each of the harmonic vectors $\boldsymbol{B}_x$, $\boldsymbol{B}_y$, $\boldsymbol{M}_x$ and $\boldsymbol{M}_y$ has a similar expression with (5-54). The fixed-point reluctivity can also be expressed in the harmonic domain:

$$\nu_{\text{FP}}(t) = dH(B)/dB = \nu_0 + \sum_{n=1}^{\infty} \{\nu_{2n-1}\sin(n\omega t) + \nu_{2n}\cos(n\omega t)\} \tag{5-55}$$

where ν_{2n-1} and ν_{2n} are the harmonic coefficients of the fixed-point reluctivity, while ν_0 is the DC component. The fixed-point harmonic-balanced equation can be obtained by substituting Equations (5-52) and (5-55) into the weak form of Equation (5-50), based on Galerkin's method:

$$\begin{aligned} \boldsymbol{S}_e\boldsymbol{A}_e + \boldsymbol{T}_e\boldsymbol{A}_e - \boldsymbol{K}_e - \boldsymbol{P}_e = & \begin{bmatrix} S_{11}\boldsymbol{D}_{\text{FP}} & S_{12}\boldsymbol{D}_{\text{FP}} & S_{13}\boldsymbol{D}_{\text{FP}} \\ S_{21}\boldsymbol{D}_{\text{FP}} & S_{22}\boldsymbol{D}_{\text{FP}} & S_{23}\boldsymbol{D}_{\text{FP}} \\ S_{31}\boldsymbol{D}_{\text{FP}} & S_{32}\boldsymbol{D}_{\text{FP}} & S_{33}\boldsymbol{D}_{\text{FP}} \end{bmatrix} \cdot \begin{bmatrix} \boldsymbol{A}_{e1} \\ \boldsymbol{A}_{e2} \\ \boldsymbol{A}_{e3} \end{bmatrix} \\ + \begin{bmatrix} T_{11}\boldsymbol{N} & T_{12}\boldsymbol{N} & T_{13}\boldsymbol{N} \\ T_{21}\boldsymbol{N} & T_{22}\boldsymbol{N} & T_{23}\boldsymbol{N} \\ T_{31}\boldsymbol{N} & T_{32}\boldsymbol{N} & T_{33}\boldsymbol{N} \end{bmatrix} & \begin{bmatrix} \boldsymbol{A}_{e1} \\ \boldsymbol{A}_{e2} \\ \boldsymbol{A}_{e3} \end{bmatrix} - \begin{bmatrix} \boldsymbol{K}_{e1} \\ \boldsymbol{K}_{e2} \\ \boldsymbol{K}_{e3} \end{bmatrix} - \begin{bmatrix} \boldsymbol{P}_{e1} \\ \boldsymbol{P}_{e2} \\ \boldsymbol{P}_{e3} \end{bmatrix} = 0 \end{aligned} \tag{5-56}$$

where $S_{i,j} = \int_{\Omega_e} \nabla N_i \cdot \nabla N_j d\Omega, T_{i,j} = \int_{\Omega_e} \sigma N_i \cdot N_j d\Omega$, and the subscripts i and j represent the node number in finite elements. N_i is the shape function on node i in the finite element region. $\boldsymbol{A}$ is the harmonic form of magnetic vector potential, and $\boldsymbol{N}$ is related to the harmonic number (and called the harmonic matrix) [46]. $\boldsymbol{D}_{\text{FP}}$ is related to the reluctivity and can be called the fixed-point reluctivity matrix, which has a similar expression with the reluctivity matrix in [33]. $\boldsymbol{P}$ is derived from the magnetization-like quantity $\boldsymbol{M}$:

$$\boldsymbol{P}_{ei} = [P_{ei,0} \ \ P_{ei,1} \ \ P_{ei,2} \ \ P_{ei,3} \ \ P_{ei,4} \ \ \cdots]^{\text{T}} \tag{5-57}$$

$$P_{ei,m} = \iint_{\Omega_e} \left(M_{y,m}\frac{\partial N_{ei}}{\partial x} - M_{x,m}\frac{\partial N_{ei}}{\partial y} \right) \mathrm{d}x\mathrm{d}y \tag{5-58}$$

5.3.4 Electromagnetic Coupling

When electromagnetic devices are excited by voltage, the coupling between external circuits and the magnetic field should be taken into account in the finite element equations, so that the magnetizing current and magnetic field can be solved simultaneously.

The electromagnetic coupling can be considered by:

$$U_{ink} = C_k\boldsymbol{A} + Z_k\boldsymbol{J}_k \tag{5-59}$$

where the subscript k represents the number of external circuits, and $\boldsymbol{U}_{\text{in}}$ and $\boldsymbol{J}$ are the harmonic vectors of the external input voltage and current density. $\boldsymbol{C}$ and $\boldsymbol{Z}$ represent the coupling matrix and impedance matrix, respectively.

The magnetic field and exciting current can be calculated by solving Equations (5-56) and (5-59) together:

$$\begin{bmatrix} \boldsymbol{F} & -\boldsymbol{G}_1 & \cdots & -\boldsymbol{G}_k & \cdots \\ \boldsymbol{C}_1 & S_{c1}\boldsymbol{Z}_1 & \boldsymbol{0} & \boldsymbol{0} & \cdots \\ \vdots & \vdots & \ddots & \boldsymbol{0} & \cdots \\ \boldsymbol{C}_k & \boldsymbol{0} & \boldsymbol{0} & S_{ck}\boldsymbol{Z}_k & \cdots \\ \vdots & \vdots & \vdots & \vdots & \ddots \end{bmatrix} \begin{bmatrix} \boldsymbol{A} \\ \boldsymbol{J}_1 \\ \vdots \\ \boldsymbol{J}_k \\ \vdots \end{bmatrix} = \begin{bmatrix} \boldsymbol{P} \\ \boldsymbol{U}_{\text{in}1} \\ \vdots \\ \boldsymbol{U}_{\text{in}k} \\ \vdots \end{bmatrix} \tag{5-60}$$

where $\boldsymbol{F} = \boldsymbol{S} + \boldsymbol{T}$, and $\boldsymbol{G}$ is related to the spatial distribution of $\boldsymbol{J}$.

The harmonic vector $\boldsymbol{P}$ nonlinearly depends on $\boldsymbol{B}$ and, hence, on $\boldsymbol{A}$. Therefore, equation (5-60) can be solved iteratively. In the first step ($p = 0$), an initial value is assigned to the vector $\boldsymbol{P}$ ($= \boldsymbol{P}_0$), $\boldsymbol{A}_1$ and $\boldsymbol{J}_1$ are computed with the given working voltage $\boldsymbol{U}_{\text{in}}$, $\boldsymbol{B}_1$ and $\boldsymbol{H}_1$ can be obtained from $\boldsymbol{A}_1$ and the single-valued function $H = F(B)$, which should be Lipschitz continuous. Consequently, the vector $\boldsymbol{P}$ can be updated by $\boldsymbol{M}$ and the computational process is performed iteratively. The stopping criterion is defined by:

$$\left\|\boldsymbol{X}^{p+1} - \boldsymbol{X}^{p}\right\| / \left\|\boldsymbol{X}^{p}\right\| < \varepsilon \tag{5-61}$$

where $\boldsymbol{X}$ represents all harmonic solutions, including $\boldsymbol{A}$ and $\boldsymbol{J}$.

5.3.5 Validation and Discussion

Due to the structural symmetry of the SLC, 1922 triangular elements and 1014 nodes are used in a quarter of the model shown in Figure 5.33 to carry out the harmonic computation.

5.3.5.1 A. Computational Results and Analysis of the Fixed-Point Harmonic-Balanced Method

The DC-biasing magnetization curve shown in Table 5.10 is employed to calculate the DC-biasing magnetic field by the fixed-point HBFEM. Different harmonics are truncated to calculate the exciting current, which is compared with the measured current in Figure 5.35. Figure 5.36 indicates the effectiveness of the locally convergent method in adapting the fixed-point reluctivity according to the changes in flux density. The waveform of flux density is raised under DC-biased excitation. The negative peak value approaches to –2.0 Tesla, which leads to the significant saturation of the ferromagnetic core.

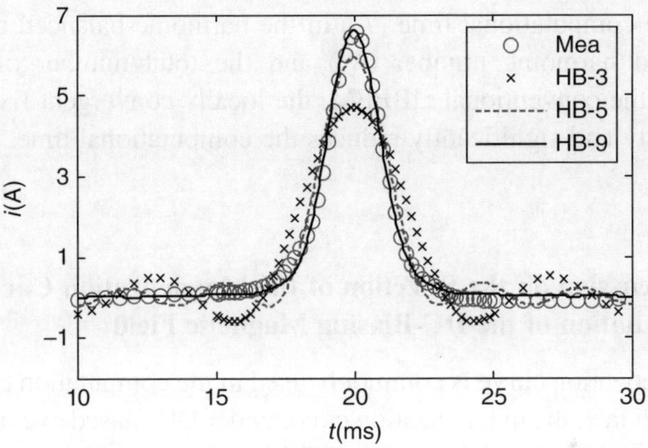

Figure 5.35 Calculated and measured exciting current (H_{dc} = 213 A/m, U_{ac} = 320 V)

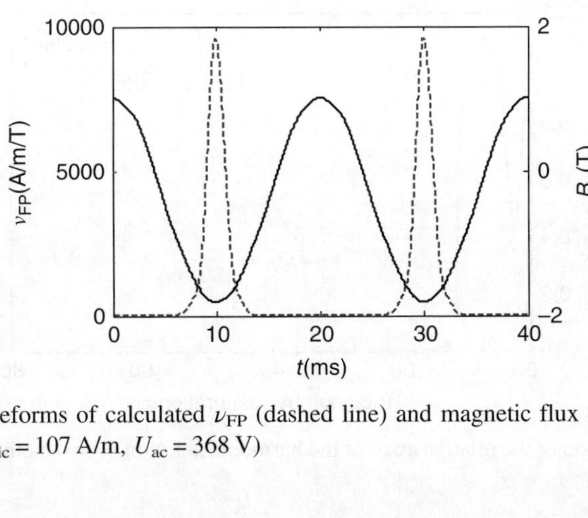

Figure 5.36 Waveforms of calculated ν_{FP} (dashed line) and magnetic flux density (solid line) in point E in SLC (H_{dc} = 107 A/m, U_{ac} = 368 V)

The convergent speed of the harmonic solutions in HBFEM depends on the method used to update magnetic reluctivity. The conventional method [33] adopts a relaxation iterative scheme to renew the reluctivity matrix, so the harmonic solutions converge slowly. In the fixed-point harmonic-balanced method, the determination of ν_{FP} plays an important role in computation. The locally convergent method determines an optimal value of ν_{FP} in a local region, which significantly speeds up the convergence of harmonic solutions.

Eleven harmonics ($N_c = 11$) are truncated in both methods to observe the evolution of the relative error of the harmonic solution in Equation (5-61). As shown in Figure 5.37, the harmonic solution converges quickly in the proposed method, but slowly in the conventional method. It is concluded that solutions in the conventional method require more iterations to reach convergence.

The two methods are compared in detail in Table 5.11. With the same memory requirement, the computational time (T_c) in the harmonic-balanced methods depends on the truncated harmonic number (N_c) and the total number of iterations (P_c). Compared with the conventional HBFEM, the locally convergent fixed-point method converges quickly and significantly reduces the computational time.

5.3.5.2 B. Discussion on the Selection of the Magnetization Curve in the Computation of the DC-Biasing Magnetic Field

The basic magnetization curve is commonly used in the computation of the DC-biasing magnetic field. In fact, the magnetization curve under DC-biased excitation differs from the basic curve under sinusoidal excitation. Table 5.11 shows that the DC-biasing magnetization curve is not symmetric in the first and third quadrants of the *B*-*H* coordinate.

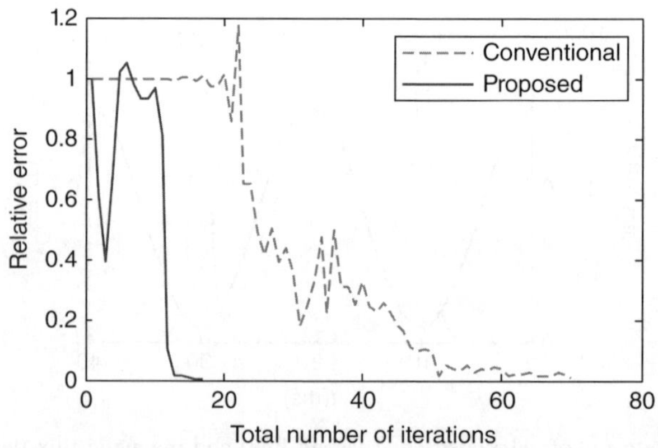

Figure 5.37 Evolution of the relative error of the harmonic solution in two methods ($H_{dc} = 107$ A/m, $U_{ac} = 368$ V, $N_c = 11$)

Table 5.11 Comparison of the two methods

HB method	H_{dc} (A/m)	U_{ac} (V)	N_c	P_c	T_c(s)
Conventional	107	290	9	53	3062.1
Proposed	107	290	9	26	1646.6
Conventional	213	420	11	74	8513.2
Proposed	213	420	11	18	2003.2

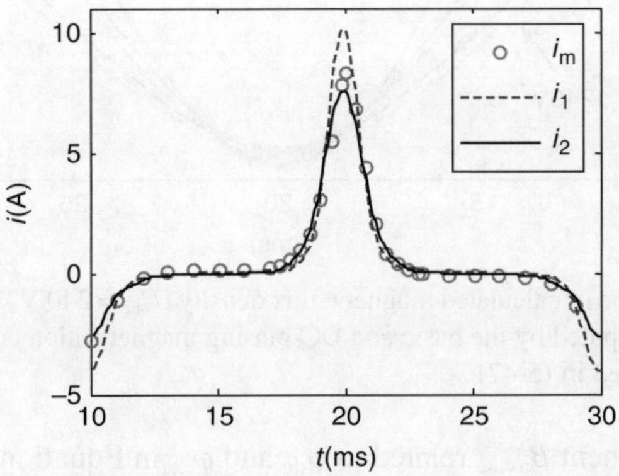

Figure 5.38 Comparison of calculated exciting currents (H_{dc} = 107 A/m, U_{ac} = 495 V); i_1 and i_2 are exciting currents calculated by basic and DC-biasing magnetization curves respectively, and i_m is the measured result

Table 5.12 Comparison of the exciting current

U_{ac}(V)	$I_{rms,m}$(A)	$I_{rms,1}$(A)	$I_{rms,2}$(A)
240	1.126 4	1.110 6	1.153 6
368	1.351 1	1.361 5	1.384 1
495	3.037 9	3.608 7	2.912 5

Therefore, the computational results obtained from the two curves must be different. Exciting currents calculated from the two different magnetization curves are compared with the measured currents in Figure 5.38, and the corresponding quantitative comparison is presented in Table 5.12. I_{rms} is the rms value of exciting current. It is obvious that the exciting current *i2* is closer to the measured result.

Point *E* in SLC is selected to observe the waveforms of the calculated magnetic flux density (B_1 and B_2) under different DC-biased magnetizations, and Figure 5.39 compares the computed results with the measured results. B_{av} is the sum of DC component

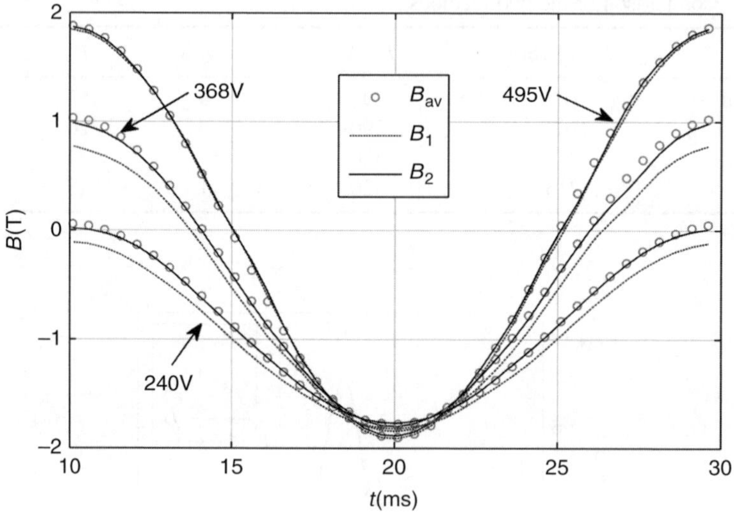

Figure 5.39 Comparison of calculated magnetic flux density (U_{ac} = 240 V, 368 V, 495 V; H_{dc} = 107 A/m). B_1 and B_2 are computed by the basic and DC-biasing magnetization curves respectively, B_{av} is the average value obtained in (5-47)

$B_{av,dc}$ and AC component $B_{av,ac}$ related to φ_{dc} and φ_{ac} in Equation (5-47), respectively. It can be seen that the waveforms of B_2 calculated by the DC-biasing magnetization curve are closer to the measured result B_{av}. The numerical data are listed in detail in Table 5.13. Compared with B_1, the AC component $B_{2,ac}$ and the DC component $B_{2,dc}$ of B_2 coincide better with $B_{av,ac}$ and $B_{av,dc}$, respectively.

Based on the comparison between computed and measured results, in terms of exciting current and magnetic flux density, the numerical results obtained from the DC-biasing magnetization curve are more accurate. It is indicated that the DC-biasing magnetization curve should be adopted in the computation of the DC-biasing magnetic field.

5.4 Hysteresis Model Based on Neural Network and Consuming Function

5.4.1 Introduction

The computation of steady-state solutions of nonlinear time-periodic magnetic fields can be performed in the time or the harmonic domain. The solutions in the time domain generally require several periods to reach a steady state, by using a time-stepping technique starting from arbitrary initial values. Variables in the electromagnetic field under steady-state excitations can be approximated by the triangular series. The electromagnetic field can be solved directly in the harmonic domain, without a long computational time.

Table 5.13 Comparison of the flux density

	B_{av}(T)		B_1(T)		B_2(T)	
U_{ac}(V)	$B_{av,dc}$	$B_{av,ac}$	$B_{1,dc}$	$B_{1,ac}$	$B_{2,dc}$	$B_{2,ac}$
240	−0.858 2	0.918 6	−0.972 7	0.851 0	−0.868 1	0.898 5
368	−0.383 0	1.416 3	−0.531 8	1.305 6	−0.410 7	1.386 5
495	−0.010 0	1.890 9	−0.033 0	1.873 3	−0.007 2	1.874 4

However, it is difficult to deal with hysteresis effects in computation of nonlinear magnetic fields. The fixed-point technique is used in the time-stepping finite element method to analyze the hysteretic characteristics and eddy current problems. A smoothing algorithm is proposed to guarantee the stability of the solution in the iterative procedure when the hysteresis model is involved in the numerical computation of the magnetic field [47].

The hysteresis loops of the iron core exhibit symmetric and asymmetric features under sinusoidal and DC-biased magnetizations respectively [43]. The open-type single sheet tester has been developed to measure the static hysteresis loops under DC-biased magnetization [38]. However, it is still difficult to obtain the DC-biasing hysteresis loops of the laminated core directly from the measurement. In addition, there are no proper hysteresis models combined with numerical methods to compute the nonlinear magnetic field under DC-bias conditions. The Preisach model requires experimental dynamic hysteresis loops, which is difficult to measure when the DC magnetic flux exists in the laminated core. The Jiles-Atherton model is not accurate enough to simulate the DC-biasing hysteresis loops [49]. The DC-biasing hysteresis loops are simulated precisely by using the neural network (NN) theory [43]. However, the NN hysteresis model may occasionally predict inaccurate magnetic fields in numerical computation, due to the limited generalization ability.

5.4.2 Hysteresis Model Based on Consuming Function

As shown in Figure 5.40, hysteretic nonlinearity of the iron core leads to the non-sinusoidal magnetizing current i, which can be resolved into two components: i_1 in-phase with the magnetic flux φ, and i_2 in-phase with the induced electromotive force e [50]. Therefore, the i-φ hysteresis model, based on consuming function, is presented as follows [51, 52],

$$\begin{cases} i(t) = i_1(\varphi) + i_2(\varphi') \\ \varphi' = d\varphi/dt \end{cases} \tag{5-62}$$

where i_2 is related to hysteresis effects.

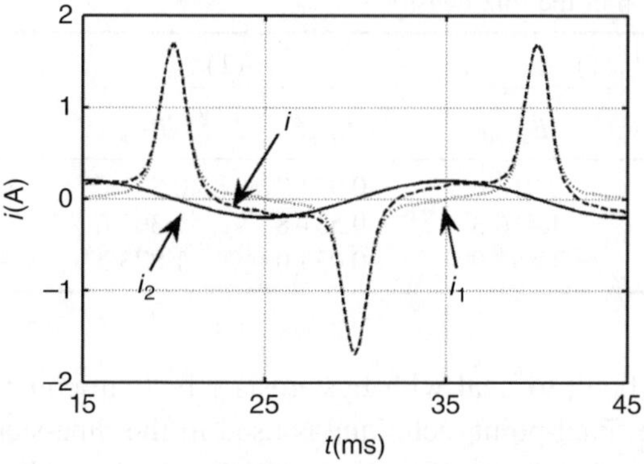

Figure 5.40 Magnetizing current under sinusoidal excitation

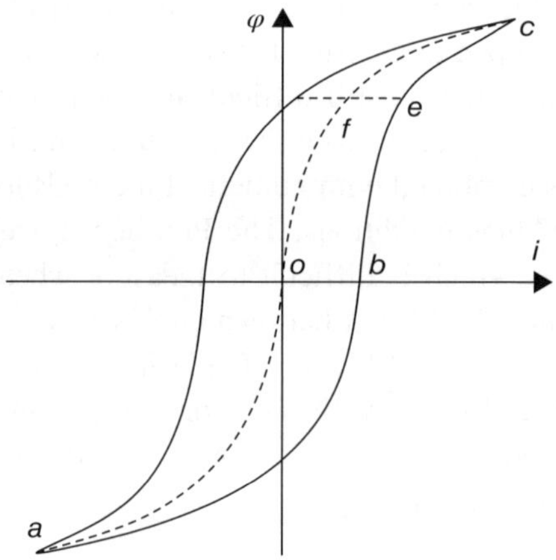

Figure 5.41 Hysteresis model based on consuming function

Figure 5.41 illustrates the hysteresis model used to simulate the static hysteresis loop of the iron core. The dotted line '*aoc*' represents the middle magnetizing curve, which is the locus of the mid-points of the hysteresis loop, and is not related to hysteresis loss. The distance between the periphery of the hysteresis loop and the middle magnetizing curve, such as '*ef*', represents hysteresis loss. The loss part corresponding to i_2 is defined as consuming function, which includes the reverse characteristic of the magnetizing curve and can be approximated by a polynomial in the derivative of the magnetic flux [51]. A simple expression of the consuming function can be assumed as follows [52]:

$$\begin{aligned} i_2(\varphi') &= D \cdot (d\varphi/dt) \\ &= D \cdot \varphi_m \cdot \omega \cdot \cos(\omega t) \\ &= I_{\mathrm{ob}} \cdot \cos(\omega t) \end{aligned} \tag{5-63}$$

where φ_{m} is the magnitude of the magnetic flux φ. I_{ob} is the coefficient of the consuming function and can be defined by the maximum distance '*ob*' between the mid-point locus and the periphery of the hysteresis loop.

5.4.3 Hysteresis Loops and Simulation

The iron core, laminated by grain-oriented silicon steel (30Q140), is used to carry out experiments according to the electric schematic diagram in Figure 5.42. The laminated core is entwined by two coils (both 312 turns). The exciting coil is fed by alternating voltage (50 Hz). The power analyzer is used to measure the magnetizing current i in the exciting coil and the no-load induced voltage U_{ac} in the search coil. The measured data are manipulated by Equations (5-64) and (5-65) to obtain hysteresis loops under sinusoidal magnetization:

$$B = \varphi/S = (1/N_{\mathrm{coil}}S)\int U_{\mathrm{ac}}dt \tag{5-64}$$

$$H = N_{\mathrm{coil}}i/L \tag{5-65}$$

where φ is the total flux in the laminated core, N_{coil} is the number of turns of the coil, S is the cross-sectional area of the core, and L is the mean length of the magnetic circuit.

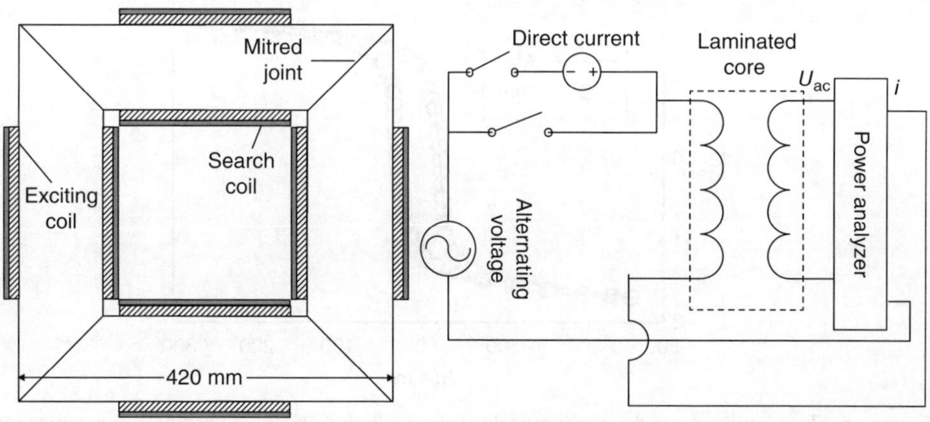

Figure 5.42 Laminated core model and electric schematic diagram

5.4.3.1 A. Hysteresis Loops Under Sinusoidal Excitation

The measured hysteresis loops of the laminated core under sinusoidal magnetization are shown in Figure 5.43(a).

According to Equations (5-62), (5-64) and (5-65), the hysteresis model based on B and H can be established by:

$$\begin{cases} B = B_{acm} \cdot \sin(\omega t) \\ H = H_1(B) + H_2(B') \end{cases} \tag{5-66}$$

where B_{acm} is the magnitude of alternating flux density, and the consuming function H_2 can be approximated simply by $H_{ob} \cdot \cos(\omega t)$, which is similar to its counterpart in

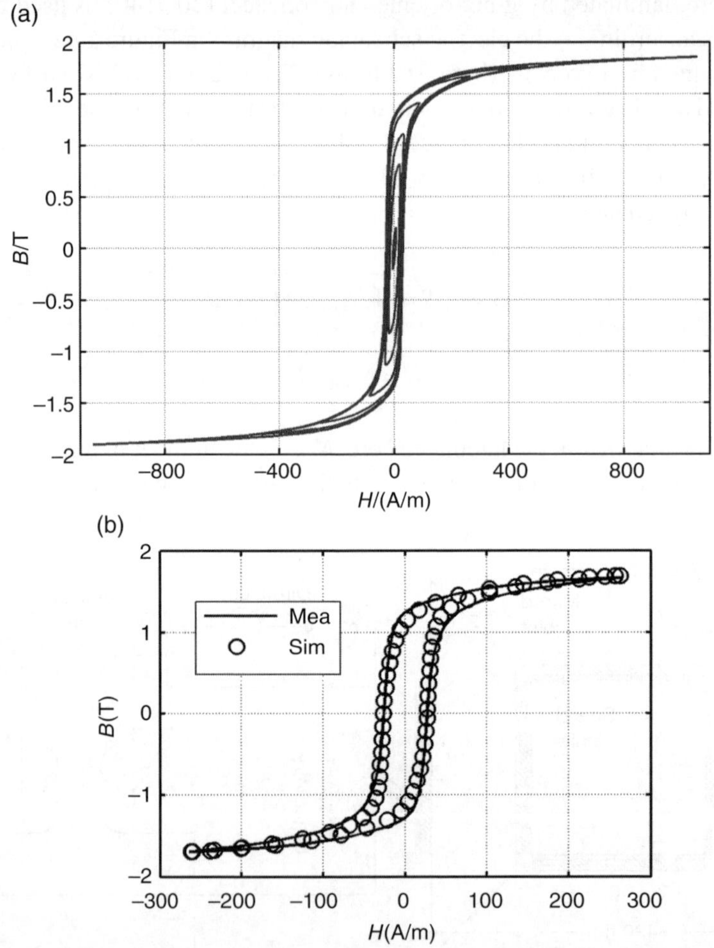

Figure 5.43 Simulated and measured hysteresis loop under sinusoidal magnetization. (a) Symmetrical hysteresis loops. (b) Simulated and measured results

Equation (5-63). The consuming coefficient H_{ob} can be obtained from the measured data in Figure 5.43(a).

Consequently, the presented model in Equation (5-66) can finally be used to simulate the hysteresis loops under sinusoidal magnetization. The simulation and measurement are compared in Figure 5.43(b).

5.4.3.2 B. DC-Biasing Hysteresis Loops

In order to simulate the real working condition of power transformers in the DC-biasing experiment, a direct current source is connected with the alternating voltage source in series, to supply the DC bias current I_{dc}. In this case, the total flux φ in (5-64) includes φ_{ac} and φ_{dc}, which represents AC and DC magnetic flux in the laminated core, respectively.

From Equations (5-64), (5-65) and (5-66), it is indicated that the hysteresis model to simulate DC-biasing hysteresis loops can be presented as follows:

$$\begin{cases} B = B_{acm} \cdot \sin(\omega t) + B_{dc} \\ H = H_{1d}(B) + H_{obd} \cdot \cos(\omega t) \end{cases} \tag{5-67}$$

where H_{1d} corresponds to the locus of mid-points of the DC-biasing hysteresis loop, and H_{obd} is the measured consuming coefficient, based on the experimental data.

φ_{ac} can be obtained from Equation (5-64), while φ_{dc} cannot be obtained directly from measurement; hence, the simple iterative method (SIM) (based on the basic magnetizing curve) has been proposed to estimate φ_{dc} iteratively [43]. The calculated I_{dc} (DC component of the magnetizing current i) and I_m (magnitude of the magnetizing current i) in each iterative step are usually compared with the measured results to ensure the accuracy of the assumed φ_{dc}. However, it is difficult to obtain accurate φ_{dc} by this method, since the computed I_{dc} and I_m cannot approach the measured results at the same time in many cases. Therefore, the modified iterative method (MIM) is presented here to predict φ_{dc} accurately, and the iterative process is as follows:

- firstly, the hysteresis model in Equation (5-62) is set up, based on the hysteresis loops under sinusoidal magnetization, which are then used to calculate i with the assumed φ_{dc} and the measured φ_{ac};
- secondly, φ_{dc} is adjusted by observing the computed I_{dc} and I_m until the accuracy of φ_{dc} is satisfied;
- finally, the DC-biasing hysteresis model is set up, based on the obtained φ_{dc}, and then used to predict new φ_{dc} more accurately following the former two steps.

Table 5.14 Comparison of the two methods ($I_{dc} = 0.426$ A)

	Results by SIM			Results by MIM		
B_{acm}/T	E_m(%)	E_{dc}(%)	B_{dc}/T	E_m(%)	E_{dc}(%)	B_{dc}/T
1.745	10.023	−10.066	0.086	2.931	−2.949	0.068
1.122	−3.766	4.103	0.648	1.224	−1.529	0.641
0.509	−6.932	7.864	1.192	−2.404	2.243	1.196

Table 5.15 Comparison of the two methods ($I_{dc} = 0.847$ A)

	Results by SIM			Results by MIM		
B_{acm}/T	E_m(%)	E_{dc}(%)	B_{dc}/T	E_m(%)	E_{dc}(%)	B_{dc}/T
1.935	24.123	−25.750	0.030	0.095	−0.653	0.023
1.816	5.130	−5.311	0.086	1.187	0.802	0.080
0.403	−5.103	5.721	1.383	1.129	−1.084	1.375

The computed results by the SIM and MIM are compared in Tables 5.14 and 5.15, which correspond to different DC-biased magnetizations. B_{dc} is the DC flux density obtained from the computed φ_{dc}. E_{dc} and E_m are the relative errors of computed I_{dc} and I_m, respectively. The computed and measured waveforms of the magnetizing current are compared in Figure 5.44 to verify the validity and accuracy of the MIM.

The DC-biasing hysteresis loops obtained by the MIM are shown in Figures 5.45(a) and 5.45(b). Therefore, the hysteresis model in Equation (5-67) can be set up to simulate the DC-biasing hysteresis loops. Figure 5.45(c) compares the simulated result with the measured result. It is verified that the consuming hysteresis model can be used to accurately simulate the DC-biasing hysteresis loops.

5.4.4 Hysteresis Model Based on a Neural Network

Neural network theory is imitative of the biological processing model to obtain the function of information intelligent processing. It deals with the pattern information, which is hard to express in a certain language by the bottom to top method, or parallel distribution formed by self-study, self-organization and non-linear dynamics. A neural network is a type of parallel and distributed information-processing network architecture. It is generally composed of massive neurons, each of which has only one output that can connect many other neurons. The reciprocity between neurons is embodied by their connected weights, and the output of neurons is also its input function. Linear function, sigmoid function and value function are common.

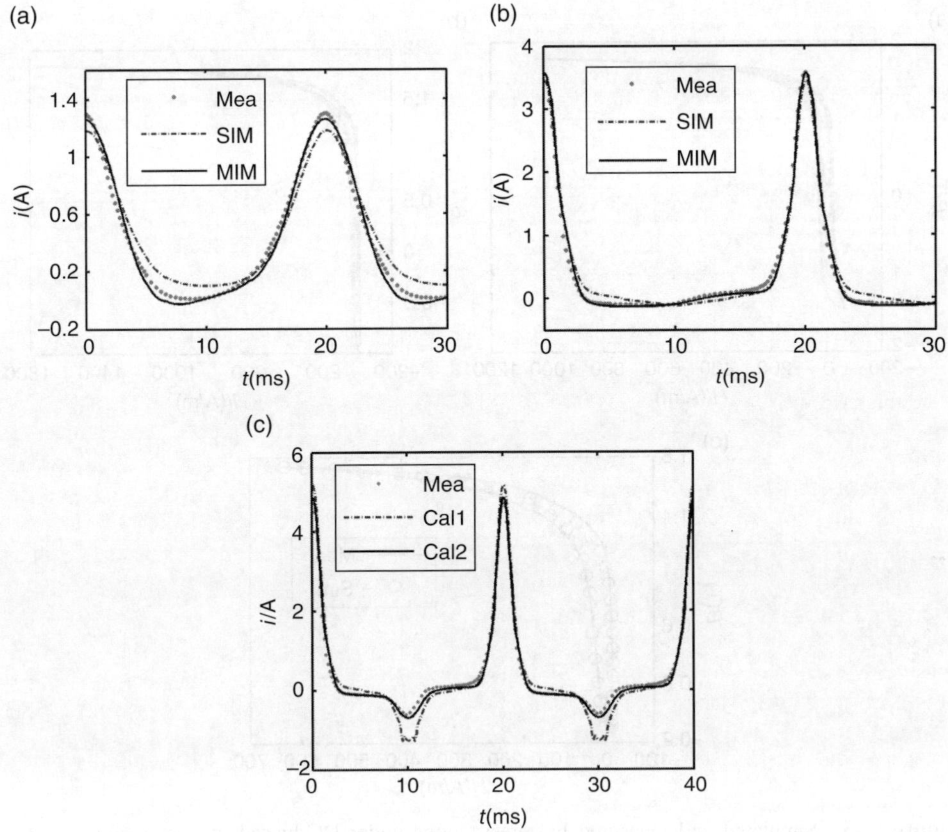

Figure 5.44 Computed and measured magnetizing currents under different DC-biased magnetizations. (a) $U_{ac} = 80$ V, $I_{dc} = 0.425$ A; (b) $U_{ac} = 370$ V, $I_{dc} = 0.425$ A; (c) $U_{ac} = 452$ V, $I_{dc} = 0.425$ A

There are two phases of positive transmission processing and error reverse transmission processing in the study processing of BP neural networks [53]. The signal inputted from outside spreads to the output layer, and provides the result through the processing layer (there is a layer of neurons in the input layer and hidden layer). If the expected output cannot be obtained from the output layer, it shifts to conversed spread processing, and the true value and error output by the network will return along with the former coupled access. The error is reduced by modifying the contacted weight value of the neurons in every layer, so that it shifts to positive spread processing and iteration until the error is smaller than the given value.

n, m, l are used to represent the number of input neurons, hidden neurons and output neurons, respectively, and can be determined according to the following equation:

$$m = \sqrt{n + l} + \alpha,$$

in which α is a constant in the range 1–10.

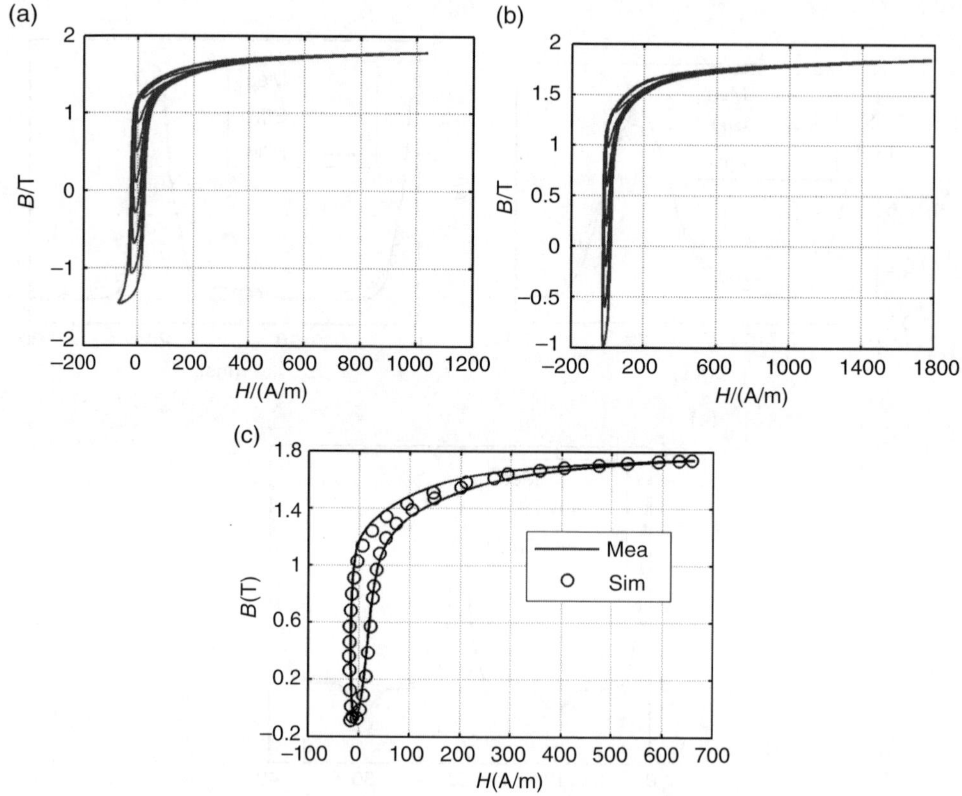

Figure 5.45 Simulated and measured hysteresis loops under DC-biased magnetization. (a) DC-biasing hysteresis loops ($H_{dc} = 107$ A/m); (b) DC-biasing hysteresis loops ($H_{dc} = 213$ A/m); (c) Simulated and measured results

The hyperbolic tangent function is an alternative transfer function, which is shown in Figure 5.46. The training performance of the neural network with different numbers of hidden layers and hidden neurons is shown in Figures 5.47 and 5.48.

5.4.5 Simulation and Validation

The numerical data of hysteresis loops under sinusoidal excitation are trained, and predict nonlinear hysteretic characteristics. Figure 5.49 compares the simulated and measured results of the hysteretic data. Table 5.16 presents the numerical data of hysteresis loops under sinusoidal excitation simulated by the neural network.

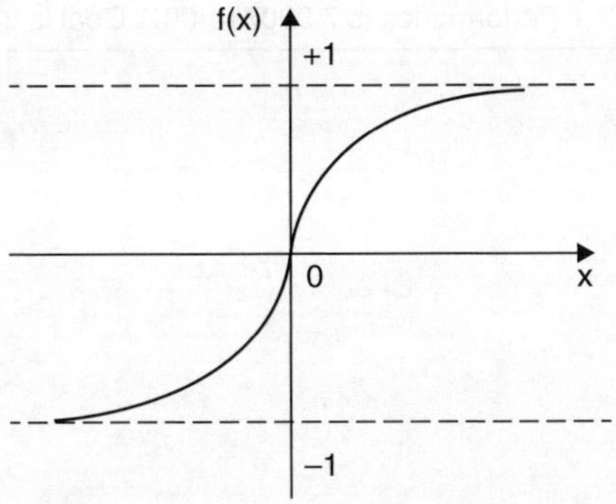

Figure 5.46 Hyperbolic tangent transfer function

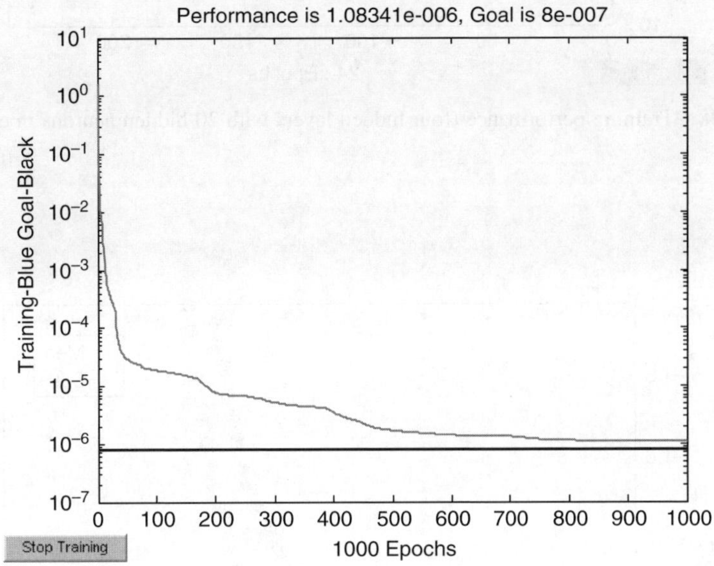

Figure 5.47 Training performance (three hidden layers with ten hidden neurons in each layer)

Figure 5.50 and Figure 5.51 present the simulated and measured hysteresis loops under DC-biased excitations. The numerical data of DC-biasing hysteresis loops are given in Table 5.17 and Table 5.18 respectively.

The hysteresis model, based on the consuming function and neural network, can be used to simulate the hysteresis effects of the ferromagnetic core under sinusoidal and DC-biased magnetizations. It can also be used to compute the hysteretic nonlinear magnetic field combined with the harmonic finite element method.

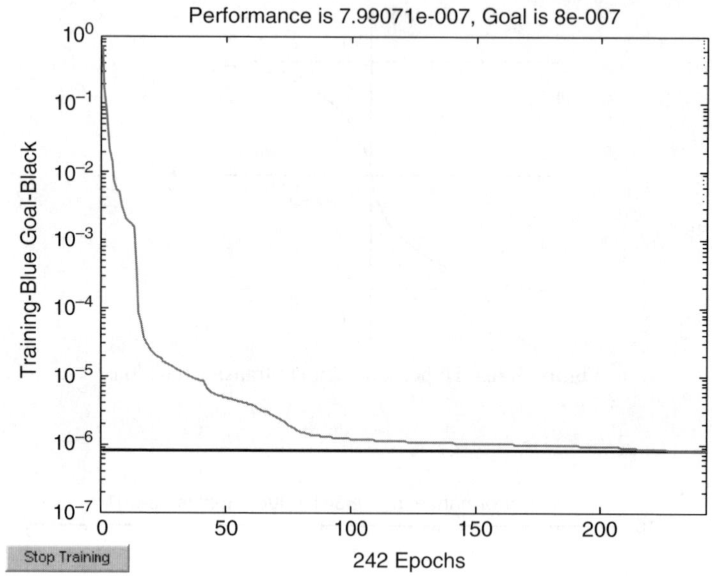

Figure 5.48 Training performance (four hidden layers with 20 hidden neurons in each layer)

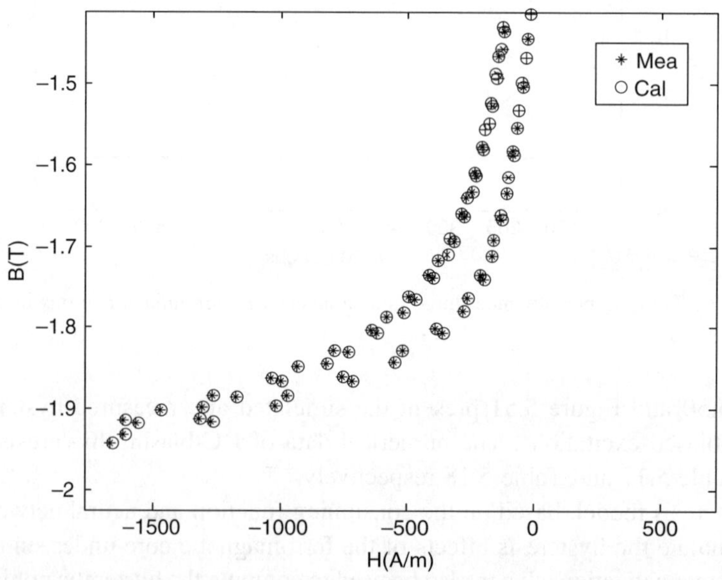

Figure 5.49 Comparison between the measured and simulated results ($I_{dc} = 0$ A)

Table 5.16 Simulated results of hysteresis loops under sinusoidal excitation ($I_{dc} = 0$ A)

Simulated results H(A/m)	Measured results H(A/m)
–3393.96626634747	–3399.41612903225
–2877.69728468093	–2887.76129032258
–2393.99020100784	–2406.01064516129
–1983.18353026803	–1994.61096774193
–1611.12974176649	–1620.9029032258
–1263.59006183157	–1271.99129032258
–969.802243469077	–976.786451612903
–710.240592177631	–716.505483870967
–512.841293537294	–518.12129032258
–357.641122188792	–362.788064516129

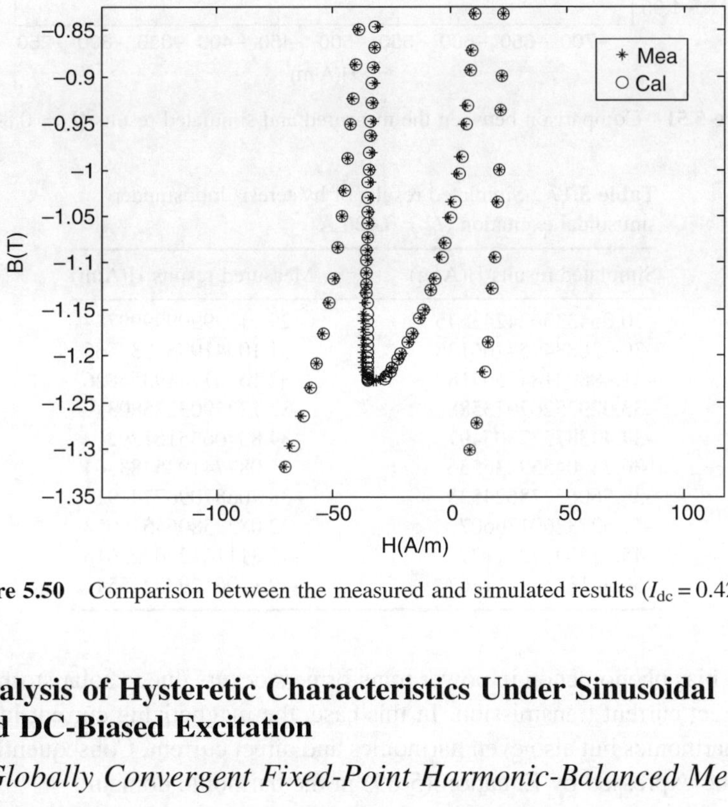

Figure 5.50 Comparison between the measured and simulated results ($I_{dc} = 0.426$ A)

5.5 Analysis of Hysteretic Characteristics Under Sinusoidal and DC-Biased Excitation

5.5.1 Globally Convergent Fixed-Point Harmonic-Balanced Method

As is shown in Equation (5-68), the magnetizing current can be termed as the sum of odd harmonics when the power transformer works under sinusoidal excitation:

$$i = \sum_{n=1,3,5\ldots}^{\infty} [i_{ns}\sin(\omega t) + i_{nc}\cos(\omega t)] \tag{5-68}$$

where n represents the harmonic number.

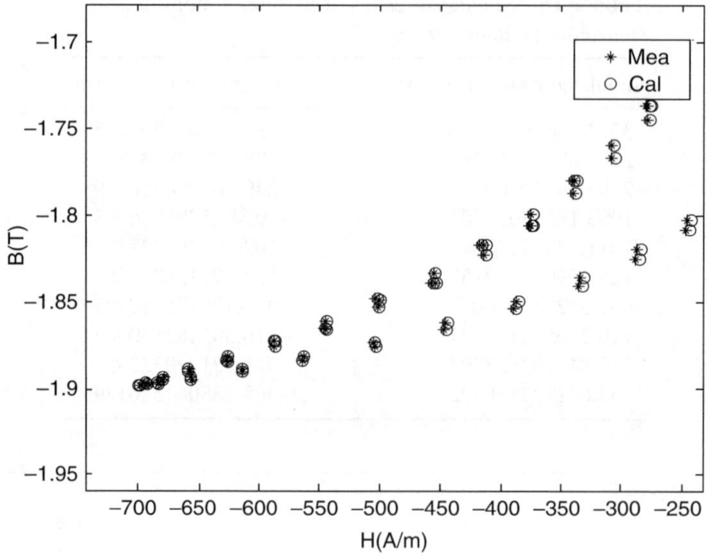

Figure 5.51 Comparison between the measured and simulated results ($I_{dc} = 0.847$ A)

Table 5.17 Simulated results of hysteresis loops under sinusoidal excitation ($I_{dc} = 0.426$ A)

Simulated results H(A/m)	Measured results H(A/m)
–30.064333624243545	–29.912999999999784
–30.926884954766138	–31.109419354838565
–31.888931847504182	–32.163677419354826
–33.02925267913588	–32.7272903225803
–34.44387577305497	–34.8370645161292
–36.24604559236536	–37.087741935483564
–38.56065278624533	–38.70687096774191
–41.5273260176607	–42.082258064516054
–45.3370977251127	–45.811161290322616
–50.34138167711467	–49.82061290322554

The DC bias phenomenon in power transformers occurs due to solar storms or high-voltage direct current transmission. In this case, the magnetizing current includes not only odd harmonics but also even harmonics and direct current. Consequently, the current i can be expressed by Equation (5-69) in the harmonic domain:

$$\boldsymbol{I} = \left[i_0 \;\; i_{1s} \;\; i_{1c} \;\; i_{2s} \;\; i_{2c} \;\; \cdots \right] \tag{5-69}$$

According to the fixed point theorem, the fixed-point reluctivity ν_{FP} is introduced to overcome the discontinuity of traditional reluctivity in the numerical computation of the

Table 5.18 Simulated results of hysteresis loops under sinusoidal excitation ($I_{dc} = 0.847$ A)

Simulated results H(A/m)	Measured results H(A/m)
713.775907309398	711.422903225806
884.992348113111	883.35
1104.40012666554	1103.41064516129
1345.18705763059	1344.5564516129
1633.42085469195	1632.99290322579
1935.53725734772	1935.05419354838
2283.28204724313	2282.58193548386
2630.21964951644	2629.35483870967
3016.73501769168	3015.83225806451
3396.60075913058	3395.89354838709

magnetic field involving hysteresis effects. Therefore, a new relationship in the harmonic domain is presented in the two-dimensional nonlinear magnetic field:

$$\begin{cases} \boldsymbol{H}_x = \nu_{\text{FP}} \boldsymbol{B}_x - \boldsymbol{M}_x \\ \boldsymbol{H}_y = \nu_{\text{FP}} \boldsymbol{B}_y - \boldsymbol{M}_y \end{cases} \tag{5-70}$$

in which a specific parameter is usually assigned to the fixed-point reluctivity ν_{FP}, and expressions of the harmonic vector $\boldsymbol{H}_x$ and $\boldsymbol{H}_y$ are similar to $\boldsymbol{I}$ in Equation (5-69).

The vector potential equation based on ν_{FP} can be used to describe the two-dimensional nonlinear magnetic field:

$$\nabla \times \nu_{\text{FP}}(\nabla \times A) + \sigma(\partial A/\partial t) = J - \nabla \times \boldsymbol{M} \tag{5-71}$$

where $\boldsymbol{M}$ is a magnetization-like quantity and σ is the electric conductivity.

When electromagnetic devices are excited by voltage, the coupling between the k-th external circuits and the magnetic field should be taken into account in the system equations, so that the magnetizing current and magnetic field can be solved simultaneously by:

$$\begin{bmatrix} \boldsymbol{Q} & \boldsymbol{G}_k \\ \boldsymbol{C}_k & \boldsymbol{Z}_k \end{bmatrix} \begin{bmatrix} \boldsymbol{A} \\ \boldsymbol{I}_k \end{bmatrix} = \begin{bmatrix} \boldsymbol{P} \\ \boldsymbol{U}_{ink} \end{bmatrix} \tag{5-72}$$

where $\boldsymbol{U}_{in}$ and $\boldsymbol{I}$ are the harmonic vectors of external input voltage and magnetizing current, $\boldsymbol{C}$ and $\boldsymbol{Z}$ represent the coupling matrix and impedance matrix, respectively, $\boldsymbol{P}$ is derived from $\boldsymbol{M}$, while $\boldsymbol{Q}$ is related to hysteretic nonlinearity and eddy current.

Harmonic solutions in Equation (5-72) can be computed iteratively, according to the following procedure (p represents the iterative step): $[\boldsymbol{A}(p), \boldsymbol{I}_k(p)] \rightarrow [\boldsymbol{B}_x(p), \boldsymbol{B}_y(p)] \rightarrow$

$[\boldsymbol{H}_x(p), \boldsymbol{H}_y(p)] \rightarrow \boldsymbol{P}(p) \rightarrow [\boldsymbol{A}(p+1), \boldsymbol{I}_k(p+1)]$. If the anisotropy of the iron core is neglected, the magnetic field intensity can be computed, based on the scalar hysteresis model and the following relationship:

$$\begin{cases} H_x \cdot B = B_x \cdot H \\ H_y \cdot B = B_y \cdot H \end{cases} \tag{5-73}$$

The convergence of harmonic solutions of magnetizing current and magnetic vector potential depends on the strategy to determine ν_{FP} in the iterative process. However, the proposed method in time-stepping finite element analysis is not applicable in harmonic computation. An alternative scheme is recommended as follows [54]: ν_{FP} is initially set to be $3\nu_0 \sim \nu_0$, and is reset in the range $\nu_0/10 \sim \nu_0/40$ after a few initial iterations. When the convergence criterion ε in Equation (5-74) is small enough, the iteration is stopped.

$$\varepsilon = \|\boldsymbol{M}(p+1) - \boldsymbol{M}(p)\| / \|\boldsymbol{M}(p)\| \tag{5-74}$$

5.5.2 *Hysteretic Characteristic Analysis of the Laminated Core*

The laminated core is computed by the fixed-point HBFEM considering hysteresis effects under different types of magnetizations. Figure 5.52 presents the geometric model of the laminated core for computation.

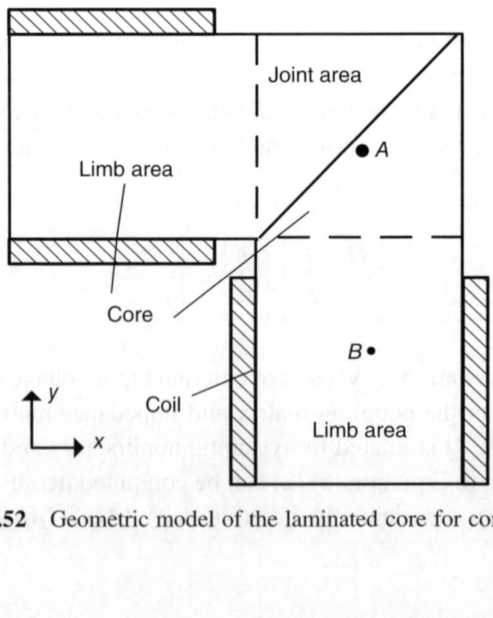

Figure 5.52 Geometric model of the laminated core for computation

The calculated and experimental results are compared in Figures 5.53 and 5.54. It is apparent that the computed magnetizing currents coincide well with the measured ones when the hysteresis effects of the laminated core are taken into account.

The flux density $\boldsymbol{B}$ and the magnetic field intensity $\boldsymbol{H}$ in each finite element is obtained from calculated harmonic vector potential $\boldsymbol{A}$. Two points (A and B) shown in Figure 5.52, are selected to observe the hysteretic characteristics in different regions of the laminated core. The computational hysteresis loops are presented in Figure 5.55-Figure 5.58.

It can be seen that the hysteretic characteristics in the limb and joint areas are different from each other. The DC bias current leads to the appearance of the DC magnetic flux in the laminated core, which makes the hysteresis loops asymmetrical. In addition, hysteretic characteristics on point A exhibit different features in the x-direction and y-direction.

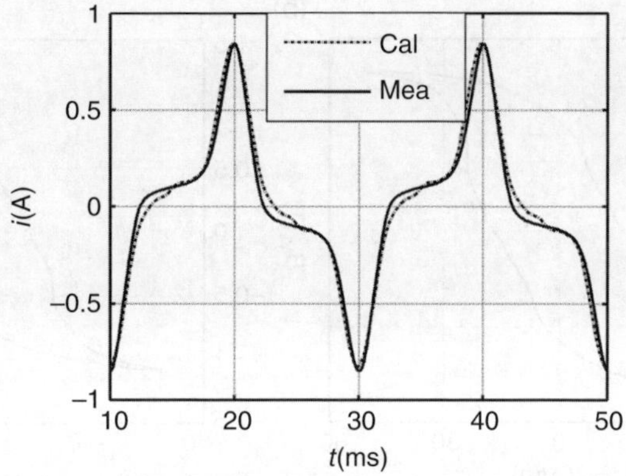

Figure 5.53 Magnetizing current under sinusoidal magnetization ($U_{ac} = 420$ V, $I_{dc} = 0$ A)

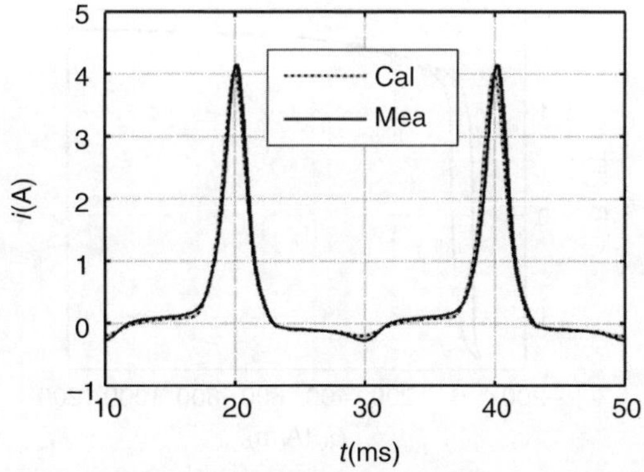

Figure 5.54 Magnetizing current under DC-biased magnetization ($U_{ac} = 420$ V, $I_{dc} = 0.426$ A)

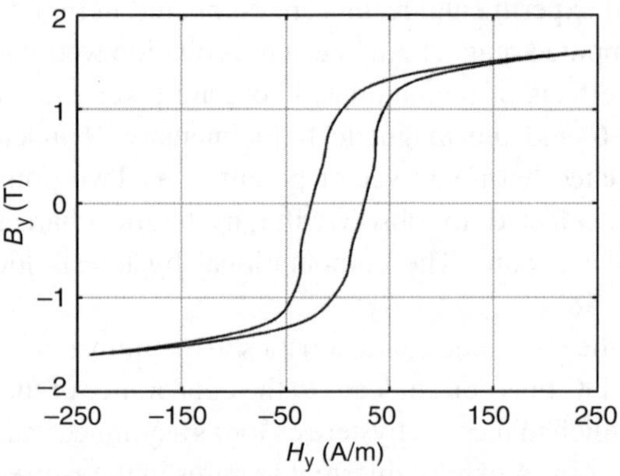

Figure 5.55 H_y-B_y hysteresis loop on point B under sinusoidal magnetization (U_{ac} = 420 V, I_{dc} = 0 A)

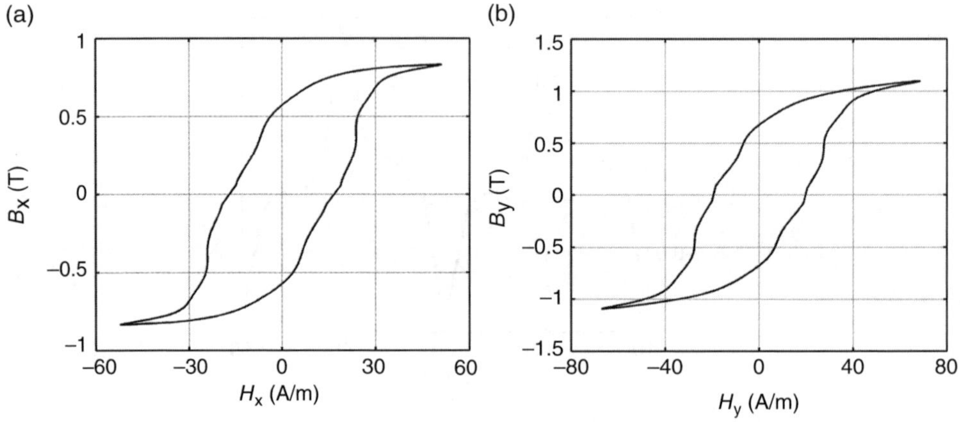

Figure 5.56 Hysteresis loop on point A under sinusoidal magnetization (U_{ac} = 420 V, I_{dc} = 0 A). (a) H_x-B_x hysteresis loop. (b) H_y-B_y hysteresis loop

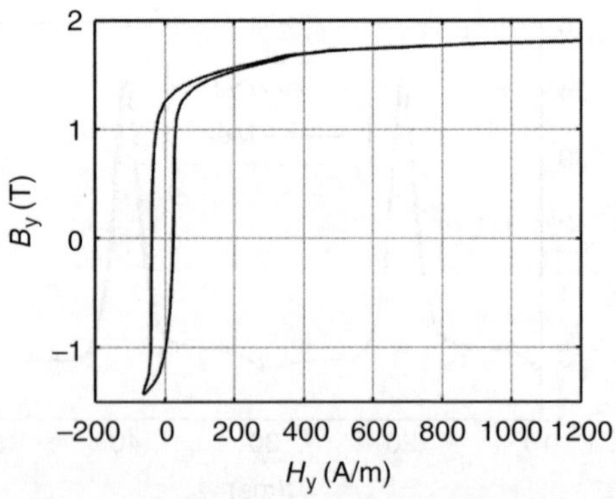

Figure 5.57 H_y-B_y hysteresis loop on point B under DC-biased magnetization (U_{ac} = 420 V, I_{dc} = 0.426 A)

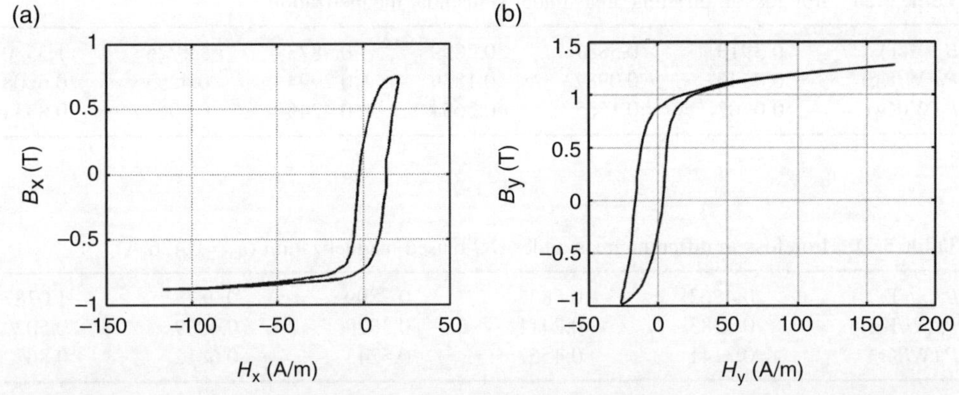

Figure 5.58 Hysteresis loop on point A under DC-biased excitation ($U_{ac} = 420$ V, $I_{dc} = 0.426$ A). (a) H_x-B_x hysteresis loop; (b) H_y-B_y hysteresis loop

Table 5.19 Iron loss under sinusoidal magnetization

B_{acm}/(T)	0.3913	0.5887	0.7899	0.9873	1.1775	1.3530
P_c/(W/Kg)	0.0614	0.1320	0.2364	0.3796	0.5501	0.7716
P_m(W/Kg)	0.0624	0.1381	0.2453	0.3785	0.5548	0.7919

With the computed $\boldsymbol{B}$ and $\boldsymbol{H}$ in each finite element, the iron loss can be calculated by the following equation [55, 56]:

$$P_t = \rho \cdot d_s \cdot N_s \cdot \sum_{i=1}^{N_e} P_i \cdot S_i \tag{5-75}$$

where P_t is the total loss of the laminated core, ρ is the density of the core material, d_s is the thickness of the electrical steel sheet, N_s is the total number of laminated sheets, S_i is the area of the i-th finite element, N_e is the number of finite elements in the computational region, and P_i is the iron loss in the i-th finite element. This is computed by the integral of the scalar product between $\boldsymbol{B}$ and $\boldsymbol{H}$ with respect to time in one period T.

The computed specific loss of the laminated core (P_c) is compared with the measured data (P_m) in Table 5.19. The computed results coincide well with the measured results.

From the calculated magnetic field on points A and B, it can be seen that flux distributions in the joint and limb areas are not the same. The non-uniformity of flux distribution may lead to the difference between the iron loss in the joint area and that in the limb area. Tables 5.20 and 5.21 present the calculated iron loss corresponding to the sinusoidal magnetization and DC-biased magnetization, respectively. P_j and P_l are the specific loss in the joint area and limb area, respectively. It is obvious that the

Table 5.20 Iron loss in different areas under sinusoidal magnetization

B_{acm}/(T)	0.3913	0.5887	0.7899	0.9873	1.1775	1.3530
P_j/(W/Kg)	0.0445	0.0987	0.1839	0.2994	0.4253	0.6208
P_l(W/Kg)	0.0707	0.1505	0.2654	0.4245	0.6063	0.8334

Table 5.21 Iron loss in different areas under DC-biased magnetization ($I_{dc} = 0.426$ A)

B_{acm}/(T)	0.4863	0.6838	0.7899	0.9873	1.0787
P_j/(W/Kg)	0.1383	0.2411	0.3029	0.4069	0.5020
P_l(W/Kg)	0.3141	0.4557	0.5245	0.6812	0.8632

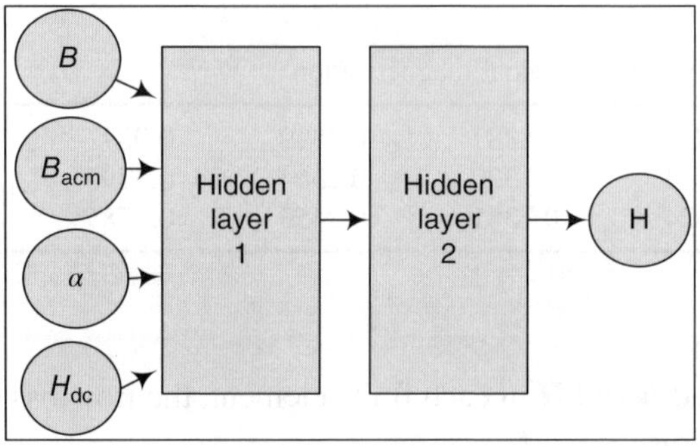

Figure 5.59 The NN DC-biasing hysteresis model in the limb-yoke region

specific loss in the limb area is larger than that in the joint area. In addition, existence of the DC magnetic flux gives rise to the significant increase in the total iron loss of the laminated core.

5.5.3 *Computation of the Nonlinear Magnetic Field Based on the Combination of the Two Hysteresis Models*

The hysteresis effects can be modeled by different methods, based on the neural network. Figure 5.59 shows the four-input and single-output NN, which is trained by the experimental data (B and H) based on the Levenberg-Marquard training algorithm. The hidden layers both have 30 neurons. α reflects the ascending or descending phase of the hysteresis loop, and B_{acm} is the magnitude of B_{ac}. The DC component of the magnetic field intensity H_{dc} can be calculated with the given direct current I_{dc} in the exciting coil.

Sixty DC-biasing hysteresis loops, corresponding to four different DC-biased magnetizations, are obtained in the experiment. All the datasets are randomly divided into 60 sets, of which 58 sets are employed to train the NN hysteresis model in Figure 5.59. The other two are used to verify the accuracy of the results predicted by the NN hysteresis model.

Figure 5.60 shows that the sample error of the trained NN is found to be within the range -8×10^{-7} to 8×10^{-7}. The measured and predicted results under different DC-biased magnetizations are compared to verify the accuracy of the DC-biasing hysteresis model based on NN. Figure 5.61 depicts the DC-biasing hysteresis loops in accordance

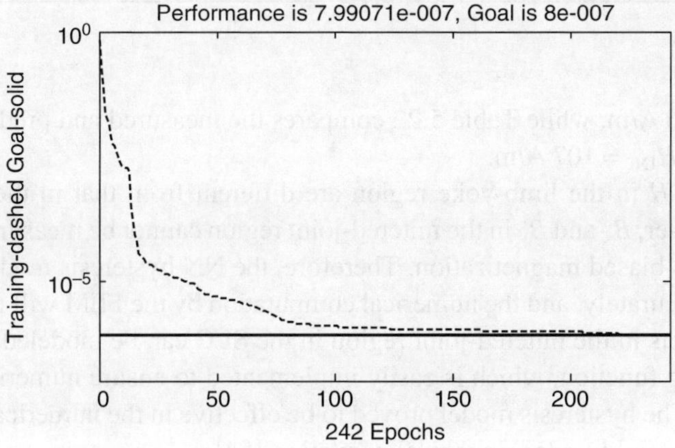

Figure 5.60 The training performance based on NN

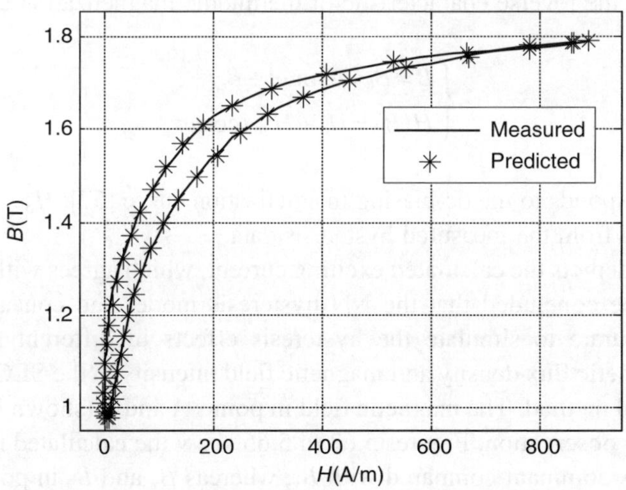

Figure 5.61 The predicted and measured DC-biasing hysteresis loops ($H_{dc} = 213$ A/m, $B_{acm} = 0.5065$ T)

Table 5.22 Comparison between measured and predicted results

B(T)	B_{acm}(T)	α	H_{dc}(A/m)	H(A/m)	
				measured	predicted
–1.8824	1.8924	–1	107	–686.8030	–687.8620
0.0159	1.8932	1	107	29.3179	29.8082
–0.2837	1.5222	–1	107	–26.6458	–26.5633
0.8016	0.8150	1	107	23.2425	23.5525
1.6539	0.3072	1	107	307.9868	308.2446
1.7659	0.8153	–1	107	604.2484	599.2828
1.9005	1.8932	1	107	2035.4480	2028.6230

with $H_{DC} = 213$ A/m, while Table 5.22 compares the measured and predicted data corresponding to $H_{DC} = 107$ A/m.

Both B and H in the limb-yoke region are different from that in the mitered-joint region. Moreover, B_x and B_y in the mitered-joint region cannot be measured experimentally under DC-biased magnetization. Therefore, the NN hysteresis model may predict H_x and H_y inaccurately, and the numerical computation by the FEM will fail. In fact, the hysteresis effects in the mitered-joint region in the SLC can be modeled by the simplified consuming function, which is easily implemented to ensure numerical stability in computation. The hysteresis model proved to be effective in the numerical computation of the magnetic field under sinusoidal excitation [54].

The distance between the periphery of the hysteresis loop and the middle magnetization curve represents hysteresis loss. The loss part is defined as the consuming function, which includes the reverse characteristic of the middle magnetization curve:

$$\begin{cases} B = B_{acm} \cdot \sin\omega t + B_{dc} \\ H(B) = H_c + H_{ob} \cdot \cos\omega t \end{cases} \tag{5-76}$$

where H_c corresponds to the dc-biasing magnetization curve [57]. H_{ob} is related to B_m and is collected from the measured hysteresis data.

Figure 5.62 depicts the calculated exciting current, which agrees with the measured result. It can be concluded that the NN hysteresis model and consuming-function model are accurate to simulate the hysteresis effects in different regions in the SLC. The magnetic flux density and magnetic field intensity in the SLC are computed by the proposed method. The magnetic field in points A and B (shown in Figure 5.52) are selected for observation. Figures 5.63 to 5.65 show the calculated magnetic field. B_y in point B is dominant compared with B_x, whereas B_x and B_y in point A are comparable. Moreover, hysteresis loops in point A and B are calculated based on different

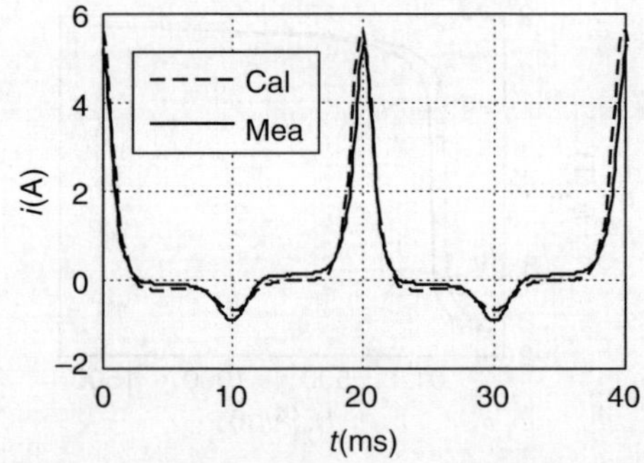

Figure 5.62 Comparison of exciting current between measurement and calculation (H_{dc} = 107 A/m, U_{ac} = 470 V)

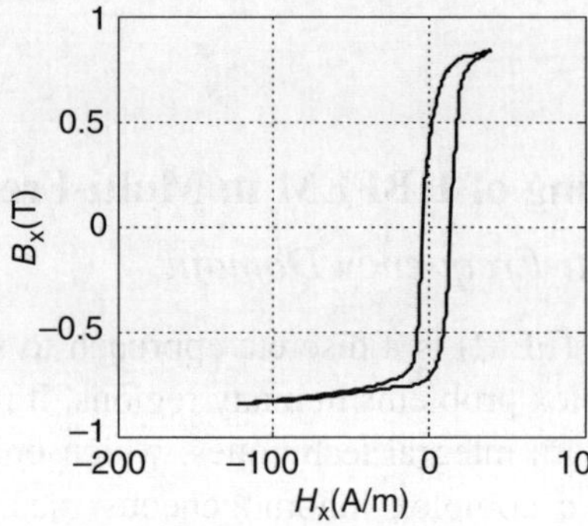

Figure 5.63 The calculated hysteresis loops in point *A* (H_{dc} = 107 A/m, U_{ac} = 470 V)

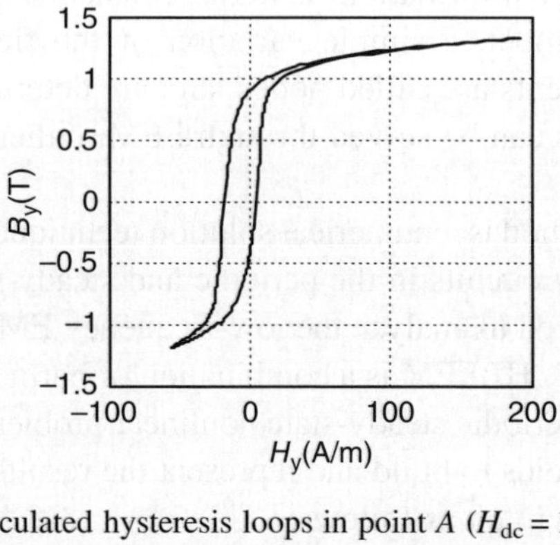

Figure 5.64 The calculated hysteresis loops in point *A* (H_{dc} = 107 A/m, U_{ac} = 470 V)

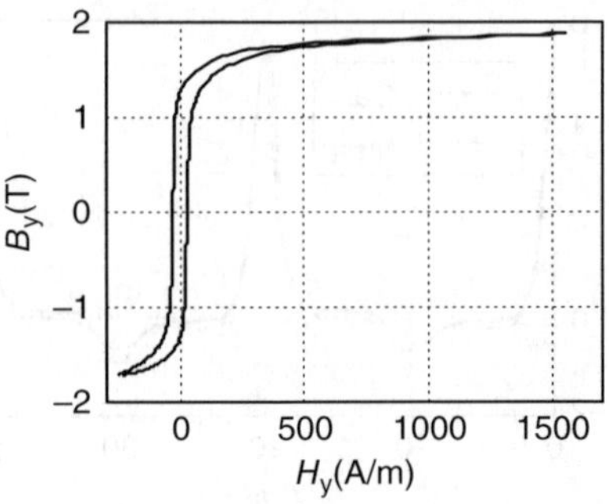

Figure 5.65 The calculated hysteresis loops in point B ($H_{dc} = 107$ A/m, $U_{ac} = 470$ V)

hysteresis models. Therefore, hysteretic characteristics in the two regions are different from one another.

5.6 Parallel Computing of HBFEM in Multi-Frequency Domain

5.6.1 HBFEM in Multi-Frequency Domain

The finite element method (FEM) is a discrete approach to solve the partial different equations (PDEs) for complex problems in many regions. It requires the entire volume to be meshed, compared with integral techniques, which only need the surface to be gridded. FEM can model a complex inhomogeneous medium where each element may have completely different material properties from neighbor elements. FEM discretizes the computing domain into a group of small sub-regions with boundary conditions.

In computational electromagnetics (CEMs), a typical electromagnetic model contains the device geometry information, material constants, excitation, and boundary constraints. In each element, a simple variation of the field quantity is assumed. The vertices of the elements are called nodes, and are determined after the FEM calculation. CEM problems can be solved through FEM, either in frequency domain or time domain.

Harmonic balance method is a numerical solution technique which has been used for nonlinear high-frequency circuits in the periodic and steady-state regime for two decades. It was first introduced to analyze the low-frequency EM field problem, as briefly introduced in Chapter 1.3. HBFEM is a combination of harmonic balance and FEM, to numerically solve time-periodic steady-state nonlinear problems. HBFEM uses the linear combination of sinusoids to build and represent the results. A generalized HBFEM matrix equation can be written as follows:

$$G^e = \frac{1}{4\Delta^e}\begin{bmatrix}(b_1b_1+c_1c_1)D & (b_1b_2+c_1c_2)D & (b_1b_3+c_1c_3)D\\(b_2b_1+c_2c_1)D & (b_2b_2+c_2c_2)D & (b_2b_3+c_2c_3)D\\(b_3b_1+c_3c_1)D & (b_3b_2+c_3c_2)D & (b_3b_3+c_3c_3)D\end{bmatrix}\begin{Bmatrix}A_1^e\\A_2^e\\A_3^e\end{Bmatrix}$$

$$+\frac{\sigma\omega\Delta^e}{12}\begin{bmatrix}2N & N & N\\N & 2N & N\\N & N & 2N\end{bmatrix}\begin{Bmatrix}A_1^e\\A_2^e\\A_3^e\end{Bmatrix}-\begin{Bmatrix}K_1^e\\K_2^e\\K_3^e\end{Bmatrix}=[S^e]\{A^e\}+[N^e][A^e]-\{K^e\} \tag{5-77}$$

The coefficients of the matrix D are determined by only the Fourier coefficients. Matrix D acts as a reluctivity, and is called the Reluctivity Matrix. On the other hand, matrix N is a constant concerned with harmonic orders, and is called the Harmonic Matrix.

The system matrix equation for current source excitation can then be written in a compact form:

$$[S]\{A\} + [M]\{A\} - \{K\} = 0 \tag{5-78}$$

where $[S]$ is the system matrix and $[M]$ is the harmonic-related matrix. All harmonic components of magnetic vector potential A can be directly obtained by solving this system matrix equation.

A significant difference between the HBFEM and conventional FEM is the size of the matrices and vectors, as shown in Figure 5.66. The conventional simulation vector contains one value for each node, while there are K values in one vector for each node in HB – that is, $K \times K$ diagonal sub-matrix for each node in the stiffness matrices.

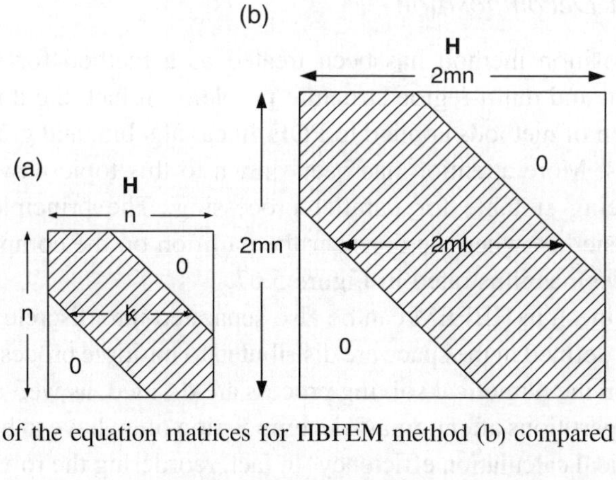

Figure 5.66 Size of the equation matrices for HBFEM method (b) compared with the traditional FEM (a)

The harmonic matrix can include DC, as well as the even and odd harmonics, depending on the application problems.

Suppose there are M harmonics considered in the application, then the size of harmonic balance matrix is $2M$ larger than that of the traditional FEM. The ratio of matrix size between HBFEM FEM is $2M : 1$.

5.6.2 Parallel Computing of HBFEM

The simulations of physical process trend to be computational-intensive, as more natural phenomena are considered. Large models take up more memory space, and solving sparse matrices equations always consume a lot of time to get accurate results. From the Krylov subspace iterative method to algebra multi-grid [58], the advanced numerical method leads to difficulty in modeling and solving the equations on the computer.

To facilitate the modern computer architecture, it is natural to divide the huge problem into a collection of small pieces, and the resulting discrete jobs can be mapped on multi-processing elements (CPUs, cores). There are two types of parallel computer on the market, according to Flynn's taxonomy: SIMD (most modern CPUs, GPU) and MIMD (multiple-processors SMP, distributed memory clusters, etc.).

The efficiency of the parallel computing relies on many factors. Generally, the overhead of communication and synchronization, granularity and load balance are points that have to be considered when designing a parallel program. The parallel computing model consists of shared memory models (OpenMP [59], POSIX threads, etc.), distributed memory/message passing models (MPI) [60] and hybrid models such as CPU-GPU CUDA [61,62], MPI-OpenMP [63], MPI-CUDA [64,65], MPI-MIC [66–68] (Intel Xeon Phi), and so on.

5.6.3 Domain Decomposition

Domain decomposition method has been treated as a method for solving the PDE boundary problem and multi-region structure problem. In fact, the domain decomposition is a collection of methods applied to PDE, linear algebra, and graph partition after decades of efforts. More attention has been drawn to this topic nowadays, due to its advantages of being suitable for parallel processing. The principle of divide-and-conquer is the basic law when we consider the partition on the computational domain (mesh and matrices), as illustrated in Figure 5.67.

The triangular mesh in HBFEM can be also separated into discrete sub-meshes, and then the elements defined in the space are distributed to multiple processors accordingly. Matrices arising in the numerical solving process are divided, as well as the operations. The independent partitions after domain decomposition often have to be transformed for reasons of numerical calculation efficiency. In fact, reordering the rows and columns of a sparse matrix is one of the most important operations in parallel implementation, both

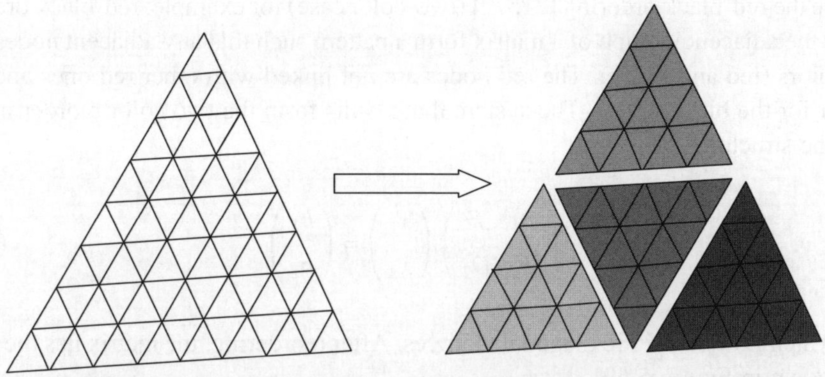

Figure 5.67 Domain decomposition approach for the FEM mesh

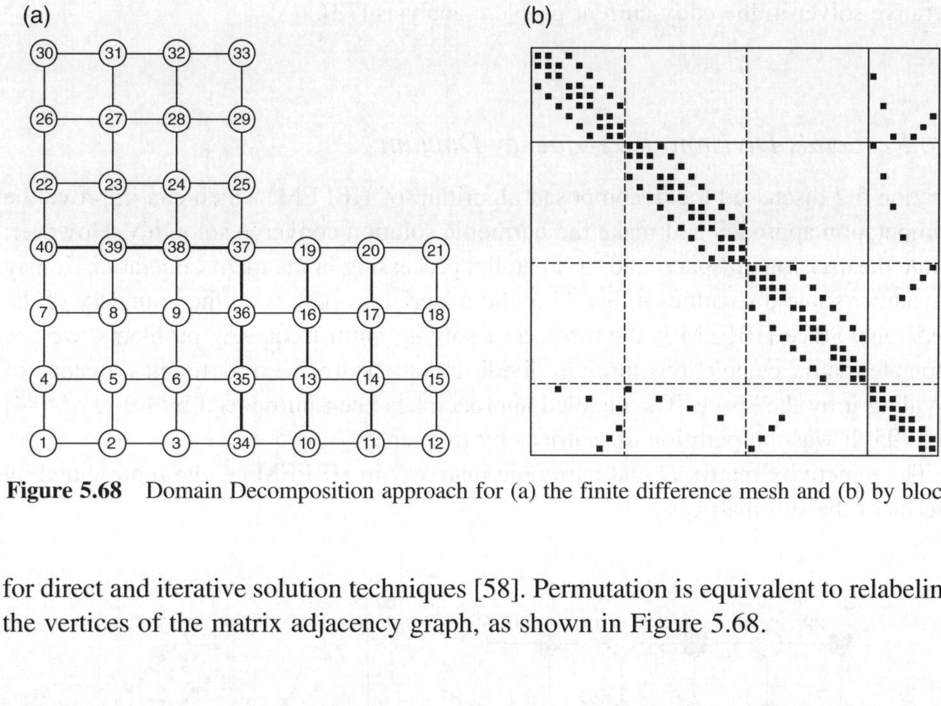

Figure 5.68 Domain Decomposition approach for (a) the finite difference mesh and (b) by block

for direct and iterative solution techniques [58]. Permutation is equivalent to relabeling the vertices of the matrix adjacency graph, as shown in Figure 5.68.

5.6.4 Reordering and Multi-Coloring

Sparse iterative methods, such as the conjugate gradient and multi-grid, are frequently used in electromagnetic simulation because of their efficiency and flexibility in many cases [69, 70]. The multi-coloring method groups the graph nodes of the mesh into subsets with different colors. It was used to improve the parallelism in iterative methods for a linear system.

Take the red-black ordering [71, 72] (two-color case) for example: red-black ordering makes the adjacency graph of a matrix form a pattern such that any adjacent nodes have two colors (red and black). The red nodes are not linked with other red ones and, it is similar for the black nodes. The system that results from the two color reordering will have the structure:

$$\begin{pmatrix} D_1 & F \\ E & D_2 \end{pmatrix} \begin{pmatrix} x_1 \\ x_2 \end{pmatrix} = \begin{pmatrix} b_1 \\ b_2 \end{pmatrix} \tag{5-79}$$

In which D1 and D2 are diagonal matrices. After reordering, the matrix has the structure shown in Figure 5.69.

A standard SSOR or ILU(0) preconditioner can be applied to the red-black reordered block matrices. The preconditioning operations can be highly parallel. The parallel algebra multi-grid preconditioning with red-black ordering was implemented as the core iterative solver in the eddy-current problem analysis [73].

5.6.5 *Loads Division in Frequency Domain*

Section 5.2 discussed the decomposed algorithm of HBFEM, which can improve the computation approach and make the harmonic solution converge smoothly. However, while the division on space leads to t parallel processing in the matrix operation, it may encounter some difficulties in handling the material properties on the boundary of the divisions. Since HBFEM is the method of solving multi-frequency problems, we can separate matrix calculations through dividing matrices by each harmonics, instead of dividing it by the space. The detailed approach has been introduced by Lu *et al.* [74] in 1995. It was the partition of matrices by frequency.

The reluctivity matrix D and harmonic matrix N in HBFEM can be represented by means of the sub-matrices:

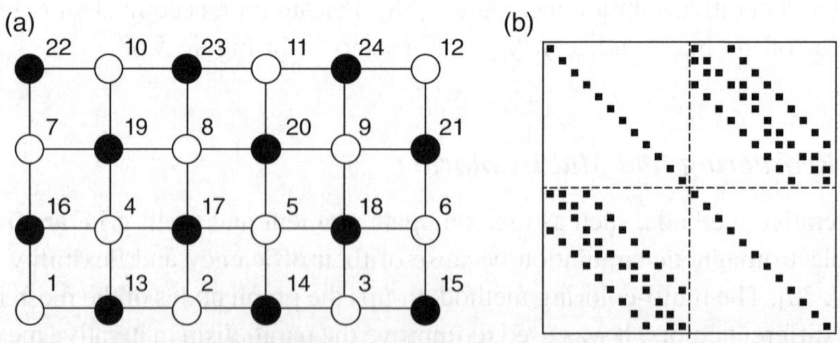

Figure 5.69 A sample of (a) Red-black ordering of a 6 × 4 grid and (b) Associated matrix

$$D = \begin{bmatrix} D_{11} & D_{13} & D_{15} & \cdots \\ D_{31} & D_{33} & D_{35} & \cdots \\ D_{51} & D_{53} & D_{55} & \cdots \\ \cdots & \cdots & \cdots & \cdots \end{bmatrix} \tag{5-80}$$

$$N = \begin{bmatrix} N_1 & 0 & 0 & \cdots \\ 0 & N_3 & 0 & \cdots \\ 0 & 0 & N_5 & \cdots \\ \cdots & \cdots & \cdots & \cdots \end{bmatrix} \tag{5-81}$$

where

$$D_{hh} = \frac{1}{2}\begin{bmatrix} 2\nu_0 - \nu_{2hc} & \nu_{2hs} \\ \nu_{2hs} & 2\nu_0 + \nu_{2hc} \end{bmatrix} \tag{5-82}$$

$$D_{h(h+2p)} = \frac{1}{2}\begin{bmatrix} \nu_{2pc} - \nu_{2(h+p)c} & -\nu_{2ps} + \nu_{2(h+p)s} \\ \nu_{2ps} + \nu_{2(h+p)S} & \nu_{2pc} + \nu_{2(h+p)c} \end{bmatrix} \tag{5-83}$$

$$N_k = \begin{bmatrix} 0 & -h \\ h & 0 \end{bmatrix} = h\begin{bmatrix} 0 & -1 \\ 1 & 0 \end{bmatrix} \tag{5-84}$$

$h = 1, 3, 5 \ldots$
$p = 1, 2, 3 \ldots$

The vectors $\{A_k^e\}$ and $\{K_k^e\}$ can be rearranged according to harmonics as follows:

$$\{A_k\}^e = \{A_{1s}{}^1 A_{1c}{}^1 A_{1s}{}^2 A_{1c}{}^2 A_{1s}{}^3 A_{1c}{}^3, A_{3s}{}^1 A_{3c}{}^1 A_{3s}{}^2 A_{3c}{}^2 A_{3s}{}^3 A_{3c}{}^3 \ldots\}^T = \begin{Bmatrix} A_1 \\ A_3 \\ A_5 \\ \vdots \end{Bmatrix} = \{A_h\}^e \tag{5-85}$$

$$\{K_k\}^e = \{K_{1s}K_{1c}K_{1s}K_{1c}K_{1s}K_{1c}, K_{3s}K_{3c}K_{3s}K_{3c}K_{3s}K_{3c}\ldots\}^T = \begin{Bmatrix} K_1 \\ K_3 \\ K_5 \\ \vdots \end{Bmatrix} = \{K_h\}^e \tag{5-86}$$

Then the matrix equation representing in harmonics is obtained:

$$\{H\}^e\{A_h\}^e = \{K_h\}^e \tag{5-87}$$

where

$$[H]^e = \begin{bmatrix} [H_{11}]^e & [H_{13}]^e & [H_{15}]^e & \cdots \\ & [H_{33}]^e & [H_{35}]^e & \cdots \\ & Symmetry & [H_{55}]^e & \cdots \\ & & & \cdots \end{bmatrix} \tag{5-88}$$

$$\begin{aligned}[H_{hh}]^e = \frac{1}{4\Delta^e}&\begin{bmatrix} (c_1c_1+d_1d_1)D_{hh} & (c_1c_2+d_1d_2)D_{hh} & (c_1c_3+d_1d_3)D_{hh} \\ (c_2c_1+d_2d_1)D_{hh} & (c_2c_2+d_2d_2)D_{hh} & (c_2c_3+d_2d_3)D_{hh} \\ (c_3c_1+d_3d_1)D_{hh} & (c_3c_2+d_3d_2)D_{hh} & (c_3c_3+d_3d_3)D_{hh} \end{bmatrix} \\ &+ \frac{\sigma\omega\Delta^e}{12}\begin{bmatrix} 2N_h & N_h & N_h \\ N_h & 2N_h & N_h \\ N_h & N_h & 2N_h \end{bmatrix}\end{aligned} \tag{5-89}$$

$$\left[H_{h(h+2p)}\right]^e = \frac{1}{4\Delta^e}\begin{bmatrix} (c_1c_1+d_1d_1)D_{h(h+2p)} & (c_1c_2+d_1d_2)D_{h(h+2p)} & (c_1c_3+d_1d_3)D_{h(h+2p)} \\ (c_2c_1+d_2d_1)D_{h(h+2p)} & (c_2c_2+d_2d_2)D_{h(h+2p)} & (c_2c_3+d_2d_3)D_{h(h+2p)} \\ (c_3c_1+d_3d_1)D_{h(h+2p)} & (c_3c_2+d_3d_2)D_{h(h+2p)} & (c_3c_3+d_3d_3)D_{h(h+2p)} \end{bmatrix} \tag{5-90}$$

By using the Gauss iteration method, the system equations for the fundamental component and the h-th harmonics ($h = 3, 5, 7, \ldots$) can be obtained as follows:

$$\begin{aligned} &[H_{11}]\{A_1\}^k = \{G_1\}^k \\ &[H_{33}]\{A_3\}^k = \{G_3\}^k \\ &\cdots \\ &[H_{hh}]\{A_h\}k = \{G_h\}^k \end{aligned} \tag{5-91}$$

In this method, the decomposition of the problem is based on harmonic components division instead of space region decomposition. A shared memory execution environment can be utilized in the frequency parallel calculation. Each computation unit has the same space structure for the finite elements, the same number of unknown variables and the same structure of the matrices. The maximum parallelism is M (numbers of

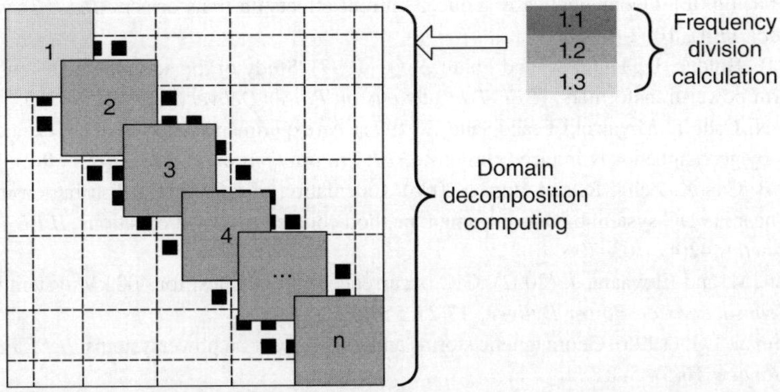

Figure 5.70 Domain decomposition and frequency domain divided for the system matrix

harmonics), and the maximum speed-up will be M time if M processors are available. This parallel computation model for nonlinear EM field analysis can be easily mapped on distributed-memory MIMD parallel computer systems, or on multiple computers connected by LANs.

5.6.6 Two Layers Hybrid Computing

A more complex design can be proposed by combining domain decomposition and the parallel frequency domain described in section 5.6.5. The harmonic matrices can be divided according to the domain decomposition results in the first layer, and then the sub-matrix is reordered and calculated independently [75]. The matrices are partitioned with the hybrid domain and frequency decomposition shown in Figure 5.70.

The two-layer hybrid scheme has the potential to be mapped on the heterogeneous computing architecture. That is, the top layer on the domain partitions can be distributed among the cluster computing nodes, and the second layer of frequency divisions are calculated through multiple threads on the massive parallel co-processors or multicore processing elements. However, there are still many factors that need to be considered in multi-layer parallel computing, such as the underlying message exchanging between partitioned domains and the global frequency sub-matrix assemble, as well as mapping on the fine-grained parallel execution elements.

References

[1] Bíró, O., Außerhoher, S., Buchgraber, G., Preis, K. and Seitlinger, W. (2007). Prediction of magnetizing current waveform in a single phase power transformer under DC bias. *IET Science, Measurement & Technology* **1**(1), 2–5.

[2] Price, P.R. (2002). Geomagnetically induced current effects on transformers. *IEEE Transactions on Magnetics* **17**(4), 1002–1008.

[3] Picher, P., Bolduc, L., Dutil, A. and Pham, V.Q. (1997). Study of the acceptable DC current limit in core-form power transformers, *IEEE Transactions on Power Delivery* **12**(1), 257–265.

[4] Takasu, N., Oshi, T., Miyawaki, F. and Saito, S. (1994). An experimental analysis of DC excitation of transformers by geomagnetically induced currents. *IEEE Transactions on Power Delivery* **9**(2), 1173–1182.

[5] Zhang, B. Cui, X. Zeng, R. and He, J. (2006). Calculation of DC current distribution in AC power system near HVDC system by using moment method coupled to circuit equations. *IEEE Transactions on Magnetics* **42**(4), 703–706.

[6] Lahtinen, M. and Elovaara, J. (2002). GIC occurrences and GIC test for 400 kV system transformer. *IEEE Transactions on Power Delivery* **17**(2), 555–561.

[7] Kappenman, J.G. (1996). Geomagnetic storms and their impact on powersystems. *IEEE Power Engineering Review* **16**(5), 5–8.

[8] Lu, S. Liu, Y. and Ree, J. (1993). Harmonics generated from a DC biased transformer. *IEEE Transactions on Power Delivery* **8**(2), 725–731.

[9] Bíró, O., Buchgraber, G. Leber, G. and Preis, K. (2008). Prediction of magnetizing current wave-forms in a three-phase power transformer under DC bias, *IEEE Transactions on Magnetics* **44**(6), 1554–1557.

[10] You, Y. Fuchs, E.F. Lin, D. and Barnes, P.R. (1996). Reactive power demand of transformers with DC bias. *IEEE Industry Application Magazine* **2**(4), 45–52.

[11] Albertson, V.D. Bozoki, B., Feero, W.E., Kappenman, J.G., Larsen, E.V., Nordell, D.E. *et al.* (1993). Geomagnetic disturbance effects on power systems. *IEEE Transactions on Power Delivery* **8**(3), 1206–1216.

[12] Yao, Y., Koh, C.S., Ni, G. and Xie, D. (2005). 3-D nonlinear transient eddy current calculation of online power transformer under DC bias. *IEEE Transactions on Magnetics* **41**(5), 1840–1843.

[13] Yamada, S. and Bessho, K. (1988). Harmonic field calculation by the combination of finite element analysis and harmonic balance method. *IEEE Transactions on Magnetics* **24**(6), 2588–2590.

[14] Lu, J., Yamada, S. and Bessho, K. (1990). Time-periodic Magnetic Field Analysis with Saturation and Hysteresis Characteristics by Harmonic Balance Finite Element Method. *IEEE Transactions on Magnetics* **26**(2), 995–998.

[15] Zhou, P., Fu, W.N., Lin, D. and Cendes, Z.J. (2004). Numerical modeling of magnetic devices. *IEEE Transactions on Magnetics* **40**(4), 1803–1809.

[16] Yamazaki, K. (2002). An efficient procedure to calculate equivalent circuit parameter of induction motor using 3-D nonlinear time-stepping finite element method. *IEEE Transactions on Magnetics* **38**(2), 1281–1283.

[17] Lu, J., Yamada, S. and Bessho, K. (1991). Harmonic Balance Finite Element Method Taking Account of External Circuits and Motion. *IEEE Transactions on Magnetics* **27**(5), 4204–4207.

[18] Cheng, Z. Takahashi, N. and Forghani, B. (2009). *Electromagnetic and Thermal Field Modeling and Application in Electrical Engineering*, pp. 386-393. Beijing: Science Press.

[19] Maas, S.A. (1988). *Nonlinear Microwave Circuits*. Artech House.

[20] Lu, J., Yamada, S. and Harrison, H.B. (1996). Application of HB-FEM in the Design of Switching Power Supplies. *IEEE Transactions on Power Electronics* **11**(2), 347–355.

[21] Bachinger, F., Langer, U. and Schoberl, J. (2006). Efficient solvers for nonlinear time-periodic eddy current problems. *Computing and. Visualization in Science* **9**(4), 197–207.

[22] Bolduc, L. (2002). GIC observations and studies in the Hydro-Quebec power system, *Journal of Atmospheric and Solar-Terrestrial Physics* **64**, 1793–1802.

[23] Yamada, S., Biringer, P. and Bessho, K. (1991), Calculation of nonlinear eddy-current problems by the harmonic balance finite element method. *IEEE Transactions on Magnetics* **27**(5), 4122–4125.

[24] Rizzoli, V., Masotti, D. and Mastri, F. (1995). General-purpose analysis of nonlinear circuits containing saturating/hysteresis inductor by the harmonic-balance technique. *IEEE Transactions on Magnetics* **31**(3), 2290–2302.

[25] Gyselinck, J., Dular, P., Geuzaine, C. and Legros, W. (2002), Harmonic-balance finite-element modeling of electromagnetic devices: A novel approach. *IEEE Transactions on Magnetics* **38**(2), 521–524.
[26] Deblecker, O. and Lobry, J. (2006). A new efficient technique for harmonic-banlance finite-element analysis of saturated electromagnetic devices. *IEEE Transactions on Magnetics* **42**(4), 535–538.
[27] Ausserhofer, S., Bíró, O. and Preis, K. (2007). An efficient harmonic balance method for nonlinear eddy current problems. *IEEE Transactions on Magnetics* **43**(4), 1229–1232.
[28] Ciric, I.R. and Hantila, F.I. (2007). An efficient harmonic method for solving nonlinear time-periodic eddy-current problems. *IEEE Transactions on Magnetics* **43**(4), 1185–1188.
[29] Yao, Y. (2005). *Research on the DC bias phenomena of large power transformers*. Shenyang: Shenyang University of Technology.
[30] Bíró, O. and Preis, K. (2006). An efficient time domain method for nonlinear periodic eddy current problems. *IEEE Transactions on Magnetics* **42**(4), 695–698.
[31] Pirjola, R. (2000). Geomagnetically induced current during magnetic storms. *IEEE Transactions on Power Apparatus and Systems* **104**(1), 212–219.
[32] Zhang, B., Cui, X., Zeng, R. and He, J. (2006). Calculation of DC current distribution in AC power system near HVDC system by using moment method coupled to circuit equations. *IEEE Transactions on Magnetics* **42**(4), 703–706.
[33] Zhao, X., Lu, J., Li, L., Cheng, Z. and Lu, T. (2011). Analysis of the DC bias phenomenon by the harmonic balance finite element method. *IEEE Transactions on Power Delivery* **26**(1), 475–485.
[34] Lombard, P. and Meunier, G. (1992). General method for electric and magnetic coupled problem in 2D and magnetodynamic domain. *IEEE Transactions on Magnetics* **28**(2), 1291–1294.
[35] Saad, Y. (2009). *Iterative methods for sparse linear systems*. Science Press, Beijing, China.
[36] Yamada, S., Biringer, P., Hirano, K. and Bessho, K. (1990). Finite element analysis of nonlinear dynamic magnetic field with DC component in the harmonic domain, *IEEE Transactions on Magnetics* **26**(5), 2199–2201.
[37] Cheng, Z., Norio, T. and Behzad, F. (2009). *Electromagnetic and Thermal Field Modeling and Application in Electrical Engineering*. Science Press, Beijing, China.
[38] Miyagi, D., Yoshida, T. and Takahashi, N. (2006). Development of measuring equipment of dc-biased magnetic properties using open-type single-sheet tester. *IEEE Transactions on Magnetics* **42**(10), 2846–2848.
[39] Koczka, G., Auberhofer, S., Bíró, O. and Preis, K. (2009). Optimal convergence of the fixed-point method for nonlinear eddy current problems. *IEEE Transactions on Magnetics* **45**(3), 948–951.
[40] Auberhofer, S., Bíró, O. and Preis, K. (2008). A strategy to improve the convergence of the fixed-point method for nonlinear eddy current problems. *IEEE Transactions on Magnetics* **44**(6), 1282–1285.
[41] Dlala, E., Belahcen, A. and Arkkio, A. (2007). Locally convergent fixed-point method for solving time-stepping nonlinear field problems. *IEEE Transactions on Magnetics* **43**(11), 3969–3975.
[42] Dlala, E. and Arkkio, A. (2008). Analysis of the convergence of the fixed-point method used for solving nonlinear rotational magnetic field problems. *IEEE Transactions on Magnetics* **44**(4), 473–478.
[43] Zhao, Z., Liu, F., Ho, S.L., Fu, W.N. and Yan, W. (2009). Modeling magnetic hysteresis under DC-Biased magnetization using the neural network. *IEEE Transactions on Magnetics* **45**(10), 3958–3961.
[44] Cheng, Z., Takahashi, N. and Forghani, B. (2011). Effect of variation of B-H properties on loss and flux inside silicon steel lamination. *IEEE Transactions on Magnetics* **47**(5), 1346–1349.
[45] Hantila, F.I. Preda, G. and Vasiliu, M. (2000). Polarization method for static field. *IEEE Transactions on Magnetics* **36**(4), 672–675.
[46] Yamada, S., Bessho, K. and Lu, J. (1989). HBFEM Applied to Nonlinear AC Magnetic Analysis. *IEEE Transactions on Magnetics* **25**(4), 2971–2973.
[47] Dlala, E., Belahcen, A. and Arkkio, A. (2008). A fast fixed-point method for solving magnetic field problems in media of hysteresis. *IEEE Transactions on Magnetics* **44**(6), 1214–1217.

[48] Enokizono, M. and Takeshima, Y. (1999). Measuring method of magnetic properties of electrical steel sheets under DC-biased magnetization with flux meter. *Journal of Magnetism and Magnetic Materials* **196–197**, 913–914.

[49] Cao, L. (2000). *Study on the influences of DC biasing current on power transformers magnetization characteristics*. Ph.D thesis, Tsinghua University, Beijing, China.

[50] Kulkarni, S. and Khaparde, S. (2004). *Transformer Engineering: Design and Practice*, pp. 26-36. New York: Marcel Dekker.

[51] Faiz, J. and Sharifian, M. (2001). Hysteresis loop modeling techniques and hysteresis loss estimation of soft materials. *COMPEL* **20**(4), 981–1001.

[52] Lin, C., Wei, J. and Huang, C. (1989). A new method for representation of hysteresis loops. *IEEE Transactions on Power Delivery* **4**(1), 413–420.

[53] Serpico, C. and Visone, C. (1998). Magnetic hysteresis modeling via feed-forward neural networks. *IEEE Transactions on Magnetics* **34**(3), 623–628.

[54] Zhao, X., Lu, J., Li, L., Cheng, Z. and Lu, T. (2010). *An efficient fixed-point harmonic-balanced method taking account of hysteresis effect based on the consuming function*. International Conference on Computational Problem-Solving, pp. 179–182.

[55] Enokizono, M. (2009). Vector magnetic property and magnetic characteristic analysis by vector magneto-hysteretic E&S model. *IEEE Transactions on Magnetics* **45**(3), 1148–1153.

[56] Zhang, Y., Liu, Y., Li, Y., Xie, D. and Bai, B. (2014). A complex-valued rotating magnetic property model and its application to iron core loss calculation. *IEEE Transactions on Magnetics* **50**(2). DOI: 10.1109/TMAG.2013.2281460

[57] Zhao, X., Lu, J., Li, L., Cheng, Z. and Lu, T. (2012). Characteristics analysis of the square laminated core under DC-biased magnetization by the fixed-point harmonic-balanced FEM. *IEEE Transactions on Magnetics* **48**(2), 747–750.

[58] Saad, Y. (2003). *Iterative methods for sparse linear systems*. Siam (Society for Industrial and Applied Mathematics).

[59] Dagum, L., and Enon, R. (1998). OpenMP: an industry standard API for shared-memory programming. *Computational Science & Engineering, IEEE* **5**(1), 46–55.

[60] Gropp, W., Lusk, E., Doss, N. and Skjellum, A. (1996). A high-performance, portable implementation of the MPI message passing interface standard. *Parallel Computing* **22**(6), 789–828.

[61] Owens, J.D., Houston, M., Luebke, D. *et al.* (2008). GPU computing. *Proceedings of the IEEE* **96**(5), 879–899.

[62] Bolz, J., Farmer, I., Grinspun, E. and Schroder, P. (2003). Sparse matrix solvers on the GPU: conjugate gradients and multigrid. *ACM Transactions on Graphics (TOG)* **22**(3), 917–924.

[63] Rabenseifner, R., Hager, G. and Jost, G. (2009). *Hybrid MPI/OpenMP parallel programming on clusters of multi-core SMP nodes*. 17th Euromicro International Conference on Parallel, Distributed and Network-based Processing.

[64] Jacobsen, D.A., Thibault, J.C. and Senocak, I. (2010). *An MPI-CUDA implementation for massively parallel incompressible flow computations on multi-GPU clusters*. 48th AIAA aerospace sciences meeting and exhibit.

[65] Yang, C.-T., Huang, C.-L. and Lin. C.-F. (2011). Hybrid CUDA, OpenMP, and MPI parallel programming on multicore GPU clusters. *Computer Physics Communications* **182**(1), 266–269.

[66] Potluri, S., Venkateshm A. Bureddy, D. Kandalla, K. and Panda, D.K. (2013). *Efficient intra-node communication on intel-mic clusters*. 13th IEEE/ACM International Symposium on Cluster, Cloud and Grid Computing (CCGrid).

[67] Meadows, L. (2012). *Experiments with WRF on Intel® many integrated core (Intel MIC) architecture*. Proceedings of the 8th international conference on OpenMP in a Heterogeneous World, pp. 130–139. Springer: Berlin, Heidelberg.

[68] Hu, Y., Li, Q., Cao, Z. and Wang, J. (2015). Parallel simulation of high-dimensional American option pricing based on CPU versus MIC. *Concurrency and Computation: Practice and Experience* **27**(5), 1110–1121.

[69] Iwashita, T. and Shimasaki, M. (2000). Parallel processing of 3-D eddy current analysis with moving conductor using parallelized ICCG solver with renumbering process. *IEEE Transactions on Magnetics* **36**(4), 1504–1509.

[70] Iwashita, T. and Shimasaki, M. (2003). Block red-black ordering: A new ordering strategy for parallelization of ICCG method. *International Journal of Parallel Programming* **31**(1), 55–75.

[71] Iwashita, T., Nakanishi, Y. and Shimasaki, M. (2005). Comparison criteria for parallel orderings in ILU preconditioning. *SIAM Journal on Scientific Computing* **26**(4), 1234–1260.

[72] Mifune, T., Iwashita, T. and Shimasaki, M. (2002). A fast solver for FEM analyses using the parallelized algebraic multigrid method. *IEEE Transactions on Magnetics* **38**(2), 369–372.

[73] Mifune, T., Iwashita, T. and Shimasaki, M. (2003). New algebraic multigrid preconditioning for iterative solvers in electromagnetic finite edge-element analyses. *IEEE Transactions on Magnetics* **39**(3), 1677–1680.

[74] Lu, J., Li, Y., Sun, C. and Yamada. S. (1995). *A parallel-computation model for nonlinear electromagnetic field analysis by harmonic balance finite element method*. Algorithms and Architectures for Parallel Processing, 1995. ICAPP 95. IEEE First ICA/sup 3/PP., IEEE First International Conference on (Volume 2).

[75] Li, W. and Lu J. (2015). High Performance Parallel Computing for HBFEM, Internal Technical Report of Griffith University, September, 2015.

6

HBFEM and Its Future Applications

6.1 HBFEM Model of Three-Phase Power Transformer

6.1.1 Three-Phase Transformer

A three-phase transformer-based magnetic frequency-tripling devices was introduced in Chapter 4.3, where the production of triple-frequency output from a three-phase fundamental frequency source was based on nonlinear magnetic saturation characteristics, as illustrated in Figure 6.2. However, a three-phase transformer used in electric power systems, including HVDC power system with different winding connections, including Y/Y, Y/Δ, Δ/Y and Δ/Δ, is also a nonlinear magnetic system [1]. In particular, the transformer is under the DC or quasi-DC biased condition, a significant nonlinear phenomenon is accrued, and a larger amount of harmonics are generated when the transformer is saturated.

Figure 6.1 shows the Δ/Y and Δ/Δ winding connection used in an electric power system. The Δ/Y connection is often used in a 10–12 kV/200–400 V distribution network, which is now widely used in solar PV and winding power-based renewable energy systems or a distributed energy resource (DER) low-voltage (LV) network.

By using the HBFEM approach, magnetic nonlinear characteristics, hysteresis losses, eddy current losses, and magnetic flux distribution for each harmonic component can be calculated and presented. HBFEM can provide magnetic flux distribution and eddy current losses at each harmonic [2].

Harmonic Balance Finite Element Method: Applications in Nonlinear Electromagnetics and Power Systems, First Edition. Junwei Lu, Xiaojun Zhao and Sotoshi Yamada.

Companion website: www.wiley.com/go/lu/HBFEM

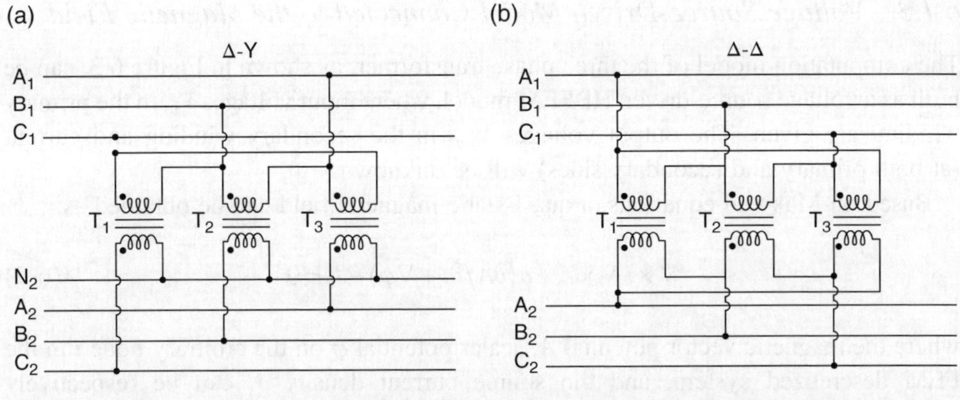

Figure 6.1 A circuit diagram of a three-phase transformer. (a) Δ/Y; (b) Δ/Δ

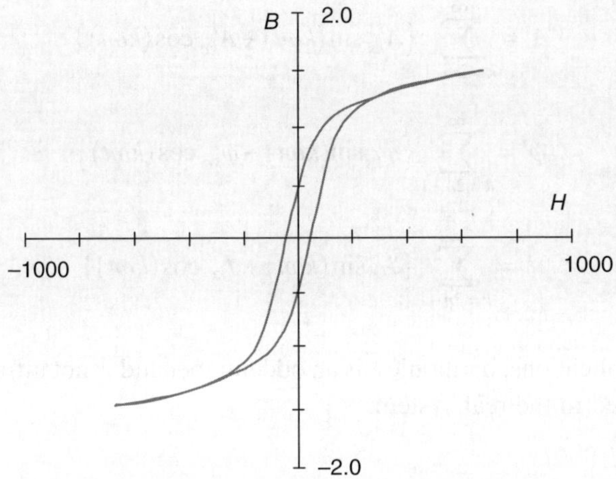

Figure 6.2 B-H curve with hysteresis

6.1.2 Nonlinear Magnetic Material and its Saturation Characteristics

As discussed in Chapter 4.3, when a nonlinear magnetic system is excited by a sinusoidal waveform, a number of harmonics will be generated in this nonlinear magnetic system. For the non-DC-biased case, only odd harmonics can be generated in the magnetic field. The magnetizing characteristics of the magnetic core used in the transformer can be also mathematically expressed by following the arbitrary function of flux density B:

$$H(B) = H_0(B) + H_e \frac{dB}{dt} = \left(\alpha B + \beta B^{2n-1}\right) + \left(\frac{1}{f_s}\frac{dB}{dt}\right)\frac{dB}{dt} \tag{6-1}$$

where the B-H curve with hysteresis can be obtained from Equation (6-1), and the curve without hysteresis characteristics is obtained from the second term of Equation (6-1).

6.1.3 Voltage Source-Driven Model Connected to the Magnetic Field

The computation model of the three phase transformer, as shown in Figure 6.3, can be built as a voltage source-driven HBFEM model, where input voltages V_{in} in the primary winding are given. The output voltages V_{out} in the secondary winding and current (at both primary and secondary sides) will be unknown.

Based on Maxwell equations, a quasi-static magnetic field can be obtained as:

$$\nabla \times \nu \nabla \times A + \sigma(\partial A/\partial t + \nabla \varphi) - J_s = 0 \tag{6-2}$$

where the magnetic vector potential A, scalar potential φ on the arbitrary node i in the FEM descretized system, and the source current density J_s can be respectively expressed as:

$$A^i = \sum_{k=2n-1}^{\infty} \{A^i_{ks} \sin(k\omega t) + A^i_{kc} \cos(k\omega t)\} \tag{6-3}$$

$$\varphi^i = \sum_{k=2n-1}^{\infty} \{\varphi^i_{ks} \sin(k\omega t) + \varphi^i_{kc} \cos(k\omega t)\} \tag{6-4}$$

$$J_s = \sum_{k=2n-1}^{\infty} \{J_{ks} \sin(k\omega t) + J_{kc} \cos(k\omega t)\} \tag{6-5}$$

In practical applications, harmonic k is an odd number and is not infinite. Only a finite number is required in the real system.

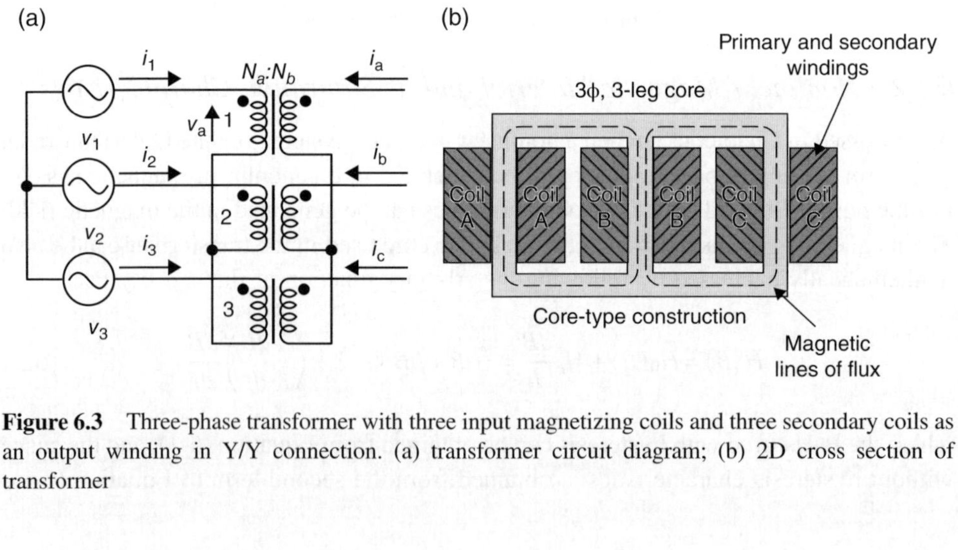

Figure 6.3 Three-phase transformer with three input magnetizing coils and three secondary coils as an output winding in Y/Y connection. (a) transformer circuit diagram; (b) 2D cross section of transformer

6.1.3.1 Magnetic Medium

The magnetic reluctivity ν corresponding to $B(t)$ can be expressed as:

$$\nu(t) = H(B(t))/B(t) = \nu_0 + \sum_{k=2n-2}^{\infty} \{\nu_{ks}\sin(k\omega t) + \nu_{kc}\cos(k\omega t)\} \tag{6-6}$$

where flux density $B(t)$ is time-dependent. This can be expressed by the B-H curve (normally provided by manufacture product catalog), including the hysteresis characteristic. ν (=1/μ) is the nonlinear magnetic reluctivity, and the Fourier coefficients are obtained from following equations:

$$\nu_0 = \frac{1}{T}\int_0^T \nu(t)dt \tag{6-7}$$

$$\nu_{ks} = \frac{2}{T}\int_0^T \nu(t)\cdot\sin(k\omega t)dt \tag{6-8}$$

$$\nu_{kc} = \frac{2}{T}\int_0^T \nu(t)\cdot\cos(k\omega t)dt \tag{6-9}$$

6.1.4 HBFEM Matrix Equations, Taking Account of Extended Circuits

Since the three phase transformer model with voltage driven source connected to the magnetic field system, is always coupled to the external circuits [2], the current in the input and output circuits are unknown. The saturation of the current waveform occurs due to the nonlinear characteristic of the magnetic core.

6.1.4.1 The Equivalent Circuit of a Three-Phase Transformer

From Kirchhoff's laws, for the equivalent circuit of a three-phase transformer with Y/Y winding connection, the equations can be obtained as:

$$\left.\begin{aligned}
&v_{1Np'} + V_{N_p'} + i_1 Z_1 = V_1 \\
&v_{2Np'} + V_{N_p'} + i_2 Z_2 = V_2 \\
&v_{3Np'} + V_{N_p'} + i_3 Z_3 = V_3 \\
&i_1 + i_2 + i_3 = 0 \\
&v_{aNs'} + V_{N_s'} + i_a Z_a = V_a \\
&v_{bNs'} + V_{N_s'} + i_b Z_b = V_b \\
&v_{cNs'} + V_{N_s'} + i_c Z_c = V_c \\
&i_a + i_b + i_c = 0
\end{aligned}\right\} \tag{6-10}$$

Considering a three-phase transformer connected in wye, a numerical simulation model with a neutral N and external circuits for both primary and secondary windings is obtained using the HBFEM technique. According to the Galerkin procedure, the system matrix equations of HBFEM for the three phase transformer can be obtained through Faraday's and Kirchhoff's laws for the external circuit [2]. The matrix equation of input voltage can be defined as:

$$\begin{aligned} V_{ink} &= V_k + R_k I_k + L_k \frac{dI_k}{dt} + \frac{1}{C_k}\int I_k dt \\ &= \left[C_k^e\right]\{A\} + S_{ck}\left[Z_k\right]\{J_k\} \end{aligned} \tag{6-11}$$

V_{ink} is the input voltage of circuit k, V_k is the corresponding induced electromotive force, S_{ck} and Z_k are the cross areas and impedance of winding k respectively, C_k and L_k are the capacitance and inductance respectively in circuit k. The induced electromotive force can be obtained based on Faraday's law:

$$\begin{aligned} V_k &= \frac{d\Psi}{dt} = \frac{d}{dt}\iint \boldsymbol{B}\cdot d\boldsymbol{S} \\ &= \frac{d}{dt}\iint \nabla \times \boldsymbol{A} d\boldsymbol{S} = \frac{d}{dt}\oint_f \boldsymbol{A} d\boldsymbol{l} \\ &= \frac{\omega N_k d_0}{3S_{ck}}[N \quad N \quad N]\begin{Bmatrix} \{A^1\} \\ \{A^2\} \\ \{A^3\} \end{Bmatrix} \end{aligned} \tag{6-12}$$

V_k can be rewritten as a compact form, as below:

$$\{V_k\} = \left[C_k^e\right]\{A\} \tag{6-13}$$

The input voltage can be defined as:

$$\{V_{ink}\} = \{V_k\} + S_{ck}[Z_k]\{J_k\} = [C_k]\{A\} + S_{ck}[Z_k]\{J_k\} \tag{6-14}$$

where matrix $[Z_k]$ is the circuit impedance, including the resistance of windings and leakage inductance corresponding to the harmonics, and S_{ck} is the area of the windings. The input voltage $\{V_{ink}\}$, including all harmonic components which have a known value, is expressed as follows:

$$\{V_{in,k}\} = \{V_{in,1sk}, V_{in,1ck}, V_{in,2sk}, V_{in,2ck}, \cdots\}^T \tag{6-15}$$

Therefore, all input and output circuits related matrix equations can be obtained as below:

$$\left.\begin{aligned}
&[C_1]\{A\}+[I]\{V_{N,p}\}+S_{cu}[Z_1]\{J_1\}=\{V_1\}\\
&[C_2]\{A\}+[I]\{V_{N,p}\}+S_{cu}[Z_1]\{J_2\}=\{V_2\}\\
&[C_3]\{A\}+[I]\{V_{N,p}\}+S_{cu}[Z_3]\{J_3\}=\{V_3\}\\
&S_{in,cu}[I]\{J_{1,in}\}+S_{in,cu}[I]\{J_{2,in}\}+S_{in,cu}[I]\{J_{3,in}\}=0\\
&[C_a]\{A\}+[I]\{V_{N,s}\}+S_{cu}[Z_a]\{J_a\}=\{V_a\}\\
&[C_b]\{A\}+[I]\{V_{N,s}\}+S_{cu}[Z_b]\{J_b\}=\{V_b\}\\
&[C_c]\{A\}+[I]\{V_{N,s}\}+S_{cu}[Z_c]\{J_c\}=\{V_c\}\\
&S_{out,cu}[I]\{J_{a,out}\}+S_{out,cu}[I]\{J_{b,out}\}+S_{out,cu}[I]\{J_{c,out}\}=0
\end{aligned}\right\} \tag{6-16}$$

where the applied three-phase voltage sources $\{V_{k,in}\}$, $\{V_{Np'}\}$, $\{V_{j,out}\}$ and $\{V_{Ns'}\}$ ($k=$ 1,2,3 and $j=a,b,c$) are the unknown voltages at the neutral point, as expressed below:

$$\{V_{k,in}\}=\{\{V_{k,1s}\}\ \{V_{k,1c}\}\ \{V_{k,3s}\}\ \{V_{k,3c}\}\ \cdots\}^T \tag{6-17}$$

$$\{V_{Np'}\}=\{\{V_{Np,1s}\}\ \{V_{Np,1c}\}\ \{V_{Np,3s}\}\ \{V_{Np,3c}\}\ \cdots\}^T \tag{6-18}$$

$$\{V_{j,out}\}=\{\{V_{j,1s}\}\ \{V_{j,1c}\}\ \{V_{j,3s}\}\ \{V_{j,3c}\}\ \cdots\}^T \tag{6-19}$$

$$\{V_{Ns'}\}=\{\{V_{Ns,1s}\}\ \{V_{Ns,1c}\}\ \{V_{Ns,3s}\}\ \{V_{Ns,3c}\}\ \cdots\}^T \tag{6-20}$$

and $[C_k]$ is obtained from:

$$[C_k]=\frac{\omega d_0\Delta}{3S_{ck}}[N\ \ N\ \ N] \tag{6-21}$$

$\{J_k\}$ and $\{J_{j,out}\}$ are the magnetizing current density and output current density, respectively. It can be expressed in a general form as below:

$$\{J^k\}=\{\{J^k_{1s}\}\ \{J^k_{1c}\}\ \{J^k_{3s}\}\ \{J^k_{3c}\}\ \cdots\}^T \tag{6-22}$$

The generalized input voltage matrix equation taking account of circuit can be expressed as:

$$\{V_{k,in}\} = \{V_k\} + S_{ck}[Z_k]\{J_k\} \tag{6-23}$$

where matrix [Zk] is the circuit impedance matrix which includes the resistance and leakage inductance of windings corresponding to the harmonics, and S_{ck} is the area of windings.

$$[\mathbf{Z}_{nk}] = [\mathbf{Z}_{Rnk}] + [\mathbf{Z}_{Lnk}] = \begin{bmatrix} R_k & -n\omega L_k \\ n\omega L_k & R_k \end{bmatrix}. \tag{6-24}$$

The input voltage $\{V_{k,in}\}$, including all harmonic components which have a known value, is expressed as follows:

$$\{V_{k,in}\} = \{V_{1sk}, V_{1ck}, V_{3sk}, V_{3ck}, \cdots\}^T \tag{6-25}$$

where $V_{3\mathrm{sk}}$ and $V_{3\mathrm{ck}}$ are zero components; only fundamental component exists in the input voltage.

Based on the Galerkin procedure, the system matrix equations of HBFEM for the three-phase transformer, including equivalent circuits, can be obtained through Faraday's and Kirchhoff's laws. Therefore, the electric circuit-related matrix and FEM system matrix can be obtained, respectively, as below:

$$[C_k]\{A\} + S_{ck}[Z_k]\{J_k\} = \{V_{k,in}\} \tag{6-26}$$

$$([S] + [N])\{A\} - [G_k]\{J_k\} = 0 \tag{6-27}$$

where $\{I_k\} = S_{ck}\{J_k\}$, and $[G_k]$ is the current density coefficient obtained from a single element of winding area, that is $[G^e] = \Delta^e/3$.

The above system matrix, including all input and output currents in the three-phase transformer windings, can be expressed as:

$$[H]\{A\} - [G_1]\{J_1\} - [G_2]\{J_2\} - [G_3]\{J_3\} \cdots - [G_{out}]\{J_{out}\} = \{0\} \tag{6-28}$$

Combining (6-16) and (6-28), the global system matrix equations for multiple input and output are obtained from following system equation:

$$\left.\begin{array}{l}[H]\{A\}-[G_1]\{J_{1,in}\}-[G_2]\{J_{2,in}\}-[G_3]\{J_{3,in}\}\\ -[G_a]\{J_{a,out}\}-[G_b]\{J_{b,out}\}-[G_c]\{J_{c,out}\}=\{0\}\\ [C_{1,in}]\{A\}+[I]\{V_{Np}\}+S_{c1}[Z_{1,in}]\{J_{1,in}\}=\{V_{1,in}\}\\ [C_{2,in}]\{A\}+[I]\{V_{Np}\}+S_{c2}[Z_{2,in}]\{J_{2,in}\}=\{V_{2,in}\}\\ [C_{3,in}]\{A\}+[I]\{V_{Np}\}+S_{c3}[Z_{3,in}]\{J_{3,in}\}=\{V_{3,in}\}\\ S_{c1}[I]\{J_{1,in}\}+S_{c2}[I]\{J_{2,in}\}+S_{c3}[I]\{J_{3,in}\}=0\\ [C_{a,out}]\{A\}+[I]\{V_{Ns}\}+S_{ca}[Z_{a,out}]\{J_{a,out}\}=\{V_{a,out}\}\\ [C_{b,out}]\{A\}+[I]\{V_{Ns}\}+S_{cb}[Z_{b,out}]\{J_{b,out}\}=\{V_{b,out}\}\\ [C_{c,out}]\{A\}+[I]\{V_{Ns}\}+S_{cc}[Z_{c,out}]\{J_{c,out}\}=\{V_{c,out}\}\\ S_{ca}[I]\{J_{a,out}\}+S_{cb}[I]\{J_{b,out}\}+S_{cc}[I]\{J_{c,out}\}=0\end{array}\right\} \quad (6\text{-}29)$$

Finally, the harmonic balance FEM matrix equations, taking into account the external and internal equivalent circuits of the transformer for voltage source excitation, can therefore be expressed as the following system matrix equation:

$$\begin{bmatrix}[H] & -[G_1] & -[G_2] & -[G_3] & 0 & -[G_a] & -[G_b] & -[G_c] & 0\\ [C_{1,in}] & S_{c1}[Z_1] & 0 & 0 & [I_p] & 0 & 0 & 0 & 0\\ [C_{2,in}] & 0 & S_{c2}[Z_2] & 0 & [I_p] & 0 & 0 & 0 & 0\\ [C_{3,in}] & 0 & 0 & S_{c3}[Z_3] & [I_p] & 0 & 0 & 0 & 0\\ 0 & S_{c1}[I_p] & S_{c2}[I_p] & S_{c3}[I_p] & 0 & 0 & 0 & 0 & 0\\ [C_{a,out}] & 0 & 0 & 0 & 0 & S_{ca}[Z_a] & 0 & 0 & [I_s]\\ [C_{b,out}] & 0 & 0 & 0 & 0 & 0 & S_{cb}[Z_b] & 0 & [I_s]\\ [C_{c,out}] & 0 & 0 & 0 & 0 & 0 & 0 & S_{cc}[Z_c] & [I_s]\\ 0 & 0 & 0 & 0 & 0 & S_{ca}[I_s] & S_{cb}[I_s] & S_{cd}[I_s] & 0\end{bmatrix}\begin{Bmatrix}A\\ J_1\\ J_2\\ J_3\\ V_{NP}\\ J_a\\ J_b\\ J_c\\ V_{NS}\end{Bmatrix}=\begin{Bmatrix}0\\ V_1\\ V_2\\ V_3\\ 0\\ 0\\ 0\\ 0\\ 0\end{Bmatrix} \quad (6\text{-}30)$$

where $\{A\}$ and $\{J_k\}$ are unknown, and can be calculated by solving the system matrix equation, $[H]$ is the matrix obtained from $([S]+[M])$, $[C_k]$ represents the geometric co-efficient related to transformer windings. $[Z_{k,nk}]$, $[Z_{j,out}]$ and $S_{c,in}$, $S_{c,out}$ are external circuit impedances and cross-sectional areas of windings, respectively, $[I]$ is unit matrix, V_{NN} is the voltage for neutral point when it is not grounded, and $[C_{in}]$ and $[C_{out}]$ are geometric coefficients related to transformer windings. Current density J_k can be presented as (no zero component and second harmonic components when transformer in DC-biased case):

$$\{J_k^e\} = \{J_{1s}\ J_{1c}\ J_{2s}\ J_{2c}\ J_{3s}\ J_{3c}\ \cdots\}^T \tag{6-31}$$

[H] is the system matrix, and the detailed definitions can be obtained from:

$$[H^e] = \frac{1}{4\Delta^e}\begin{bmatrix}(b_1b_1+c_1c_1)D & (b_1b_2+c_1c_2)D & (b_1b_3+c_1c_3)D \\ (b_2b_1+c_2c_1)D & (b_2b_2+c_2c_2)D & (b_2b_3+c_2c_3)D \\ (b_3b_1+c_3c_1)D & (b_3b_2+c_3c_2)D & (b_3b_3+c_3c_3)D\end{bmatrix}\begin{Bmatrix}A_1^e \\ A_2^e \\ A_3^e\end{Bmatrix} + \frac{\sigma\omega\Delta^e}{12}\begin{bmatrix}2N & N & N \\ N & 2N & N \\ N & N & 2N\end{bmatrix}\begin{Bmatrix}A_1^e \\ A_2^e \\ A_3^e\end{Bmatrix} \tag{6-32}$$

where matrices D and N are:

$$D = \begin{bmatrix}d_{11} & d_{12} & d_{13} & d_{14} & \cdots \\ d_{21} & d_{22} & d_{23} & d_{24} & \cdots \\ d_{31} & d_{32} & d_{33} & d_{34} & \cdots \\ d_{41} & d_{42} & d_{43} & d_{44} & \cdots \\ \cdots & \cdots & \cdots & \cdots & \ddots\end{bmatrix} = \frac{1}{2}\begin{bmatrix}2\nu_0-\nu_{2c} & \nu_{2s} & \nu_{2c}-\nu_{4c} & -\nu_{2s}+\nu_{4s} & \cdots \\ & 2\nu_0+\nu_{2c} & \nu_{2s}+\nu_{4s} & \nu_{2c}+\nu_{4c} & \cdots \\ & & 2\nu_0-\nu_{6c} & \nu_{6s} & \cdots \\ & Symmetry & & 2\nu_0+\nu_{6c} & \cdots \\ & & & & \ddots\end{bmatrix} \tag{6-33}$$

and

$$N^e = \begin{bmatrix}0 & -1 & 0 & 0 & 0 & 0 & \cdots \\ 1 & 0 & 0 & 0 & 0 & 0 & \cdots \\ 0 & 0 & 0 & -3 & 0 & 0 & \cdots \\ 0 & 0 & 3 & 0 & 0 & 0 & \cdots \\ 0 & 0 & 0 & 0 & 0 & -5 & \cdots \\ 0 & 0 & 0 & 0 & 5 & 0 & \cdots \\ \cdots & \cdots & \cdots & \cdots & \cdots & \cdots & \ddots\end{bmatrix} \tag{6-34}$$

The compact system matrix can be expressed as below:

$$\begin{bmatrix} [H] & [G_k] & 0 \\ [C_k] & S_{ck}[Z_k] & [I] \\ 0 & S_{ck}[I] & 0 \end{bmatrix} \begin{Bmatrix} A_k \\ J_k \\ V_N \end{Bmatrix} = \begin{Bmatrix} o \\ V_k \\ 0 \end{Bmatrix} \tag{6-35}$$

where $[H]$ is the FEM system matrix, $[G_k]$ is the current density related coefficient matrix, $[C_k]$ is the voltage-related coefficient matrix, and $[Z_k]$ is the impedance matrix.

$\{A_k\}$, $\{J_k\}$ and $\{V_N\}$ are unknown, and can be calculated by solving the HBFEM system matrix equation, $[H]$ is the matrix obtained from $([S]+[N])$, and:

$$[C_k] = \frac{\omega d_0 \Delta}{3S_{ck}}[N \;\; N \;\; N] \tag{6-36}$$

is the geometric coefficient related to transformer windings.

6.2 Magnetic Model of a Single-Phase Transformer and a Magnetically Controlled Shunt Reactor

6.2.1 Electromagnetic Coupling Model of a Single-Phase Transformer

The calculation method of the electromagnetic coupling is implemented by combining the electric circuit, which reflects characteristics of the ports, with the magnetic circuit and the internal magnetic field's distribution. Applying this method not only reflects the intrinsic attributes of transformers more accurately, but also meet the needs of system analysis [3]. The electric power system wiring diagram, including a single-phase two-limb transformer under no-load conditions, is shown in Figure 6.4.

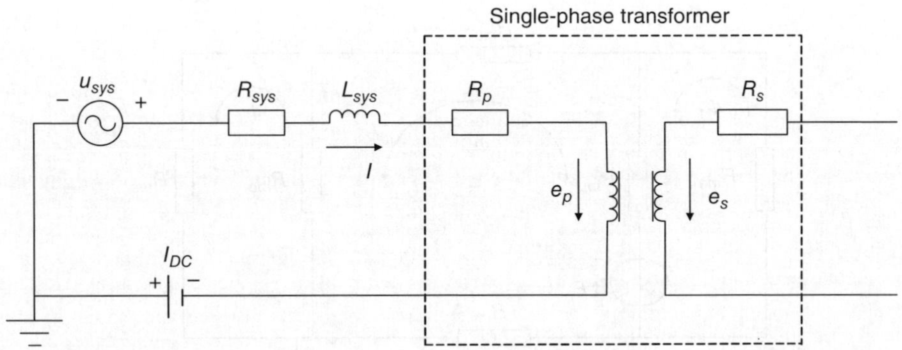

Figure 6.4 A simplified power system wiring diagram under no-load conditions

U_{sys}, R_{sys} and L_{sys} are the alternating voltage, resistance and inductance of the AC system, respectively; I_{DC} and I are the DC current source and the current in the transformer's primary winding respectively; R_p and R_s are the primary and secondary winding resistances of the transformer; and e_p and e_s are the induced electromotive forces of the transformer's primary and secondary windings, respectively. The electric circuit can be described by the following equation:

$$U_{sys} = I\left(R_{sys} + R_p\right) + L_{sys}\frac{dI}{dt} + \frac{d\Psi}{dt} - U_{dc} \tag{6-37}$$

Ψ represents the total flux linkage and U_{dc} is the product of I_{DC} and the resistance of the DC current source. Although single-phase two-limb transformer windings are not generally wound separately on different limbs in practical projects, each winding can be separated in the magnetic equivalent circuit. An equivalent magnetic circuit model structure is shown in Figure 6.5.

R_{mp} and R_{my} represent the reluctances of transformer winding limbs, the upper and lower yokes reluctance, respectively. R_{ma} and R_{ms} represent the leakage reluctances of transformer winding limbs, obtained from the open and short-circuit testing method [4]. F_p is the magnetic motive force; Φ_1, Φ_2, Φ_3 represent the loop magnetic flux; and N_{coil} is the number of turns of the windings. Combined with the magnetic circuit structure, a magnetic circuit equation can be obtained:

$$[\boldsymbol{R}_m(\boldsymbol{\Phi})][\boldsymbol{\Phi}] = [\boldsymbol{N}_{coil}][I] \tag{6-38}$$

I is the exciting current, and the expressions of $\boldsymbol{R}_m$, $\boldsymbol{\Phi}$ and $\boldsymbol{N}_{coil}$ are given by Equations (6-39), (6-40) and (6-41) respectively,

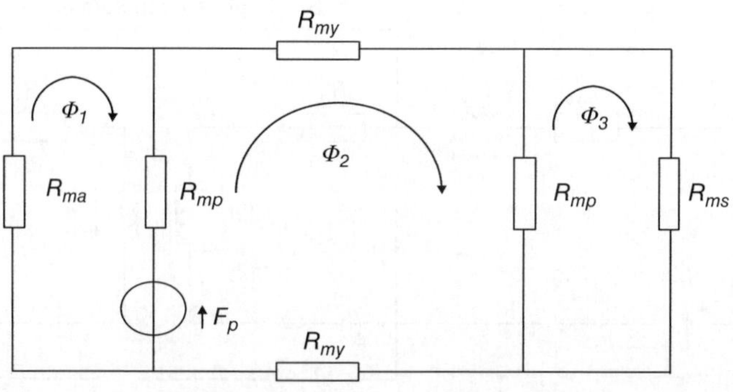

Figure 6.5 Equivalent magnetic circuit model structure

$$[\boldsymbol{R}_m(\boldsymbol{\Phi})] = \begin{bmatrix} R_{ma} + R_{mp} & -R_{mp} & 0 \\ -R_{mp} & 2(R_{my} + R_{mp}) & -R_{ms} \\ 0 & -R_{mp} & R_{ms} + R_{mp} \end{bmatrix} \tag{6-39}$$

$$[\boldsymbol{\Phi}] = [\Phi_1, \Phi_2, \Phi_3]^T \tag{6-40}$$

$$[\boldsymbol{N}_{coil}] = [-N_{coil}, -N_{coil}, N_{coil}]^T \tag{6-41}$$

According to Figure 6.5 the total flux linkage $\Psi = N_{coil}(\Phi_2 - \Phi_1 - \Phi_3)$. The resistance and the inductance of the AC source can be ignored, in view of the simplified calculation for solving the equation. The electric circuit equation can be written as:

$$U_{sys} = IR_p + N_{coil}\frac{d(\Phi_2 - \Phi_1 - \Phi_3)}{dt} - U_{dc} \tag{6-42}$$

Φ_2, Φ_1, Φ_3 and I can be obtained by combining Equation (6-38) with Equation (6-42). However, the equations are difficult to solve directly in the time domain, due to the differential equations and nonlinear reluctance. The time-step method can be applied to solve the equations, which requires discretization in the time domain and complex iterative computation in the time-stepping solution process. Considering the periodic characteristics of electromagnetic fields under steady-state excitation, a limited Fourier series is used to represent the periodic behavior of the steady-state variables, and solving equations in the time domain can be avoided. Therefore, the harmonic balance method (carried out in the frequency domain) is proposed to solve equations.

6.2.2 *Solutions of the Nonlinear Magnetic Circuit Model by the Harmonic Balance Method*

The harmonic balance method is the most effective method for obtaining the approximate periodic solution of a nonlinear differential equation. It is an effective and efficient method of analyzing the harmonic problem of alternating electromagnetic fields [5]. In the electromagnetic coupling model, the electric circuit equation is a linear differential equation, and the magnetic circuit equation is a nonlinear equation. According to the harmonic balance method, all variables, such as the magnetic flux Φ, the magnetic reluctance R and the excitation current I, can be given as follows:

$$\Phi_i = \Phi_{i.0} + \sum_{n=1}^{\infty}\{\Phi_{i.ns}sin(n\omega t) + \Phi_{i.nc}cos(n\omega t)\} \tag{6-43}$$

$$R = R_0 + \sum_{n=1}^{\infty} \{R_{ns} \sin(n\omega t) + R_{nc} \cos(n\omega t)\} \tag{6-44}$$

$$I = I_0 + \sum_{n=1}^{\infty} \{I_{ns} \sin(n\omega t) + I_{nc} \cos(n\omega t)\} \tag{6-45}$$

The subscripts i and n indicate circuit number and harmonic number respectively. ω represents the angular frequency. Combined with the above formulas, the equation (6-42) can be rewritten as,

$$[\boldsymbol{C}][\boldsymbol{\phi}] + [\boldsymbol{R_p}][\boldsymbol{I}] = [\boldsymbol{U}] \tag{6-46}$$

$\boldsymbol{\Phi}$, $\boldsymbol{U}$ and $\boldsymbol{I}$ are the harmonic vectors of the magnetic flux, the input voltage and the excitation current respectively. $\boldsymbol{C}$ is the coupling matrix and $\boldsymbol{R_p}$ is the resistance matrix. The matrix $\boldsymbol{N}$ is related to the harmonic number and is called the harmonic matrix. They can be given as follows,

$$\boldsymbol{N} = \begin{bmatrix} 0 & 0 & 0 & 0 & 0 & \cdots \\ 0 & 0 & -1 & 0 & 0 & \cdots \\ 0 & 1 & 0 & 0 & 0 & \cdots \\ 0 & 0 & 0 & 0 & -2 & \cdots \\ 0 & 0 & 0 & 2 & 0 & \cdots \\ \cdots & \cdots & \cdots & \cdots & \cdots & \ddots \end{bmatrix} \tag{6-47}$$

$$\boldsymbol{C} = -N_{coil}\omega[\boldsymbol{N}, -\boldsymbol{N}, \boldsymbol{N}] \tag{6-48}$$

$$\boldsymbol{R_p} = diag\left[R_p, R_p, \ldots R_p\right] \tag{6-49}$$

$$\boldsymbol{I} = [I_0, I_{1s}, I_{1c} \ldots I_{ns}, I_{nc}]^T \tag{6-50}$$

$$\boldsymbol{\Phi} = [\boldsymbol{\Phi}_1, \boldsymbol{\Phi}_2, \boldsymbol{\Phi}_3]^T \tag{6-51}$$

$$\boldsymbol{\Phi}_i = [\Phi_{i.0}, \Phi_{i.1s}, \Phi_{i.1c} \ldots \Phi_{i.ns}, \Phi_{i.nc}]^T, i = 1,2,3 \tag{6-52}$$

The new magnetic circuit equation can be obtained by combining Equation (6-38) with Equations (6-43), (6-44) and (6-45),

$$[\boldsymbol{R}][\boldsymbol{\phi}] + [\boldsymbol{G}][\boldsymbol{I}] = [\boldsymbol{0}] \tag{6-53}$$

$\boldsymbol{R}$ and $\boldsymbol{G}$ represent the reluctance matrix and the winding number matrix, respectively:

$$\boldsymbol{R}=\frac{1}{2}\begin{bmatrix} 2R_0 & R_{1s} & R_{1c} & R_{2s} & R_{2c} & \cdots \\ 2R_{1s} & 2R_0-R_{2c} & R_{2s} & R_{1c}-R_{3c} & -R_{1s}+R_{3s} & \cdots \\ 2R_{1c} & & 2R_0+R_{2c} & R_{1s}+R_{3s} & R_{1c}+R_{3c} & \cdots \\ 2R_{2s} & & & 2R_0-R_{4c} & R_{4s} & \cdots \\ 2R_{2c} & & Symmetry & & 2R_0+R_{4c} & \cdots \\ \cdots & & & & & \ddots \end{bmatrix} \tag{6-54}$$

$$\boldsymbol{G}=\mathrm{diag}[N_{coil},N_{coil}\ldots N_{coil}] \tag{6-55}$$

The new system matrix equation can be obtained by combining Equations (6-46) and (6-53):

$$\begin{bmatrix} \boldsymbol{R} & \boldsymbol{G} \\ \boldsymbol{C} & \boldsymbol{R_p} \end{bmatrix}\begin{bmatrix} \boldsymbol{\Phi} \\ \boldsymbol{I} \end{bmatrix}=\begin{bmatrix} \boldsymbol{0} \\ U \end{bmatrix} \tag{6-56}$$

An algebraic set of nonlinear and linear equations in the Fourier coefficients can be obtained. All harmonic components of the exciting current and the magnetic flux can be calculated directly and quickly in the frequency domain.

6.2.3 Magnetically Controlled Shunt Reactor

The magnetically controlled shunt reactor (MCSR) is an important device in the high-voltage direct current transmission system. Figure 6.6 shows the scheme circuit for an

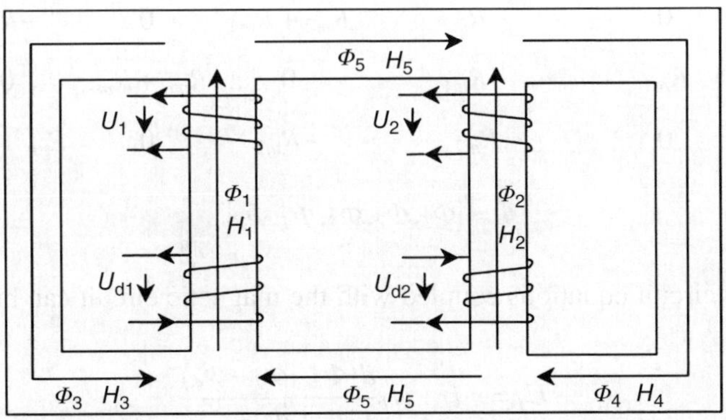

Figure 6.6 Scheme circuit for iron core structure and coil arrangement

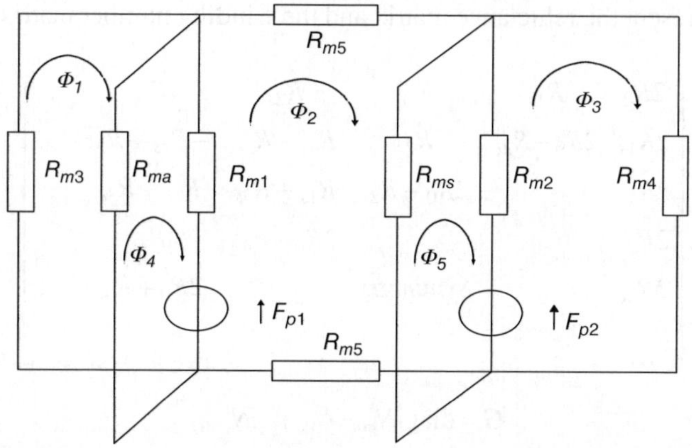

Figure 6.7 Scheme circuit for iron core structure and coil arrangement

iron core structure and coil arrangement of a single phase MCSR. U_1 and U_2 represent the terminals connected to the same phase of the voltage source. U_{d1} and U_{d2} are connected to the direct voltage source. It is clear that DC and AC magnetic flux will co-exist in the magnetic core when the MCSR is operating.

The magnetic circuit model coupled with electric circuits, as illustrated in Figure 6.7, can be established, based on the proposed method in the above section.

Therefore, the nonlinear equation of the magnetic circuit model is presented as follows,

$$[\boldsymbol{R}_m(\boldsymbol{\phi})][\boldsymbol{\phi}] = [\boldsymbol{N}_a][I_a] + [\boldsymbol{N}_d][I_d] \tag{6-57}$$

$$[\boldsymbol{R_m}(\boldsymbol{\Phi})] = \begin{bmatrix} R_{m1}+R_{m3} & -R_{m2} & 0 & R_{m1} & 0 \\ -R_{m1} & R_{m1}+2R_{m5}+R_{m2} & -R_{m2} & -R_{m1} & R_{m2} \\ 0 & -R_{m2} & R_{m2}+R_{m4} & 0 & -R_{m5} \\ R_{m1} & -R_{m1} & 0 & R_{ma}+R_{m1} & 0 \\ 0 & R_{m2} & -R_{m2} & 0 & R_{ms}+R_{m2} \end{bmatrix} \tag{6-58}$$

$$\boldsymbol{\Phi} = [\Phi_1, \Phi_2, \Phi_3, \Phi_4, \Phi_5]^T \tag{6-59}$$

The electric circuit equations coupled with the magnetic circuit can be obtained by:

$$U_1 = r_1 i_1 + N_1 \frac{d(\Phi_2 - \Phi_1 - \Phi_4)}{dt} \tag{6-60}$$

$$U_2 = r_2 i_2 + N_2 \frac{d(\Phi_3 - \Phi_2 - \Phi_5)}{dt} \tag{6-61}$$

where r_1 and r_2 are the resistance of the coil, N_1 and N_2 are the turns of the coil excited by alternating voltage.

Finally, the harmonic solutions of magnetic flux in the iron core and exciting currents in the coil can be computed simultaneously by the harmonic balance method.

6.2.4 *Experiment and Computation*

A laminated core model for a DC-biased test is shown in Figure 6.8. The iron core is made up of silicon steel, of which the model number is 30Q140. The exciting coil and the measuring coil are uniform and thick over the iron core. The effective cross-sectional area of the core is 2770 mm^2; the mean length of the magnetic circuit is 1.24 m.

The laminated core model is driven by both AC and DC currents simultaneously, as shown in Figure 6.9, where I is the exciting current which flows through the exciting coil; U is the alternating voltage; and U_1 is the induced voltage. They are measured by using the Precision Power Analyzer, WT-3000, Yokogawa, Japan.

The peak value of rated no-load magnetizing current I_{dc} without a DC bias is selected as a reference. The value of the current reference I_{dc} measured on the laminated core model is 1.68 A. Different DC bias currents, such as 25% I_{dc} and 50% I_{dc}, are applied to the core model, respectively. Different AC voltages and DC currents are applied, and the induced voltage of the search coil and the exciting current I are measured simultaneously by using the Precision Power Analyzer.

Figure 6.8 The laminated core model

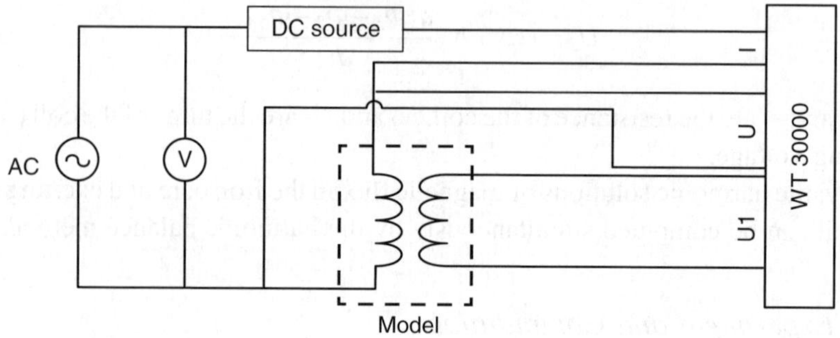

Figure 6.9 Experimental circuit of the DC-biased test

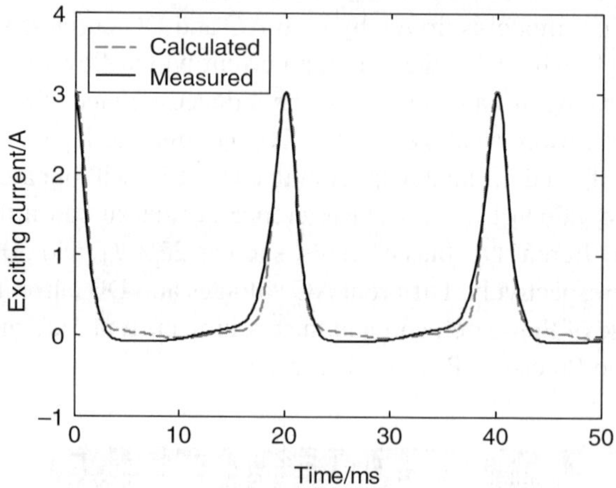

Figure 6.10 Exciting current under 25% I_{dc} (U = 293 V)

The DC bias is applied in different proportions of the current reference I_{dc}: 25%, 50%. The comparisons of calculation results and experimental data are partly shown in Figures 6.10 and 6.11.

By comparing calculated values with experimental data under DC-biased magnetization, it is observed that the calculated results coincide well with the measured results, which proves the effectiveness of the harmonic balance method. It is clear that when the alternating voltage is constant, the peak value of the exciting current increases rapidly with the growth of the DC bias.

The magnetic flux density is also analyzed in different DC-biased excitations. Two waveforms of the magnetic flux density, corresponding to 25% I_{dc} and 50% I_{dc} (under the condition that the alternating voltage is 293 V), are shown in Figure 6.12.

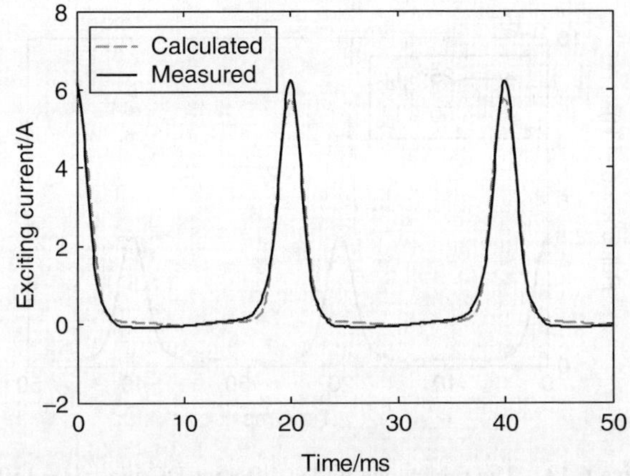

Figure 6.11 Exciting current under 50% I_{dc} (U = 293 V)

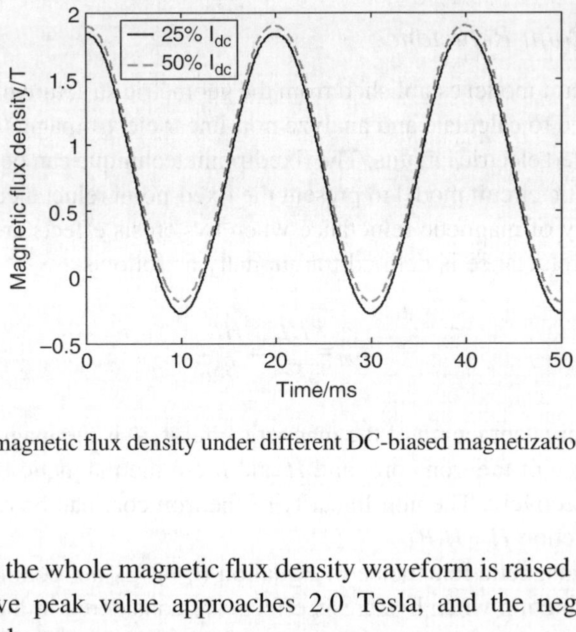

Figure 6.12 The magnetic flux density under different DC-biased magnetization with an alternating voltage of 293 V

It is found that the whole magnetic flux density waveform is raised up, due to the DC bias. The positive peak value approaches 2.0 Tesla, and the negative peak value approaches 0 Tesla.

The nonlinear reluctance is an important factor affecting the convergence of the nonlinear magnetic field in computation. The nonlinear reluctance calculated by the proposed method is shown in Figure 6.13.

As shown in Figure 6.13, the effect of the different DC bias on the waveform of the nonlinear reluctance can be observed when the alternating voltage is constant. It is obvious that when the voltage is constant, the nonlinear reluctance increases rapidly with the growth of the DC bias.

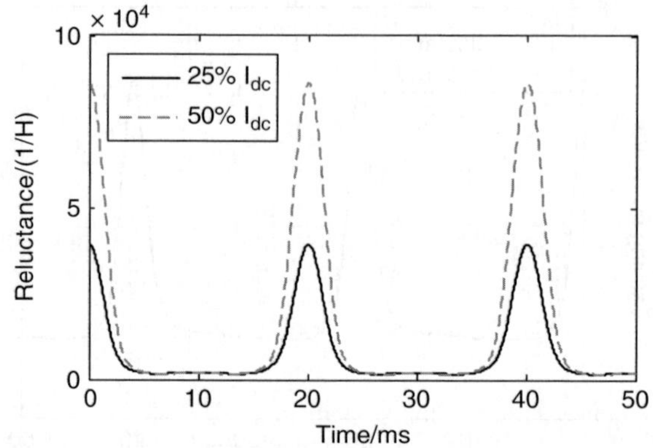

Figure 6.13 The reluctance under different DC bias (U = 159 V)

6.3 Computation Taking Account of Hysteresis Effects Based on Fixed-Point Reluctance

6.3.1 Fixed-Point Reluctance

The magnetic circuit model established from the geometric structure of the power transformer can be used to calculate and analyze non-linear electromagnetic problems, considering the coupled electric circuits. The fixed-point technique can be introduced in the non-linear magnetic circuit model to present the fixed-point reluctance, which can indicate the continuity of magnetic reluctance when hysteresis effects are involved.

The magnetic reluctance is defined traditionally as follows:

$$R_m = \frac{HL}{\Phi} = \frac{HL}{BS} \tag{6-62}$$

where L is the equivalent length of the magnetic circuit, Φ is the magnetic flux, S is the cross-sectional area of the iron core, and H and B are the magnetic field intensity and flux density, respectively. The non-linearity of the iron core can be represented by the single-valued function $H = H(B)$.

However, the magnetic reluctance in Equation (6-62) is not continuous, due to the non-monotonic function when hysteresis effects are considered in the numerical computation. That may lead to the unstable convergence in computing the non-linear magnetic field.

The widely used fixed point technique in solving non-linear systems can be introduced in the non-linear magnetic circuit model:

$$\begin{cases} F = \Re_{Fp}\Phi - M(\Phi) \\ F = HL \end{cases} \tag{6-63}$$

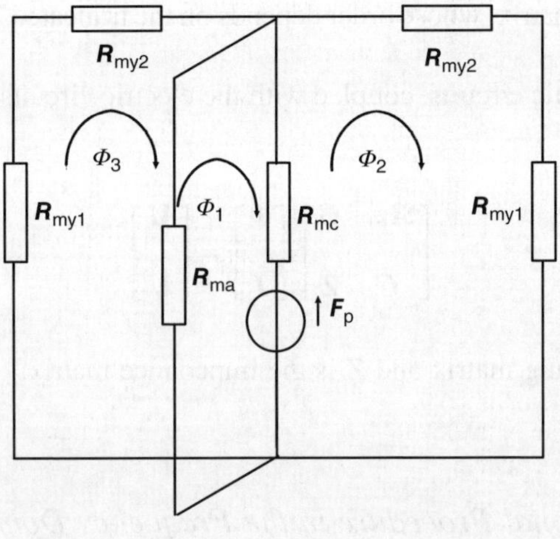

Figure 6.14 Magnetic circuit model of a single-phase three limb transformer

where F is the magnetomotive force and M is a non-linear quantity independent of the magnetic flux.

The magnetic circuit model of a single-phase three-limb transformer is presented in Figure 6.14, and the general non-linear system equations can be obtained in Equations (6-64) to (6-66).

$$[\Re_m(\Phi)][\boldsymbol{\Phi}] = [N_{coil}][\boldsymbol{I}] \tag{6-64}$$

$$[\boldsymbol{\Phi}] = [\Phi_1, \Phi_2, \Phi_3]^T \tag{6-65}$$

$$\boldsymbol{R}_{\mathrm{m}} = \begin{bmatrix} \boldsymbol{R}_{mc} + \boldsymbol{R}_{\mathrm{ma}} & -\boldsymbol{R}_{mc} & \boldsymbol{R}_{mc} \\ -\boldsymbol{R}_{mc} & \boldsymbol{R}_{my1} + \boldsymbol{R}_{\mathrm{my2}} + \boldsymbol{R}_{mc} & -\boldsymbol{R}_{mc} \\ \boldsymbol{R}_{mc} & -\boldsymbol{R}_{mc} & \boldsymbol{R}_{my1} + \boldsymbol{R}_{\mathrm{my2}} + \boldsymbol{R}_{mc} \end{bmatrix} \tag{6-66}$$

The fixed-point harmonic-balanced equation can be obtained by substituting Equation (6-63) into Equation (6-64), based on the harmonic balance method:

$$[\Re_{FP}][\boldsymbol{\Phi}] + [G][I] = [\boldsymbol{M}] \tag{6-67}$$

$$\Re_{FP} = \begin{bmatrix} \boldsymbol{R}_{ma} + \boldsymbol{R}_{FP1} & -\boldsymbol{R}_{FP1} & 0 \\ -\boldsymbol{R}_{FP1} & 2(\boldsymbol{R}_{FP1} + \boldsymbol{R}_{FP2}) & -\boldsymbol{R}_{FP1} \\ 0 & -\boldsymbol{R}_{FP1} & \boldsymbol{R}_{ma} + \boldsymbol{R}_{FP1} \end{bmatrix} \tag{6-68}$$

$$\boldsymbol{R}_{FPi} = R_{FPi}\boldsymbol{I}_{2n+1} \tag{6-69}$$

where $\boldsymbol{I}_{2n+1}$ is a unit matrix, whose order depends on the truncated harmonic number n in the computation.

Finally, the magnetic circuits, coupled with the electric circuits, are presented by the following equations:

$$\begin{bmatrix} \mathfrak{R}_{FP} & \boldsymbol{G} \\ \boldsymbol{C} & \boldsymbol{Z} \end{bmatrix} \begin{bmatrix} \boldsymbol{\Phi} \\ \boldsymbol{I} \end{bmatrix} = \begin{bmatrix} \boldsymbol{M} \\ \boldsymbol{U} \end{bmatrix} \tag{6-70}$$

where $\boldsymbol{C}$ is the coupling matrix and $\boldsymbol{Z}$ is the impedance matrix.

6.3.2 Computational Procedure in the Frequency Domain

The convergent performance of the harmonic solutions depends on the determination of the fixed-point reluctance. An optional scheme to determine the fixed-point reluctance can be selected according to the method proposed by Hantila [6],

$$R_{FP} = \frac{R_{FPmax} + R_{FPmin}}{2} \tag{6-71}$$

where R_{FPmax} and R_{FPmin} correspond to the maximum and minimum values of differential reluctance, which is related to the differential reluctivity.

Harmonic solutions of the magnetic flux and exciting currents in Equation (6-71) are computed according to the following procedure:

a. Equation (6-71) is solved to obtain the harmonic solutions of the magnetic flux $\boldsymbol{\Phi}_0$ and exciting current $\boldsymbol{I}_0$, with an initial value of $\boldsymbol{M}_0$ and a properly determined $\boldsymbol{R}_{\mathrm{FP}}$ in the right hand side of the equation.
b. The flux density $\boldsymbol{B}_0$ can be obtained from the computed $\boldsymbol{\Phi}_0$, and the magnetic field intensity $\boldsymbol{H}_0$ can be calculated by the hysteretic function $H = H(B)$. The static hysteresis loops of the iron core are simulated by the consuming function hysteresis model. Therefore, the non-linear quantity $\boldsymbol{M}_1$ can be updated by Equation (6-63).
c. The harmonic vector $\boldsymbol{\Phi}_1$ and $\boldsymbol{I}_1$ can be computed by the updated M_1 and U, according to the above procedure, until the criterion of relative error is satisfied:

$$\left\| \frac{\boldsymbol{M}_{p+1} - \boldsymbol{M}_p}{\boldsymbol{M}_p} \right\| < \varepsilon$$

where the subscript represents the number of iterative steps.

6.3.3 *Computational Results and Analysis*

The exciting current in the winding of a single-phase three-limb transformer is calculated, and the depicted waveforms under different excitations are presented in Figures 6.15 to 6.18.

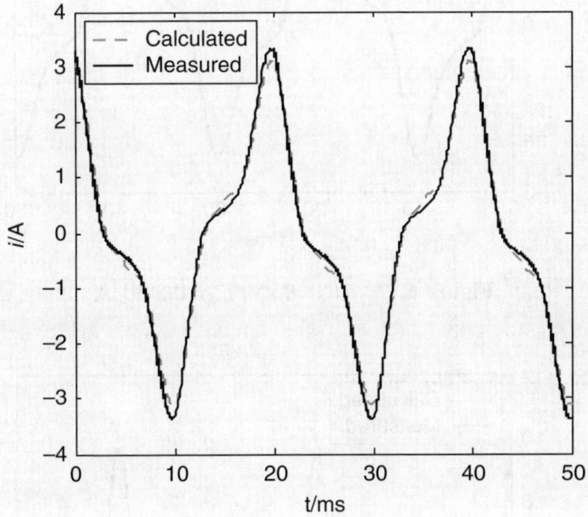

Figure 6.15 $U_{ac} = 95$ V, $I_{dc} = 0$

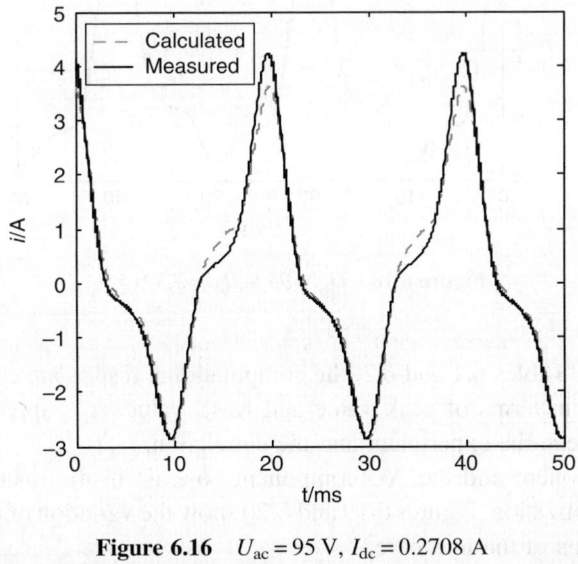

Figure 6.16 $U_{ac} = 95$ V, $I_{dc} = 0.2708$ A

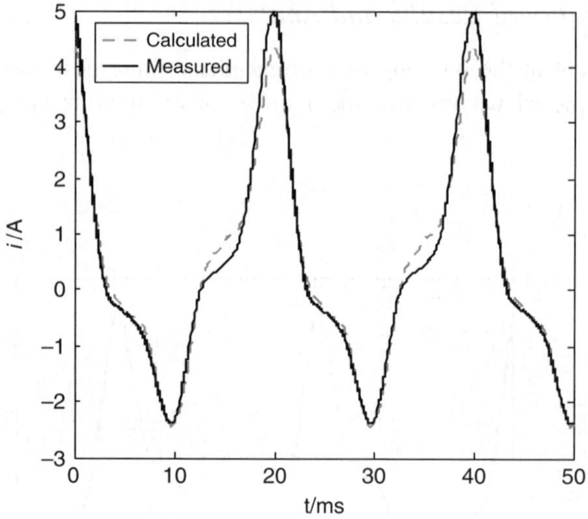

Figure 6.17 $U_{ac} = 95$ V, $I_{dc} = 0.5091$ A

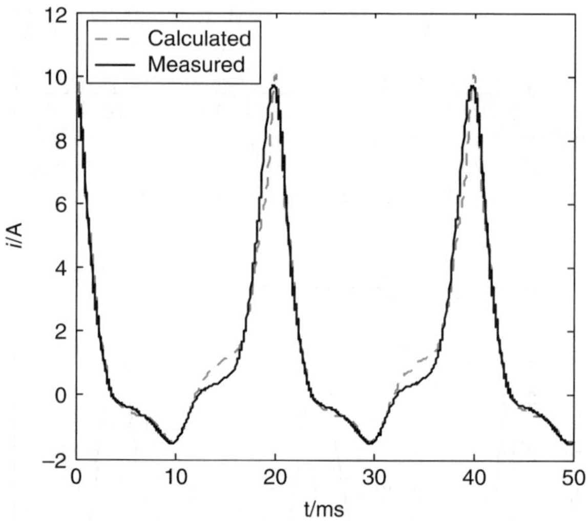

Figure 6.18 $U_{ac} = 95$ V, $I_{dc} = 1.5214$ A

As is shown in Tables 6.1 and 6.2, the computational results are compared with the measured results in terms of peak value and RMS value. It is apparent that there is consistency between the experiment and the simulation.

The DC component and the AC component co-exist in the magnetic core under DC-biased magnetization. Figures 6.19 and 6.20 show the variation of the magnetic flux in different regions of the iron core.

Table 6.1 Comparison of the RMS value of the magnetizing currents under different DC-biased conditions

I_{dc}/A	0	0.2708	0.5091	1.5214
Computed	1.6991	1.7838	1.9153	3.3483
Measured	1.7662	1.8065	2.0437	3.5056

Table 6.2 Comparison of the peak value of magnetizing currents under different DC-biased conditions

I_{dc}/A	0	0.2708	0.5091	1.5214
Computed	3.111	3.6070	4.3443	10.0483
Measured	3.500	4.2300	4.9900	9.7400

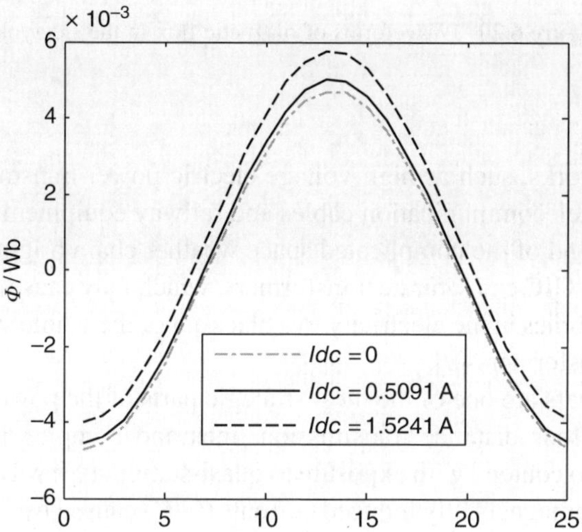

Figure 6.19 Waveforms of magnetic flux in the central core

6.4 HBFEM Modeling of the DC-Biased Transformer in GIC Event

6.4.1 GIC Effects on the Transformer

During a space weather storm, which is a result of solar activity, the magnetosphere-ionosphere system becomes disturbed, with intense and rapidly changing currents. At the Earth's surface, varying space currents manifest themselves as disturbances or storms in the geomagnetic field. As expressed by Faraday's law of induction, a geo-electric field also exists in connection with a geomagnetic storm. The electric field produces currents, known as "geomagnetically induced currents" (GIC), in ground-based

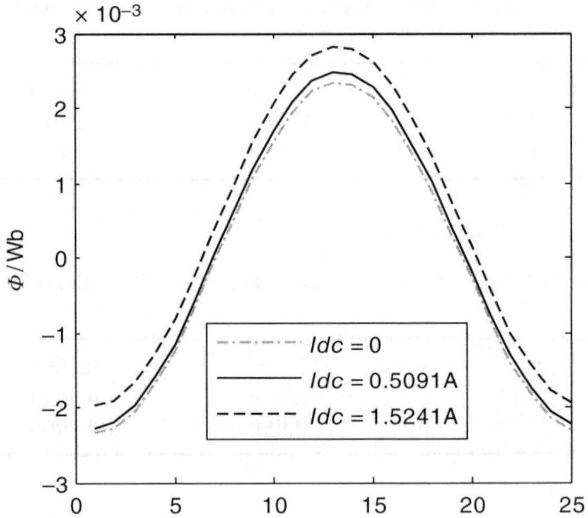

Figure 6.20 Waveforms of magnetic flux in the side yoke

technological networks, such as high-voltage electric power transmission systems, oil and gas pipelines, telecommunication cables and railway equipment [7]. GIC thus constitute the ground end of the complicated space weather chain originating from the sun [8]. In power grids, GIC can saturate transformers, which may cause different problems, ranging from harmonics in the electricity to a blackout of the whole system and permanent damage of transformers [9].

Power transformers are one of the most strategic parts of the power system. They are in operation with long-distance transmission lines and complex intermeshed power grids, which have to contend with exposure to quasi-stationary low DC currents, or even high-amplitude geomagnetically induced currents (GIC) caused by recurrent solar magnetic disturbances. Although transformers are generally designed for operation under sinusoidal waves (including the harmonics), in reality they may be subjected to superimposed DC current excitation with varying levels. These levels may reach up to a few hundreds amps. These DC currents may be of external origin, as GIC or HVDC ground return mode stray currents.

Depending on their magnitude, the DC bias currents may have a detrimental effect on the integrity of the power transformers or their long-term performance, which may affect the power system's reliability. With respect to this, user specifications relating to concern with superimposed DC excitations are generally clear enough regarding expected levels and possible durations. On the other hand, a good understanding of the behavior of power transformers under combined AC and DC excitations, as well as comprehensive modeling tools, are essential to enable the design of power transformers that fit these requirements [11].

To solve and analyze the GIC-related problems in transformers, several GIC modeling techniques are developed, including transmission line models, magnetic equivalent models, FEM-based numerical models and so forth [7, 11–13]. Among these, the equivalent circuit network based model is the most common method used to analyze GIC flow (from grounded transformer neutrals, through the HV windings into the transmission lines, with some impacts on the transformer). For structural problems with the transformer, the conventional finite element method has often been used to solve GIC-related DC biased transformer problems [12–14]. However, the well-established and accurate HBFEM approach has not been used in GIC modeling and simulation, due to the lack of technology transferring across disciplines and the complexity of HBFEM theory. As introduced and discussed in previous chapters, the HBFEM-based full-wave computational electromagnetic method has been employed in HVDC transformer analyses in the DC biased condition. The computer modeling techniques are well established [15, 16].

Since the transformer is a nonlinear magnetic structure with an arbitrary size, a full-wave computational electromagnetic technique is required. Within the various full-wave techniques, the solution can be found in either the frequency or time domain. Each technique has its own strengths and weaknesses. Which technique is most appropriate for a particular problem depends upon the problem. There is no single numerical technique that will solve all GIC problems that an electrical engineer is likely to model. Different numerical techniques can be used to solve different application problems. For GIC effects on the transformer in combination with the grid, transient electromagnetic network models are required. However, a time domain technique will be more appropriate for a transient problem during the GIC event. In this case. the transformer is modeled as a magnetic subsystem of the total grid, to allow determination of the reactive power consumption and the harmonics of the current [7, 14].

It is common practice to design power transformers operated under sinusoidal conditions. However, in reality, some power transformers need to be able to maintain normal operations under increasingly complex excitation waveforms. The most common case of this is where DC currents are superimposed on the normal AC excitation. Users located in areas where GIC may occur often specify that the transformers or shunt reactors must be able to withstand the DC currents, superimposed to the rated AC current in the windings, without damage or excessive hot spot temperature rises. The currents are specified in absolute numerical values by the user, and may be as high as 100 A, having a duration that varies from a few seconds to several minutes.

The capability of the transformers or shunt reactors to withstand DC currents is normally demonstrated in the "Design Review" [10]. Since the main requests in user specifications are related to transformers/shunt reactors prone to combined DC/AC excitation, the impact on transformer design and design verification processes needs to be investigated. HBFEM-based numerical modeling techniques will play an important role to support the design verification process, in order to mitigate the problems caused by the combined AC and DC excitations.

6.4.2 GIC Modeling and Harmonic Analysis

During a GIC event, the transformer will be operated under the magnetic saturating state when it is in a DC-biased case, as shown in Figure 6.21. The harmonics will be generated in the magnetic core, due to nonlinear magnetizing characteristics [17]. The HBFEM numerical computation model can be used for both GIC and HVDC transformer modeling. In the case of DC biased transformers, a DC current in the line neutral is often referred to, especially in ground return mode, as shown in Figure 6.21(b).

By perturbing Earth's magnetic field, geomagnetic disturbances (GMDs) can induce quasi-DC earth surface potential. This results in variation in the Earth's magnetic field of low (quasi-stationary) frequency (0.1 to 0.001 Hz), geomagnetically induced currents (GICs) that flow in the loops of transmission lines, power transformer windings with grounded neutrals, and the grounding grid of substations [18]. During GMDs, GICs in power systems cause half-cycle saturation of transformers. This produces harmonics and increases reactive power consumption and transformer overheating issues, which

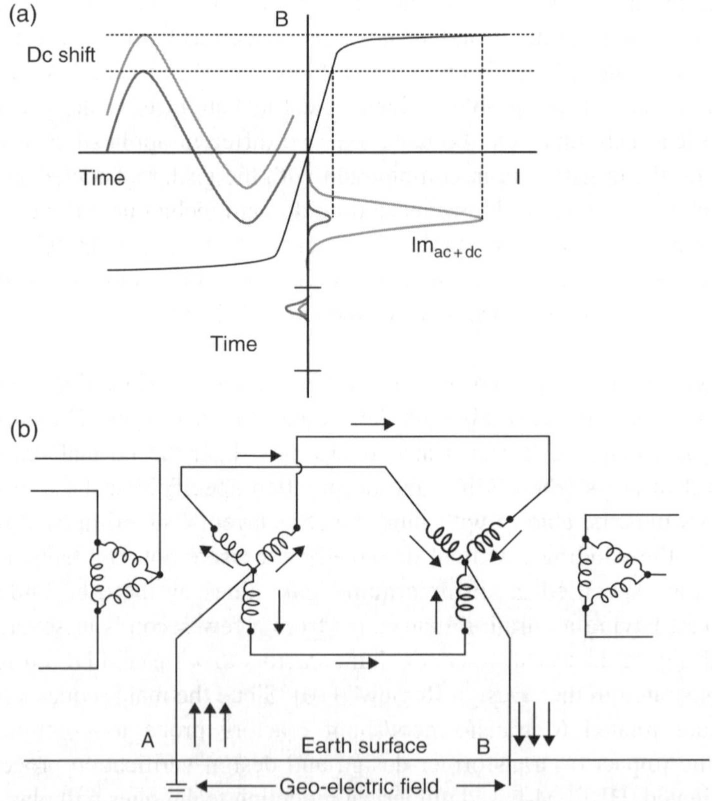

Figure 6.21 (a) DC-biased voltage and current waveform against magnetizing curve; (b) the problems caused by combined DC and AC excitation

can lead to transformer damage, voltage dips, relay misoperation, overload of the reactive power-compensating capacitor and system stability problems [19]. High-voltage overhead transmission lines, earth return paths and grounded high-voltage HV windings are linked to geomagnetic field changes (dB/dt is likely to be 200 to 400 nT/minute). The following parameters should be considered in the GIC calculations [14].

a. The magnitude of GIC depends on:
- induced voltage (up to 2–10 V/mile);
- network topology;
- resistivity of soil;
- high voltage transmission line resistances;
- angle between HV transmission line and magnetic field.

b. Back effects from the transformer on the grid:
- higher magnetizing current – reactive power consumption up to 50 MVA;
- harmonics occur.

c. GIC effects on transformers:
- additional noise (15 to 20% enhancement) with characteristic spectrum: odd multiple frequencies of 60 Hz in spectra … 300 Hz; 420 Hz). Example: Noise increase with 2 A DC/Phase is 15 dB(A);
- tank vibrations;
- critical eddy loss heating in specific metallic parts;
- tie bars near core are most vulnerable (even at low DC levels);
- clamping plate, tank cover, and windings are vulnerable at extremely high DC levels only.

6.4.3 GIC Modeling Using HBFEM Model

Figure 6.22 shows a DC-biased transformer model in GIC or HVDC events, where the internal equivalent circuit of the transformer (e.g., leakage inductance and winding resistance) and external equivalent circuits (e.g., transmission line impedance and ground resistance) should be included in the HBFEM simulation model. The DC bias transformer model can be excited as voltage source-driven or current source-driven, depending on the availability of the calculated or observed excitation data. If the induced quasi-DC voltage is obtained, then the excitation voltage source applied to the transmission line will include both DC voltage and sinusoidal voltage, with fundamental frequency (50/60 Hz). Based on the voltage source-driven HBFEM model discussed in Chapter 3.5, the unknown current magnitude can be calculated through this HBFEM model.

Transformer (A) is connected to transformer (B) through a long-distance transmission line and excited by the voltage source, including AC from the generator and DC induced by GIC. The third harmonic magnetizing current, caused by the non-linear magnetic core B-H characteristic during GIC or HVDC events, remains trapped inside the

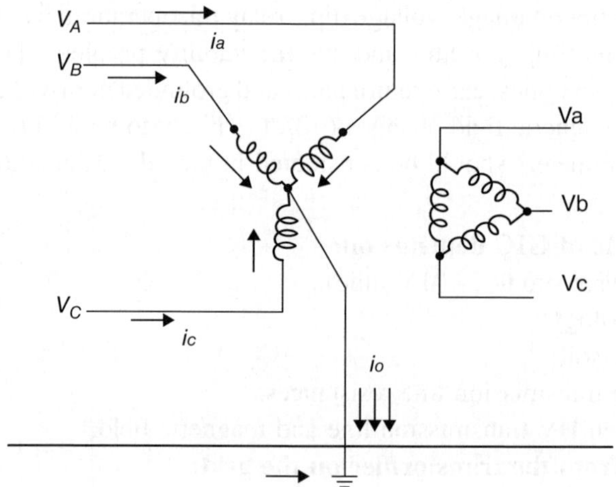

Figure 6.22 GIC or HVDC DC-biased transformer, including internal equivalent circuit (leakage inductance and winding resistance) and external circuits (transmission line impedance)

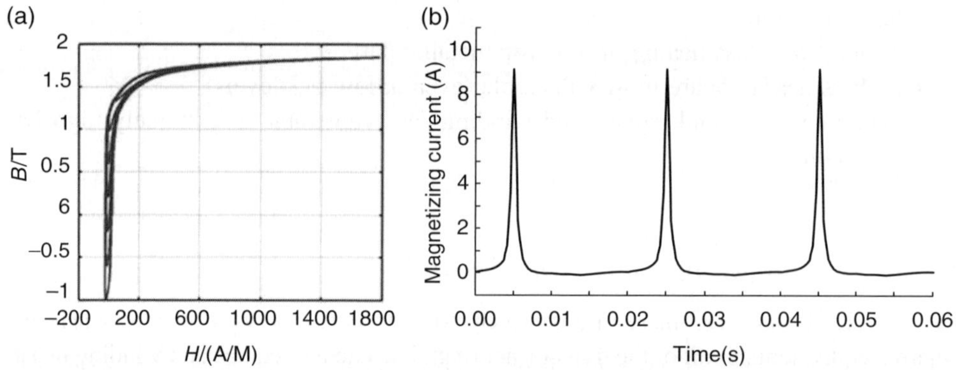

Figure 6.23 (a) B-H curve with hysteresis characteristics and DC-biased condition; (b) excitation current during DC bias

secondary Δ winding. Third harmonic currents are zero-sequence currents, which cannot enter or leave a Δ connection, but can flow within the Δ.

In the DC-biased case [15], the B-H curve of the transformer with hysteresis characteristics is illustrated as Figure 6.23. The mathematical expression of H(B), as presented in Chapter 3, can be obtained from a B-H curve data table.

If the secondary windings are in the open circuit condition, the HBFEM numerical model (including internal and external impedances) will be derived as follows:

$$\left.\begin{aligned}&[H]\{A\}-[G_A]\{J_A\}-[G_B]\{J_B\}-[G_C]\{J_C\}=\{0\}\\ &[C_{A,in}]\{A\}+[I]\{V_{Np}\}+S_{A,in}[Z_{A,in}]\{J_A\}=\{V_A\}\\ &[C_{B,in}]\{A\}+[I]\{V_{Np}\}+S_{B,in}[Z_{B,in}]\{J_B\}=\{V_B\}\\ &[C_{C,in}]\{A\}+[I]\{V_{Np}\}+S_{C,in}[Z_{C,in}]\{J_C\}=\{V_C\}\\ &S_{in,cA}[I]\{J_A\}+S_{in,cB}[I]\{J_B\}+S_{in,cC}[I]\{J_c\}=\{V_{Np}\}\end{aligned}\right\} \tag{6-72}$$

where $[H]$ (= $[S]+[M]$) is the system coefficient matrix, $[G_k]$ is the geometric matrix of the transformer windings, $[Z_{k,in}]$ and $S_{k,in}$ are external circuit impedances and cross-sectional areas of windings, respectively, $[I]$ is the unit matrix and V_{Np} is the voltage at the neutral point which can be obtained through DC equivalent resistance $Z_0 = V_{Np}/J_0$ (DC equivalent resistance obtained from DC equivalent circuits including ground resistance, transmission line DC resistance and transformer winding DC resistance), and $[C_{k,in}]$ refers to geometric coefficients related to transformer windings.

With the given input or excitation voltage sources $\{V_{k,in}\}$:

$$\{V_{k,in}\}=\{V_0\ V_{1s}\ V_{1c}]^T \tag{6-73}$$

where V_0 is the quasi-DC voltage (DC component) caused by GIC.

The unknown magnetic vector potential $\{A_n\}$ and the unknown current density $\{J_{k,n}\}$ in the system equation (6-72) can be obtained as follows:

$$\{A_n\}=\{\,A_0\ A_{1s}\ A_{1c}\ A_{2s}\ A_{2c}\ A_{3s}\ A_{3c}\ A_{4s}\ A_{4c}\ \cdots\ A_{ns}\ A_{nc}\,\}^T \tag{6-74}$$

$$\{J_{k,n}\}=\{\,J_0\ J_{1s}\ J_{1c}\ J_{2s}\ J_{2c}\ J_{3s}\ J_{3c}\ J_{4s}\ J_{4c}\cdots J_{ns}\ J_{ns}\,\}^T \tag{6-75}$$

where n is the harmonic number and k is the transformer winding number or phase number.

Using the HBFEM approach, the single element matrix $[H]$ and the detailed definitions of each matrix can be expressed from:

$$\begin{aligned}[H^e]=&\frac{1}{4\Delta^e}\begin{bmatrix}(b_1b_1+c_1c_1)D & (b_1b_2+c_1c_2)D & (b_1b_3+c_1c_3)D\\ (b_2b_1+c_2c_1)D & (b_2b_2+c_2c_2)D & (b_2b_3+c_2c_3)D\\ (b_3b_1+c_3c_1)D & (b_3b_2+c_3c_2)D & (b_3b_3+c_3c_3)D\end{bmatrix}\begin{Bmatrix}A_1^e\\ A_2^e\\ A_3^e\end{Bmatrix}\\ &+\frac{\sigma\omega\Delta^e}{12}\begin{bmatrix}2N & N & N\\ N & 2N & N\\ N & N & 2N\end{bmatrix}\begin{Bmatrix}A_1^e\\ A_2^e\\ A_3^e\end{Bmatrix}\end{aligned} \tag{6-76}$$

where matrices D and N (including the DC component) are expressed as follows:

$$D = \frac{1}{2}\begin{bmatrix} 2\nu_0 & \nu_{1s} & \nu_{1c} & \nu_{2s} & \nu_{2c} & \nu_{3s} & \nu_{3c} & \cdots \\ 2\nu_{1s} & 2\nu_0 - \nu_{2c} & \nu_{2s} & \nu_{1c} - \nu_{3c} & -\nu_{1s} + \nu_{3s} & \nu_{2c} - \nu_{4c} & -\nu_{2s} + \nu_{4s} & \cdots \\ 2\nu_{1c} & & 2\nu_0 + \nu_{2c} & \nu_{1s} + \nu_{3s} & \nu_{1c} + \nu_{3c} & \nu_{2s} + \nu_{4s} & \nu_{2c} + \nu_{4c} & \cdots \\ 2\nu_{2s} & & & 2\nu_0 - \nu_{4c} & \nu_{4s} & \nu_{1c} - \nu_{5c} & -\nu_{1s} + \nu_{5s} & \cdots \\ 2\nu_{2c} & & & & 2\nu_0 + \nu_{4c} & \nu_{2s} + \nu_{4s} & \nu_{1c} + \nu_{5c} & \cdots \\ 2\nu_{3s} & & \text{Symmetry} & & & 2\nu_0 - \nu_{6c} & \nu_{6s} & \cdots \\ 2\nu_{3c} & & & & & & 2\nu_0 + \nu_{6c} & \cdots \\ \vdots & \vdots & \vdots & \vdots & \vdots & \vdots & \vdots & \ddots \end{bmatrix} \tag{6-77}$$

where the magnetic reluctance can be obtained from equation (6-78):

$$\nu(t) = H(B(t))/B(t) = \nu_0 + \sum_{k=2n-2}^{\infty} \{\nu_{ks}\sin(k\omega t) + \nu_{kc}\cos(k\omega t)\} \tag{6-78}$$

$$N^e = \begin{bmatrix} 0 & 0 & 0 & 0 & 0 & 0 & 0 & \cdots \\ 0 & 0 & -1 & 0 & 0 & 0 & 0 & \cdots \\ 0 & 1 & 0 & 0 & 0 & 0 & 0 & \cdots \\ 0 & 0 & 0 & 0 & -2 & 0 & 0 & \cdots \\ 0 & 0 & 0 & 2 & 0 & 0 & 0 & \cdots \\ 0 & 0 & 0 & 0 & 0 & 0 & -3 & \cdots \\ 0 & 0 & 0 & 0 & 0 & 3 & 0 & \cdots \\ \vdots & \vdots & \vdots & \vdots & \vdots & \vdots & \vdots & \ddots \end{bmatrix} \tag{6-79}$$

The HBFEM system matrix equations can be obtained as:

$$\begin{bmatrix} [H] & -[G_A] & -[G_B] & -[G_C] & 0 \\ [C_A] & S_A[Z_A] & 0 & 0 & [I] \\ [C_B] & 0 & S_B[Z_B] & 0 & [I] \\ [C_C] & 0 & 0 & S_C[Z_C] & [I] \\ 0 & S_A[I] & S_B[I] & S_C[I] & 0 \end{bmatrix} \begin{Bmatrix} \{A_n\} \\ \{J_A\} \\ \{J_B\} \\ \{J_C\} \\ \{V_{Np}\} \end{Bmatrix} = \begin{Bmatrix} 0 \\ \{V_A\} \\ \{V_B\} \\ \{V_C\} \\ 0 \end{Bmatrix} \tag{6-80}$$

where $\{V_{Np}\} = \{3 \times J_0 \times Z_0\}$.

By substituting the following input voltages and current:

$$\{V_{k,in}\} = \{V_{k,A}, V_{k,B}, V_{k,C}\}^T \tag{6-81}$$

$$\{J_{k,in}\} = \{J_{k,A}, J_{k,B}, J_{k,C}\}^T \tag{6-82}$$

the HBFEM system matrix equation in a compact form can be expressed as:

$$\begin{bmatrix} [H] & [G_k] & 0 \\ [C_k] & S_{ck}[Z_k] & [I] \\ 0 & S_{ck}[I] & 0 \end{bmatrix} \begin{Bmatrix} A_k \\ J_k \\ V_{Np} \end{Bmatrix} = \begin{Bmatrix} o \\ V_k \\ 0 \end{Bmatrix} \tag{6-83}$$

With given voltage excitation sources V_k and, by solving the above HBFEM matrix equations, the magnetic flux density B and current i (including all harmonic components) can be obtained.

6.5 HBFEM Used in Renewable Energy Systems and Microgrids

6.5.1 Harmonics in Renewable Energy Systems and Microgrids

The uses of nonlinear loads connected to electric power systems include static power converters, arc discharge devices, saturated magnetic devices and, to a lesser degree, rotating machines. Static converters of electric power are the largest nonlinear loads, and are used in industry for a variety of purposes, such as electrochemical power supplies, adjustable speed drives, uninterruptible power supplies, the EV charging system and vehicle to grid (V2G), and energy storage system used in Renewable Energy Systems and Microgrid. The harmonic currents injected by DC/AC inverters (power electronic devices), as used for renewable energy systems and microgrids, might be able to cause harmonics. The magnitude and the order depend on the technology of the inverter and mode of operation, as described in IEEE Std. 519-2014 [20].

Nonlinear loads change the sinusoidal nature of the AC power current (and consequently the AC voltage drop), thereby resulting in the flow of harmonic currents in the AC power system that can cause interference with communication circuits and other types of equipment. These harmonic currents also lead to increased losses and heating in numerous electromagnetic devices (motors, transformers, etc.). When reactive power compensation is used, in the form of power factor improvement capacitors, resonant conditions can occur, which may result in high levels of harmonic voltage and current distortion when the resonant condition occurs at a harmonic associated with nonlinear loads.

In order to analyze harmonic problems, the network models that are often used for system harmonic analyses in existing works include impedance of the lines and cables,

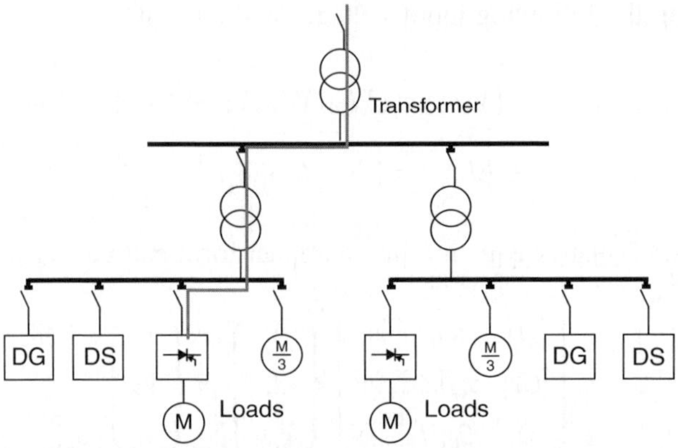

Figure 6.24 Block diagram of harmonics current flowing through a transformer in the microgrid with nonlinear load and DG and DS

transformers, and filter components. The inverter is modeled as either a simple harmonic source, or a harmonic source with output impedance in the form of a Thevenin or Norton equivalent circuit [7, 14]. However, as discussed in Chapter 2.4 (in the microgrid with a grid connected case), the current harmonics generated by power electronics and drives will be injected into the main grid through a PCC to the transformer, as shown in Figure 6.24. The voltage waveform applied to the transformer at the PCC, and current waveforms from the inverter, include a larger number of harmonics if poor DC/AC inverters are used.

Generally, current harmonics and voltage-frequency imbalances increase losses in the LV transformer and AC power lines. The main power quality parameters, including voltage, frequency, and harmonic distortion, have been improved by adopting Statcom, Static VAR Compensator (SVC) and the Shunt Active Power Filter (SAPF) in [23]. The voltage and frequency control mode is adopted in the microgrid under grid tied or islanded, in order to regulate voltage and frequency in the microgrid [24, 25].

6.5.2 Harmonic Analysis of the Transformer in Renewable Energy Systems and Microgrids

In renewable energy systems and microgrids, the transformer windings always use a Y connection at the LV side, and Δ connection at the distribution network side or high-voltage side, with the voltage around 10–12 kV, as shown in Figure 6.25. Therefore, the harmonic analysis model of the transformer can be considered as either a voltage source-driven or current source-driven HBFEM model from the LV side.

The waveforms of voltage and current in the LV side contain harmonics due to the transformer being connected to the DC/AC inverters or power electronic drives, are illustrated in Figure 6.26.

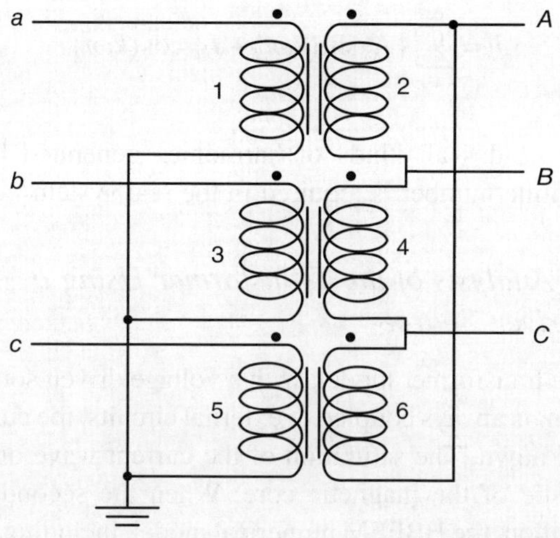

Figure 6.25 Transformer with Y/Δ connected to the grid

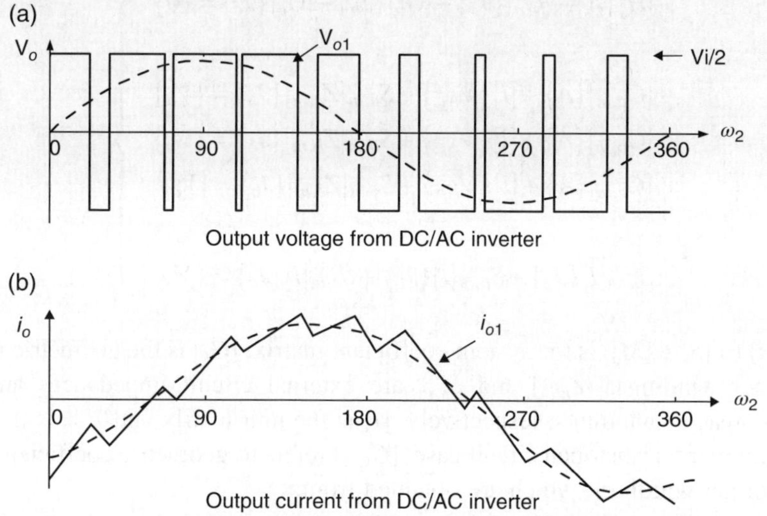

Figure 6.26 Waveforms of voltage and current in the LV side due to power electronic devices

Based on Maxwell equations and a quasi-static magnetic field expressed in Equation (6-2), the magnetic vector potential A on the arbitrary node i in the HBFEM descretized system and the source current density J_s can be respectively expressed as:

$$A^i = \sum_{k=n}^{\infty} \{A^i_{ks} \sin(k\omega t) + A^i_{kc} \cos(k\omega t)\} \tag{6-84}$$

$$J_s = \sum_{k=n}^{\infty} \{J_{ks} \sin(k\omega t) + J_{kc} \cos(k\omega t)\} \tag{6-85}$$

where harmonic k includes all kinds of harmonics generated by power electronic devices, but only a finite number is required in the real system.

6.5.3 Harmonic Analysis of the Transformer Using a Voltage Driven Source

Since the three-phase transformer model, with a voltage driven source connected to the magnetic field system, is always coupled to external circuits, the current in the input and output circuits is unknown. The saturation of the current waveform occurs due to the nonlinear characteristic of the magnetic core. When the secondary windings are in the open circuit condition, the HBFEM numerical model, including internal and external impedances, will be derived as follows:

$$\left.\begin{array}{l}
[H]\{A\} - [G_a]\{J_a\} - [G_b]\{J_b\} - [G_c]\{J_c\} = \{0\} \\
\\
[C_{a,in}]\{A\} + [I]\{V_{Np}\} + S_{a,in}[Z_{a,in}]\{J_a\} = \{V_a\} \\
[C_{b,in}]\{A\} + [I]\{V_{Np}\} + S_{b,in}[Z_{b,in}]\{J_b\} = \{V_b\} \\
[C_{c,in}]\{A\} + [I]\{V_{Np}\} + S_{c,in}[Z_{c,in}]\{J_c\} = \{V_c\} \\
\\
S_{a,in}[I]\{J_a\} + S_{b,in}[I]\{J_b\} + S_{c,in}[I]\{J_c\} = \{V_{Np}\}
\end{array}\right\} \tag{6-86}$$

where $[H]$ (= $[S] + [M]$) is the system coefficient matrix, $[G_k]$ is the geometric matrix of transformer windings, $[Z_{k,in}]$ and $S_{k,in}$ are external circuit impedances and cross-sectional areas of windings, respectively, $[I]$ is the unit matrix and V_{Np} is the voltage at a neutral point of the open circuit case. $[C_{k,in}]$ refers to geometric coefficients related to transformer windings, which are obtained from:

$$[C_k] = \frac{\omega d_0 \Delta}{3S_{ck}}[N \;\; N \;\; N] \tag{6-87}$$

The applied three-phase voltage sources $\{V_{k,in}\}$, $\{V_{Np'}\}$ and $\{V_{j,out}\}$ ($k = 1,2,3$ and $j = a,b,c$) are the unknown voltage at the neutral point, as expressed as below:

$$\{V_{k,in}\} = \{\{V_{k,1s}\} \;\; \{V_{k,1c}\} \;\; \{V_{k,2s}\} \;\; \{V_{k,2c}\} \;\; \cdots\}^T \tag{6-88}$$

$$\{V_{Np'}\} = \{\{V_{Np,1s}\} \;\; \{V_{Np,1c}\} \;\; \{V_{Np,2s}\} \;\; \{V_{Np,2c}\} \;\; \cdots\}^T \tag{6-89}$$

$$\{V_{j,out}\} = \{\{V_{j,1s}\} \ \{V_{j,1c}\} \ \{V_{j,2s}\} \ \{V_{j,2c}\} \ \cdots\}^T \tag{6-90}$$

The unknown magnetic vector potential $\{A_n\}$ and the unknown current density $\{J_{k,n}\}$ in the system equation (6-86) can be presented as follows:

$$\{A_n\} = \{A_{1s} \ A_{1c} \ A_{2s} \ A_{2c} \ A_{3s} \ A_{3c} \ A_{4s} \ A_{4c} \ \cdots \ A_{ns} \ A_{nc}\}^T \tag{6-91}$$

$$\{J_{k,n}\} = \{J_{1s} \ J_{1c} \ J_{2s} \ J_{2c} \ J_{3s} \ J_{3c} \ J_{4s} \ J_{4c} \cdots J_{ns} \ J_{nc}\}^T \tag{6-92}$$

where n is the harmonic number and k is the transformer winding number or phase number.

$\{J_k\}$ is the magnetizing current density, which can be expressed in a general form as follows:

$$\{J^k\} = \{\{J^k_{1s}\} \ \{J^k_{1c}\} \ \{J^k_{2s}\} \ \{J^k_{2c}\} \ \cdots\}^T \tag{6-93}$$

The generalized input voltage matrix equation taking account of circuits can be expressed as:

$$\{V_{k,\,in}\} = \{V_k\} + S_{ck}[Z_k]\{J_k\} \tag{6-94}$$

where matrix $[Z_k]$ is the circuit impedance matrix, which includes the resistance and leakage inductance of windings corresponding to the harmonics, and S_{ck} is the area of the windings.

$$[Z_{nk}] = [Z_{Rnk}] + [Z_{Lnk}] = \begin{bmatrix} R+n\omega L & 0 \\ 0 & R+n\omega L \end{bmatrix} \tag{6-95}$$

The HBFEM system matrix equations can be obtained as:

$$\begin{bmatrix} [H] & -[G_a] & -[G_b] & -[G_c] & 0 \\ [C_a] & S_a[Z_a] & 0 & 0 & [I] \\ [C_b] & 0 & S_b[Z_b] & 0 & [I] \\ [C_c] & 0 & 0 & S_c[Z_c] & [I] \\ 0 & S_a[I] & S_c[I] & S_b[I] & 0 \end{bmatrix} \begin{Bmatrix} \{A_n\} \\ \{J_a\} \\ \{J_b\} \\ \{J_c\} \\ \{V_{Np}\} \end{Bmatrix} = \begin{Bmatrix} 0 \\ \{V_a\} \\ \{V_b\} \\ \{V_c\} \\ 0 \end{Bmatrix} \tag{6-96}$$

By substituting the following input voltages and current into (6-96):

$$\{V_{k,in}\} = \{V_{k,a}, V_{k,b}, V_{k,c}\}^T \tag{6-97}$$

$$\{J_{k,in}\} = \{J_{k,a}, J_{k,b}, J_{k,c}\}^T \tag{6-98}$$

the HBFEM system matrix equation in a compact form can be expressed as:

$$\begin{bmatrix} [H] & [G_k] & 0 \\ [C_k] & S_{ck}[Z_k] & [I] \\ 0 & S_{ck}[I] & 0 \end{bmatrix} \begin{Bmatrix} A_k \\ J_k \\ V_{Np} \end{Bmatrix} = \begin{Bmatrix} o \\ V_k \\ 0 \end{Bmatrix} \tag{6-99}$$

Given voltage excitation sources V_k, and by solving the above HBFEM matrix equations, the magnetic flux density B and current i (including all harmonic components) can be obtained.

6.5.4 *Harmonic Analysis of the Transformer Using a Current-Driven Source*

Assuming $\nabla\varphi = 0$ in the two-dimensional case, and the governing equation (6-2), two-dimensional problems can be written as:

$$\nabla \times \nu \nabla \times A + \sigma(\partial A/\partial t) - J_s \tag{6-100}$$

By using Galerkin's method to discretize the governing equation (6-100) for two-dimensional problems, the equation can be written in integral form that is given as:

$$G = \iint_S \left\{ \frac{\partial N_i}{\partial x}\nu\frac{\partial A}{\partial x} + \frac{\partial N_i}{\partial y}\nu\frac{\partial A}{\partial y} \right\} dxdy - \iint_S \left\{ J_s - \sigma\frac{\partial A}{\partial t} \right\} N_i dxdy = 0 \tag{6-101}$$

Considering the orthogonal characteristic of trigonometric functions, substituting magnetic vector A, interpolation functions N_i, magnetic reluctivity ν and current density J_s (and using Galerkin's method, combined with harmonic balance techniques, to discretize), the following harmonic balance matrix equation (considering only odd harmonics for a single element) can be obtained as:

$$\begin{aligned} G^e = {} & \frac{1}{4\Delta^e} \begin{bmatrix} (b_1b_1 + c_1c_1)D & (b_1b_2 + c_1c_2)D & (b_1b_3 + c_1c_3)D \\ (b_2b_1 + c_2c_1)D & (b_2b_2 + c_2c_2)D & (b_2b_3 + c_2c_3)D \\ (b_3b_1 + c_3c_1)D & (b_3b_2 + c_3c_2)D & (b_3b_3 + c_3c_3)D \end{bmatrix} \begin{Bmatrix} A_1^e \\ A_2^e \\ A_3^e \end{Bmatrix} \\ & + \frac{\sigma\omega\Delta^e}{12} \begin{bmatrix} 2N & N & N \\ N & 2N & N \\ N & N & 2N \end{bmatrix} \begin{Bmatrix} A_1^e \\ A_2^e \\ A_3^e \end{Bmatrix} - \begin{Bmatrix} K_1^e \\ K_2^e \\ K_3^e \end{Bmatrix} = [S^e]\{A^e\} + [N^e][A^e] - \{K^e\} \end{aligned} \tag{6-102}$$

where the vector potential $\{A^e\}$ is called the frequency-domain representation, Δ is the area of the triangular elements, and ω indicates the fundamental frequency. b and c are obtained from x and y coordinates ($b_{ie} = x_{je}y_{ke} - x_{ke}y_{je}$, $c_{ie} = y_{je} - y_{ke}$). The D and N matrix blocks represent nonlinear magnetic material related to harmonics and harmonic frequencies, respectively.

The unknown magnetic vector potential $\{A_n\}$ and the given excitation current density $\{J_{k,n}\}$ in equation (6-101) can be presented as follows:

$$\{A_n\} = \{\, A_0 \;\; A_{1s} \;\; A_{1c} \;\; A_{2s} \;\; A_{2c} \;\; A_{3s} \;\; A_{3c} \;\; A_{4s} \;\; A_{4c} \;\; \cdots \;\; A_{ns} \;\; A_{nc} \,\}^T \tag{6-103}$$

$$\{J_{k,n}\} = \{\, J_0 \;\; J_{1s} \;\; J_{1c} \;\; J_{2s} \;\; J_{2c} \;\; J_{3s} \;\; J_{3c} \;\; J_{4s} \;\; J_{4c} \;\; \cdots \;\; J_{ns} \;\; J_{nc} \,\}^T \tag{6-104}$$

where n is the harmonic number and k is the transformer winding number or phase number.

Using the HBFEM approach, the single element matrix $[H]$ and the detailed definitions of each matrix can be expressed from:

$$[H^e] = \frac{1}{4\Delta^e}\begin{bmatrix} (b_1b_1+c_1c_1)D & (b_1b_2+c_1c_2)D & (b_1b_3+c_1c_3)D \\ (b_2b_1+c_2c_1)D & (b_2b_2+c_2c_2)D & (b_2b_3+c_2c_3)D \\ (b_3b_1+c_3c_1)D & (b_3b_2+c_3c_2)D & (b_3b_3+c_3c_3)D \end{bmatrix}\begin{Bmatrix} A_1^e \\ A_2^e \\ A_3^e \end{Bmatrix} + \frac{\sigma\omega\Delta^e}{12}\begin{bmatrix} 2N & N & N \\ N & 2N & N \\ N & N & 2N \end{bmatrix}\begin{Bmatrix} A_1^e \\ A_2^e \\ A_3^e \end{Bmatrix} \tag{6-105}$$

where matrices D and N, including DC components, are expressed as follows:

$$D = \frac{1}{2}\begin{bmatrix} 2\nu_0 & \nu_{1s} & \nu_{1c} & \nu_{2s} & \nu_{2c} & \nu_{3s} & \nu_{3c} & \cdots \\ 2\nu_{1s} & 2\nu_0 - \nu_{2c} & \nu_{2s} & \nu_{1c} - \nu_{3c} & -\nu_{1s} + \nu_{3s} & \nu_{2c} - \nu_{4c} & -\nu_{2s} + \nu_{4s} & \cdots \\ 2\nu_{1c} & & 2\nu_0 + \nu_{2c} & \nu_{1s} + \nu_{3s} & \nu_{1c} + \nu_{3c} & \nu_{2s} + \nu_{4s} & \nu_{2c} + \nu_{4c} & \cdots \\ 2\nu_{2s} & & & 2\nu_0 - \nu_{4c} & \nu_{4s} & \nu_{1c} - \nu_{5c} & -\nu_{1s} + \nu_{5s} & \cdots \\ 2\nu_{2c} & & & & 2\nu_0 + \nu_{4c} & \nu_{2s} + \nu_{4s} & \nu_{1c} + \nu_{5c} & \cdots \\ 2\nu_{3s} & & \text{Symmetry} & & & 2\nu_0 - \nu_{6c} & \nu_{6s} & \cdots \\ 2\nu_{3c} & & & & & & 2\nu_0 + \nu_{6c} & \cdots \\ \vdots & \vdots & \vdots & \vdots & \vdots & \vdots & \vdots & \ddots \end{bmatrix} \tag{6-106}$$

where the magnetic reluctance can be obtained from equation (6-78) and the harmonic number can be expressed as:

$$N^e = \begin{bmatrix} 0 & 0 & 0 & 0 & 0 & 0 & 0 & \cdots \\ 0 & 0 & -1 & 0 & 0 & 0 & 0 & \cdots \\ 0 & 1 & 0 & 0 & 0 & 0 & 0 & \cdots \\ 0 & 0 & 0 & 0 & -2 & 0 & 0 & \cdots \\ 0 & 0 & 0 & 2 & 0 & 0 & 0 & \cdots \\ 0 & 0 & 0 & 0 & 0 & 0 & -3 & \cdots \\ 0 & 0 & 0 & 0 & 0 & 3 & 0 & \cdots \\ \vdots & \vdots & \vdots & \vdots & \vdots & \vdots & \vdots & \ddots \end{bmatrix} \tag{6-107}$$

Finally, $\{K_i^e\}$ is expressed as:

$$\{K_i^e\} = \frac{\Delta^e}{3}\{J_0 \;\; J_{1s} \;\; J_{1c} \;\; J_{2s} \;\; J_{2c} \;\; \cdots, \;\; \cdots\}^T \tag{6-108}$$

The system matrix equation for current source excitation can be then written as:

$$[S]\{A\} + [N]\{A\} - \{K\} = 0 \tag{6-109}$$

All the harmonic components of A can be obtained by solving this matrix equation:

$$\left.\begin{array}{l} [H]\{A\} - [G_a]\{J_a\} - [G_b]\{J_b\} - [G_c]\{J_c\} = \{0\} \\ \\ [C_{a,in}]\{A\} - \{V_a\} + [I]\{V_{Np}\} = -S_{a,in}[Z_{a,in}]\{J_a\} \\ [C_{b,in}]\{A\} - \{V_b\} + [I]\{V_{Np}\} = -S_{b,in}[Z_{b,in}]\{J_b\} \\ [C_{c,in}]\{A\} - \{V_c\} + [I]\{V_{Np}\} = -S_{c,in}[Z_{c,in}]\{J_c\} \\ \\ S_{a,in}[I]\{J_a\} + S_{b,in}[I]\{J_b\} + S_{c,in}[I]\{J_c\} = \{V_{Np}\} \end{array}\right\} \tag{6-110}$$

$\{J_k\}$ is the given input current density, including all harmonics for three phases, and V_k is the induced voltage at three-phase transformer terminals.

The HBFEM system matrix equations for current source driven can be obtained as:

$$\begin{bmatrix} [H] & -[G_A] & -[G_B] & -[G_C] & 0 \\ [C_A] & -1 & 0 & 0 & [I] \\ [C_B] & 0 & -1 & 0 & [I] \\ [C_C] & 0 & 0 & -1 & [I] \\ 0 & S_{A,in}[I] & S_{B,in}[I] & S_{C,in}[I] & 0 \end{bmatrix} \begin{Bmatrix} \{A_n\} \\ \{V_A\} \\ \{V_B\} \\ \{V_C\} \\ \{V_{Np}\} \end{Bmatrix} = \begin{Bmatrix} 0 \\ \{J^*{}_A\} \\ \{J^*{}_B\} \\ \{J^*{}_C\} \\ 0 \end{Bmatrix} \tag{6-111}$$

where $\{J_k^*\}$ can be expressed as follows:

$$\{J_k^*\} = S_{k,in}[Z_{k,in}]\{J_k\} \tag{6-112}$$

References

[1] Glover, J.D., Sarma, M.S. and Overbye, T.J. (2011). *Power System Analysis and Design, Fifth Edition*. Cengage Learning.

[2] Lu, J., Yamada, S. and Bessho, K. (1990). Development and Application of Harmonic Balance Finite Element Method in Electromagnetic Field. *International Journal of Applied Electromagnetics in Materials* **1**(2–4), 305–316.

[3] Yacamini, R. and Bronzeado, H. (1994). Transformer inrush calculations using a coupled electromagnetic model. *IEE Proceedings – Science, Measurement and Technology* **141**(6), 491–498.

[4] Enringt, W.G. (1996). Transformer models for electromagnetic transient studies with particular reference to HV DC transmission. D. Eng. dissertation, University of Canterbury, New Zealand.

[5] Wang, J., Zhang, K. and Wang, X. (1999). Finite element method using harmonic balance, *Journal of Xi'an Jiaotong University* **33**(12), 5–10.

[6] Hantila, F., Preda, G. and Vasiliu, M. (2000). Polarization method for static fields. *IEEE Transactions on Magnetics* **36**(4), 672–675.

[7] NERC (2013). Appliaction Guide – Computing Geomagnetically-Induced Current in the Bulk-Power System.

[8] Pirjola, R. (2007). Calculation of geomagnetically induced currents (GIC) in a high-voltage electric power transmission system and estimation of effects of overhead shield wires on GIC modelling. *Journal of Atmospheric and Solar-Terrestrial Physics* **69**, 1305–1311.

[9] Kappenman, J.G. and Albertson, V.D. (1990). Bracing for the Geomagnetic Storm. *IEEE Spectrum* March, 27–33.

[10] ABB (no date). *SolidGroundTM GIC grid stability and harmonics mitigation system, Geomagnetic Storm Induced Current (GIC) and Electromagnetic Pulse (EMP) protection*. Available at: www.abb.us/highvoltage.

[11] Ngnegueu, T., Marketos, F., Devaux, F. *et al.* (2012). Behaviour of transformers under DC/GIC excitation: Phenomenon, Impact on design/design evaluation process and Modelling aspects in support of Design, A2-303. *CIGRE* **21**. Available at: http://www.cigre.org

[12] Lu, S. and Liu, Y. (1993). FEM analysis of DC saturation to assess transformer susceptibility to geomagnetically induced currents. *IEEE Transactions on Power Delivery* **8**(3), 1367–1376.

[13] Mousavi, S.A. (2015). Electromagnetic Modelling of Power Transformers for Study and Mitigation of Effects of GICs, Chapter 5. *Numerical modelling and calculation of power Transformers*. PhD thesis, Royal Institute of Technology (KTH).

[14] Siemens AG Energy Sector (2014). *Designed to Withstand-Siemense GIC-safe transformers*. Published by and copyright © 2014: Siemens AG Energy Sector Freyeslebenstrasse 1 91058 Erlangen, Germany. Available online at: siemens.com/energy/transformers

[15] Zhao, X., Lu, J., Li, L., Cheng, Z. and Lu, T. (2011). Analysis of the DC bias phenomenon by the harmonic balance finite element method. *IEEE Transactions on Power Delivery* **26**(1), 475–485.

[16] Bíró, O., Buchgraber, G. Leber, G. and Preis, K. (2008). Prediction of magnetizing current wave-forms in a three-phase power transformer under DC bias, *IEEE Transactions on Magnetics* **44**(6), 1554–1557.

[17] Power IT Lab (no date). *GIC Impact for Transformer Harmonics and Reactive Power*. The University of Tennessee. Available at: http://powerit.utk.edu/GIC_impact.html

[18] Pan, Z. and Liu, Y. (2014). *Harmonics current and reactive power consumption of threephase seven-limb transformer during geomagnetically disturbance*. Technical report.

[19] Li, X., Wen, X., Markham, P. and Liu, Y. (2010). Analysis of nonlinear characteristics for a three phase, Five limb transformer under DC bias. *IEEE Transactions on Power Delivery* **25**(4), 2504–2510.

[20] IEEE Std. 519-2014 (2014: revision of IEEE Std. 519, 1992). *IEEE Recommended Practices and Requirements for Harmonic Control in Electric Power Systems*. Approved 27 March 2014.

[21] García-Gracia, M., El Halabi, N., Alonso, A. and Paz Comech, M. (2011). *Harmonic Distortion in Renewable Energy Systems: Capacitive Couplings*. Available online at www.intechopen.com.

[22] Bouloumpasis, I., Vovos, P., Georgakas, K. and Vovos, N.A. (2015). Current Harmonics Compensation in Microgrids Exploiting the Power Electronics Interfaces of Renewable Energy Sources. *Energies* **8**(4), 2295–2311.

[23] Gao Xiaozhi, Li Linchuan and Chen Wengan (2011). Power quality improvement for microgrid in islanded mode. *Procedia Engineering* **23**, 174–179.

[24] Wang Chengshen, Xiao Zhaosia and Wang Shouxiong (2008). Synthetical control and analysis of microgrid. *Industrial Automation of Electric Power Systems* **32**(7), 98–103.

[25] Blaabjerg, F., Trodorecer, R., Liserre, M. and Timbas, A.V. (2006) Overview of control and grid synchronization for distributed power generation systems. *IEEE Transactions on Industrial Electronics* **53**, 1398–1409.

Appendix

Appendix I & II

Matlab Program and the Laminated Core Model for Computation

The laminated core model is made up of silicon steel lamination, of which the model number is 30Q140. Due to the symmetrical structure, one quadrant can be used to compute the two-dimensional magnetic field, which is shown in Figure A1.1. The detailed

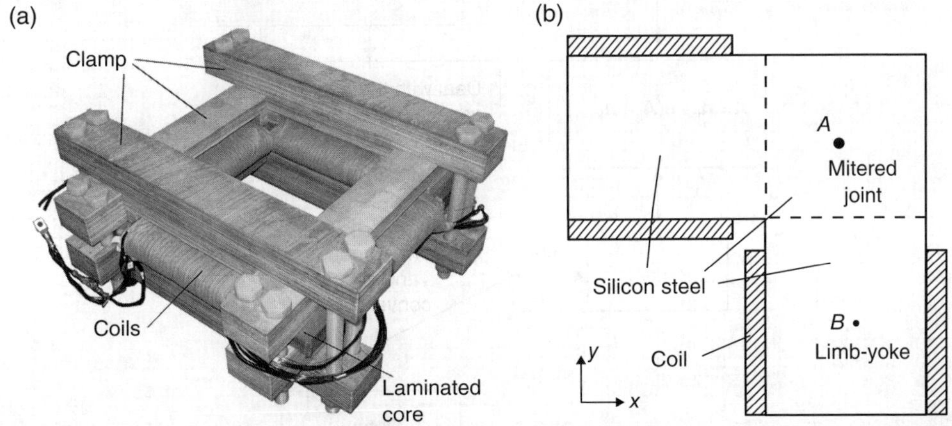

Figure A1.1 (a) The laminated core model; (b) geometric model for numerical computation

Harmonic Balance Finite Element Method: Applications in Nonlinear Electromagnetics and Power Systems, First Edition. Junwei Lu, Xiaojun Zhao and Sotoshi Yamada.

Companion website: www.wiley.com/go/lu/HBFEM

formulations and results related to this example program can be found in Chapter 5-1, HBFEM for DC-biased problems in HVDC power transformers.

The program flowchart for numerical computation is presented below in Figure A1.2:

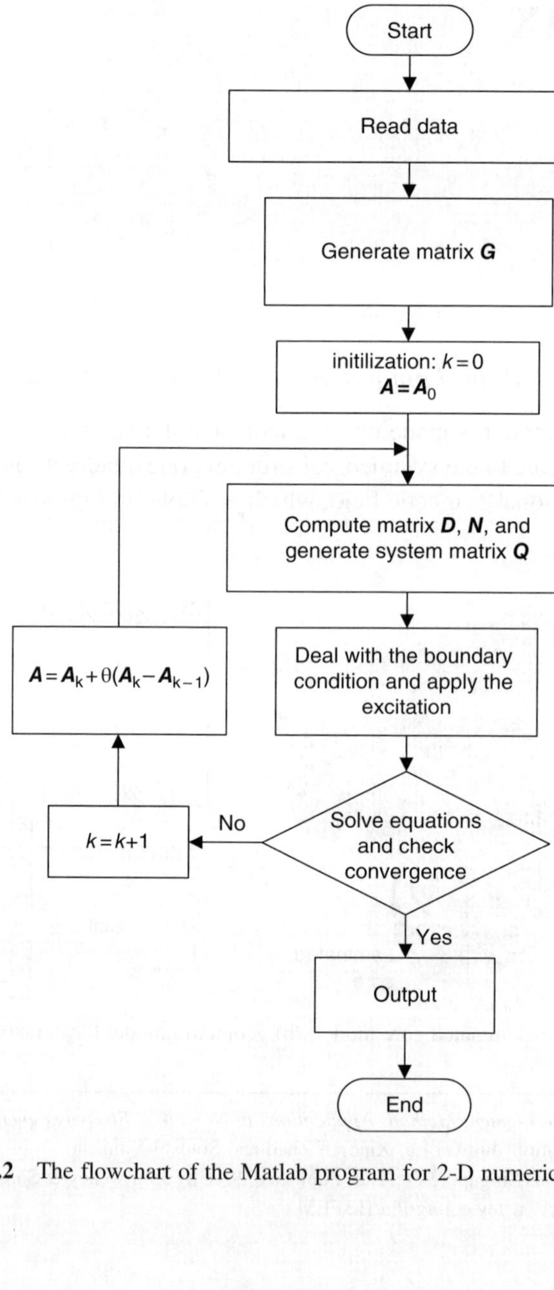

Figure A1.2 The flowchart of the Matlab program for 2-D numerical computation

Appendix III

FORTRAN-Based 3D Axi-Symmetrical Transformer with DC-Biased Excitation

A typical port core transformer has an axi-symmetrical structure and a B-H hysteresis curve, as shown in Figure A3.1. The transformer is excited by quasi-sinusoidal waveform, which includes AC fundamental, harmonics, and DC components. When the magnetic core becomes saturated, the waveforms will be distorted, and harmonics will be generated in the magnetic field and circuits. The hysteresis loss in winding is also increased, due to the effect of high-frequency harmonics.

Numerical simulation of 3-D axi-symmetrical structure can be described in Figure A3.2 as below, where the 3-D structure model can be reduced to a 2-D axi-symmetrical structure model in the **z** and **r** domain.

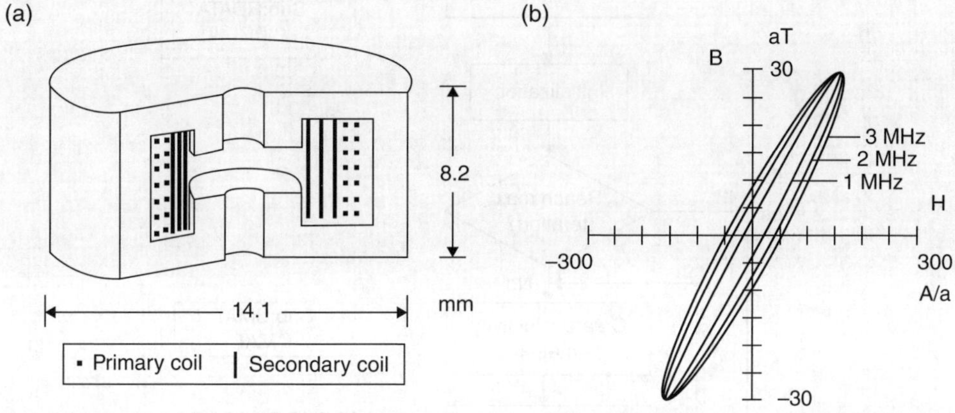

Figure A3.1 (a) Simulation model of a switching mode transformer; (b) B-H curve of the magnetic core

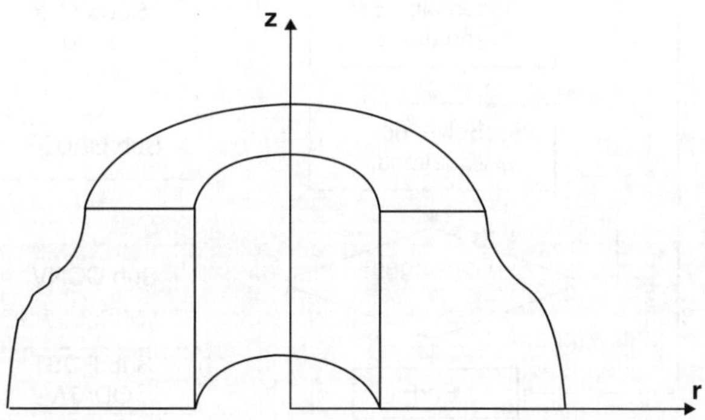

Figure A3.2 Axi-symmetrical structure in the **z** and **r** domain

Since the 3-D port core transformer has an axi-symmetrical structure, the numerical model can be reduced to two dimensional axi-symmetrical structure problems. The detailed numerical computation procedure related to this example program can be found from Chapter 4-5, Three-dimensional axi-symmetrical transformer with DC-biased excitation.

As illustrated in Figure A3.3, the example program flowchart of a three-dimensional axi-symmetrical transformer with DC-biased excitation is presented below:

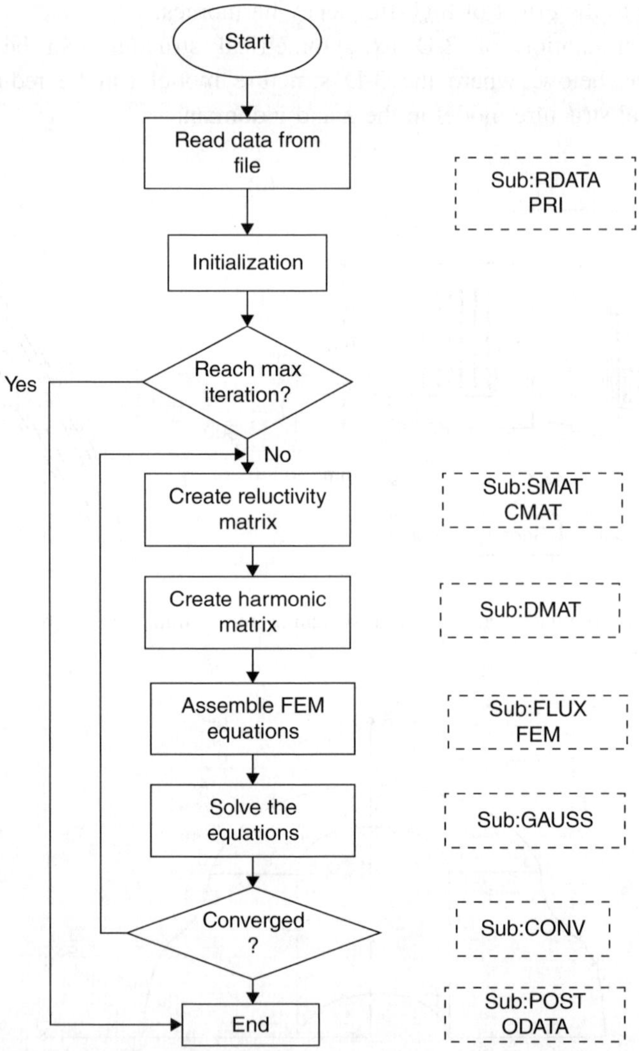

Figure A3.3 Example program flowchart of a three-dimensional axi-symmetrical transformer with DC-biased excitation

Index

Harmonic Balance Finite Element Method: Applications in Nonlinear Electromagnetics and Power Systems, First Edition. Junwei Lu, Xiaojun Zhao and Sotoshi Yamada.

Companion website: www.wiley.com/go/lu/HBFEM